A series of student texts in

CONTEMPORARY BIOLOGY

General Editors:
Professor E. J. W. Barrington, F.R.S.
Professor Arthur J. Willis

Nitrogen Metabolism in Plants

Leonard Beevers

Professor of Plant Physiology, University of Oklahoma

Edward Arnold

First published 1976
by Edward Arnold (Publishers) Ltd.,
25 Hill Street, London W1X 8LL

Boards edition ISBN: 0 7131 2514 4
Paper edition ISBN: 0 7131 2515 2

Printed in Great Britain by
William Clowes & Sons, Limited
London, Beccles and Colchester

Preface

As the most recent textbook on Nitrogen Metabolism in Plants had been written in 1959 it seemed a very worthwhile and timely endeavour when I was asked to prepare a text on this topic for the Contemporary Biology Series. Recent introductory Plant Physiology and Plant Biochemistry textbooks mention nitrogen metabolism; however, they are unable to cover adequately the increased understanding of the subject which has resulted from the unprecedented information explosion since 1959.

Consequently the aim of this text is to provide a comprehensive, up to date coverage of our knowledge of nitrogen metabolism in higher plants. In the preparation of the manuscript attention has been given to emphasizing the understanding of nitrogen metabolism in plants as opposed to the more usually cited bacterial and animal systems. The assembly of the information demonstrating the functioning of a potential metabolic pathway in plants frequently involves the utilization of data from a range of plant species and tissues. I have attempted to assemble this widely dispersed information into a condensed coherent form. The source of the accumulated material is indicated by the extensive literature citations. Thus the text in addition to satisfying the requirements for a course in nitrogen metabolism will provide a ready reference source for researchers in specific areas.

The text can be divided into two parts. The initial six chapters are devoted to a study of nitrogen metabolism, biosynthetic and degradative processes, etc., and the final three chapters stress the importance of these metabolic processes in plant growth and development. The introductory chapter describes the interconversions of the inorganic nitrogenous components of the soil and discusses the sources of nitrogen available to the plant. A description of the uptake and conversion of inorganic nitrogen into organic form is followed by a chapter on amino acid biosynthesis. Subsequently the metabolism of amino acids to other cellular constituents such as alkaloids, cyanogenic glycosides, growth hormones, porphyrins and cofactors is described. A chapter discussing the metabolism of the purine and pyrimidine nucleosides and nucleotides is followed by a study of nucleic acid metabolism. The sixth chapter discusses protein synthesis and its regulation.

Where appropriate, in the chapters on metabolism, an attempt has been made to document the occurrence of a particular component in the plant. The biosynthesis of the component is discussed, followed by a description of its degradation or secondary metabolism. Attention is drawn to those areas of metabolism which have not been investigated or which would merit further study. The final three chapters describe nitrogen metabolism in seeds, in the whole plant and in fruit ripening and senescence. In these chapters I have tried to indicate how an increased knowledge of nitrogen metabolism has advanced our understanding of plant growth and development.

Many people have helped in the preparation of the manuscript. Over the years I have appreciated the support of the National Science Foundation research grants which have allowed me to conduct much of my own research cited in the text. I am indebted to the many authors and publishers who have allowed me to use information for the figures and text illustration. I am particularly thankful to Dr. H. H. Mollenhauer and Professor J. W. Bradbeer for kindly providing previously unpublished electron micrographs. I would like to thank Mrs. Cynthia Collins who prepared several of the original illustrations, and Mrs. Irene Hackett who had the unenviable task of transcribing my illegible handwriting into the final typescript. I have appreciated the patience and tolerance of the staff of Edward Arnold and special thanks must go to Professor A. J. Willis for his editorial supervision of the manuscript. Professor Willis, and his former colleagues at Bristol, in no small measure contributed to my embarkation on a career in nitrogen metabolism. A two-day visit to the Botany Department at the University of Bristol in 1959 provided me with the then up to date techniques which allowed me to complete my Ph.D. studies at the University College of Wales, Aberystwyth. Coincidentally the weekend before the visit to Bristol I met Patricia, now my wife. I thank her and our sons Richard and Kevin for their constant source of encouragement and help.

Oklahoma 1975 L.B.

Table of Contents

I

Nitrogen Nutrition

NITROGEN SOURCE

Most plants derive the nitrogen required for their metabolism from the soil solution. The majority of the soil nitrogen is derived predominantly by biological activities associated with the 'nitrogen cycle' (Fig. 1.1).

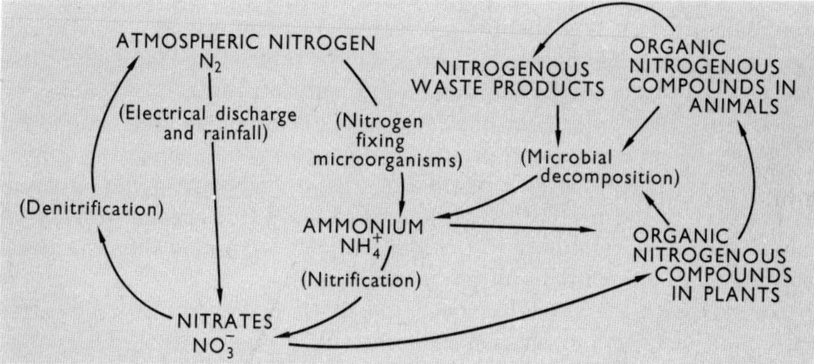

Fig. 1.1 The nitrogen cycle.

The details of this cycle are discussed at length later, but it is clear that all of the soil nitrogen is ultimately generated from atmospheric nitrogen. The soil nitrogen status is a function of the processes of nitrogen fixation, microbial decomposition, nitrification and denitrification. In view of the many reactions involved it is not surprising that great diversity exists both in the total nitrogen content of the soil and the composition of the soil nitrogen. Current soil analysts divide the soil nitrogen into inorganic and organic nitrogen fractions and as indicated in Table 1.1 the majority of the

Table 1.1 Analysis of soils showing the diversity of total nitrogen content and composition of nitrogen fractions.

Soil	Total N (%dry wt)	Inorganic nitrogen (ppm) E.A.N.	Ni.N	FiA.N	Organic N (ppm)
Iowa					
Clyde*	0.402	6	31	124	3859
Webster*	0.233	3	6	117	2204
Thurman*	0.056	0	10	28	522
Edina*	0.164	7	10	126	1497
Hayden*	0.164	8	5	61	1566
Oklahoma					
Tall Grass Prairie†	—	5.1	1.2	—	—
Post Oak, Blackjack-Oak Forest†	—	4.2	0.9	—	—
Oak–Pine Forest†	—	5.5	1.0	—	—

E.A.N., Exchangeable ammonia N ; Ni.N, Nitrate N ; FiA.N., Fixed ammonia N ; Organic N, Calculated by difference.
*Data from Nelson & Bremner.[456]
†E. L. Rice personal communication ; average calculated from 6 bimonthly samplings.

soil nitrogen is in the organic form. The inorganic nitrogen is further subdivided into fixed ammonia, exchangeable ammonia, nitrate and nitrite; however, in most soils nitrite is usually negligible. The fixed ammonia results from the replacement by ammonia of the inter-layer cations Ca^{2+} and Mg^{2+} in the expanded lattice of clay minerals. This ammonia nitrogen is probably unavailable as a plant nutrient. In contrast the exchangeable ammonia and nitrate can be utilized by plants and these serve as the prime source of nitrogen in plant nutrition.

In many soils nitrate is present in greater amounts than ammonia. However, in some grassland prairies and hardwood and coniferous forests ammonia appears to be the principal form of available inorganic nitrogen. It has been suggested[522] that certain grasses and tree species growing in such habitats are inhibitory to the nitrification process and thus the usual conversion of ammonia to nitrate is retarded.

An extensive literature has been published documenting the relative suitability of nitrate or ammonia for sustaining plant growth. Of course, comparisons conducted in soil culture are extremely difficult to interpret owing to the fact that the nitrogen source applied may be effectively converted into the other form by soil microorganisms.

In order to overcome this criticism, attempts have been made to compare the efficiency of the two nitrogen sources using hydroponic nutrient culture techniques. However, although many comparisons of this type have been made it is often difficult to interpret the results. As the nitrate or ammonium ions are removed from the nutrient medium by the growing plant there is a change of the pH of the nutrient medium and it thus becomes difficult to determine if the relative difference in growth in the two nitrogen sources is due to a variation in nitrogen availability or due to a tolerance of different solution pH. Depletion of nitrate nitrogen from the nutrient solution causes a drift toward alkalinity; conversely with an ammonium medium plant growth causes an acid drift. These pH drifts are due to differential rates of cation and anion absorption. In the presence of ammonium, cation absorption predominates; in the presence of nitrate, anion absorption predominates. Analyses of the voluminous literature indicates that both forms of nitrogen can be utilized by the plant. However, the uptake and assimilation of these ions are affected by carbohydrate status of the plant, species of plant and plant age.[606]

In studies with *Spirodela*[183] and wheat (*Triticum aestivum*)[430] it appears that ammonium ions repress the uptake and metabolism of nitrate. However, tissue analyses showing nitrate to be present in plant materials clearly indicate that nitrate is readily taken up from the soil solution. The uptake of nitrate appears to be dependent on the development of some carrier mechanism which facilitates nitrate entry into the tissues.[302]

Although tissue culture experiments have clearly demonstrated the capacity of plant cells in the absence of contaminating bacteria to assimilate nitrogen from such organic sources as amino acids, amides and urea, it is unlikely that organic nitrogenous compounds in the soil represent a major source of nitrogen for higher plant nutrition. On the other hand many fungi probably depend extensively on such nitrogen sources.

In parasitic plants such as the mistletoe (*Viscum* sp., *Phoradendron* sp.), nitrogen will be supplied by way of the xylem stream and this source will potentially contain both organic and inorganic forms. In the semi-parasites which are partially parasitic on the roots of other host species and produce haustorial connections with the host roots it is also possible that host-produced organic nitrogenous constituents can serve to meet the nitrogen requirements. Additionally, however, there is the potential of uptake of soil inorganic nitrogen by the roots of the semi-parasite. In the obligate parasites the nitrogen nutrition is dependent solely on the nitrogen supplied by the host plant and thus a considerable proportion, if not all, of the nitrogen will be assimilated from low molecular weight organic components present in the vascular system of the host plant.

In carnivorous (insectivorous) plants nitrogen nutrition is not restricted to the utilization of low molecular weight components, but since such plants can hydrolyse proteins their source of nitrogenous nutrition is extended to macromolecular nitrogenous compounds.

As discussed later, certain plants can derive an appreciable quantity of nitrogen from their nitrogen-fixing symbionts. This nitrogen may be in the form of ammonia produced in the fixation process or organic nitrogen such as amino acids or amide produced as a result of metabolic activity by the symbiont.

NITROGEN FIXATION

It has been estimated that 100 million tons of atmospheric nitrogen are fixed annually on this planet.[597] This conversion of atmospheric nitrogen to ammonia is achieved by the activities of free-living bacteria and blue-green algae and by symbiotic associations of these microorganisms with plants.

Originally the detection of organisms capable of nitrogen fixation was a somewhat insensitive and laborious procedure based upon a chemical estimation of the ammonia produced in the fixation process or the utilization of the stable isotope $^{15}N_2$ of nitrogen.[701] However, in 1966 it was demonstrated that organisms which fix nitrogen also have the capacity of reducing acetylene to ethylene.[150] The ethylene produced can be readily detected by sensitive techniques of gas chromatography and the procedure has been used extensively since that time. As well as expanding knowledge of organisms capable of carrying out nitrogen fixation, the technique has allowed for the assay of N_2-fixing activity of plants, soil, water, etc., so that there is now the possibility of assessing the nitrogen-fixing potential of the biosphere components. It has been estimated[597] that symbiotic legumes account for the fixation of 100–200 lb N_2 fixed/acre/year (112–224 kg/ha/year), symbiotic non-legumes 50–100 lb N_2 fixed /acre/year (56–112 kg/ha/year), blue-green algae 10–70 lb/acre/rice crop (11–78 kg/ha/rice crop).

In a recent review, Stewart[598] lists 15 genera of bacteria which have been shown to fix nitrogen. All the photosynthetic bacteria possess the capability in addition to various facultative aerobes and obligate anaerobes. Most of the detailed studies have been performed on *Azotobacter* and *Clostridium*.

Among the blue-green algae are many species capable of nitrogen fixation. Significantly most of the free-living nitrogen-fixing members of this group produce heterocysts (Fig. 1.2a, 1.2b) and are members of the

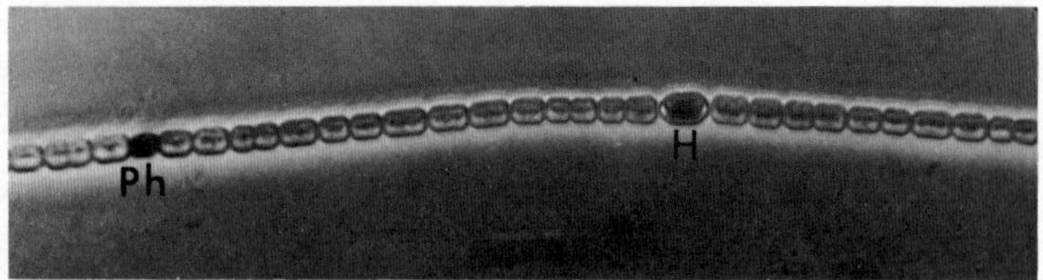

Fig. 1.2a Filament of *Anabaena cylindrica* showing mature heterocyst (H) and a proheterocyst (Ph). Phase contrast light micrography (×1200). (From Kulasooriya, Lang and Fay,[347] Fig. 6, Plate 27. Reproduced by permission of The Royal Society.)

Nostocales and Stigonematales. The production of the heterocysts by these blue-green algae has been found to be related to their nutrition and thus

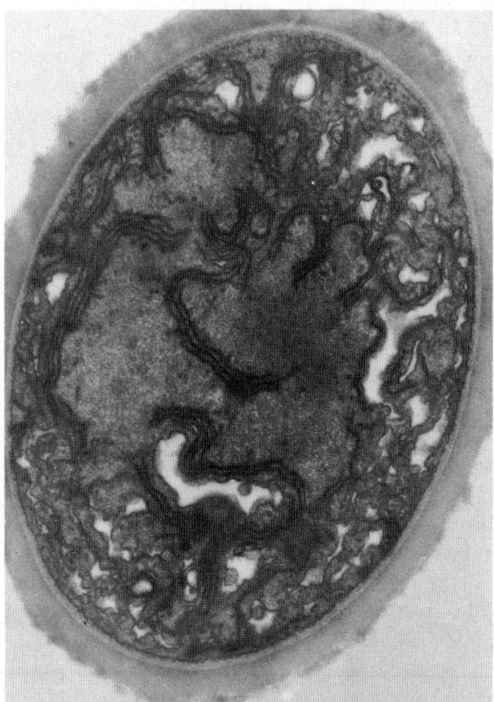

Fig. 1.2b Heterocyst of *Anabaena cylindrica* showing multilayered envelope surrounding the cell and extensive membrane deposition (×18 200). (From Kulasooriya, Lang and Fay,[347] Fig. 16, Plate 29. Reproduced by permission of The Royal Society.)

possession of heterocysts can no longer be considered as a morphological criterion for the classification of these organisms. More reliably, the presence of heterocysts is indicative of nitrogen-fixing ability in the free-living form and there is outstanding evidence to indicate that the heterocyst is the site of nitrogen fixation.[347] In the symbiotic association of blue-green algae with fungi, as occurs in the lichens however, the formation of the heterocyst is not a prerequisite for nitrogen fixation.[425]

The greatest amount of nitrogen fixation occurs in the symbiotic relationship of microorganisms with higher plants. This phenomenon is classically characterized by the nodule-bearing legumes. The nodules represent the sites of infection of the legumes by rhizobial bacteria.

The free-living bacteria occur in the soil rhizosphere. Entry of the bacterial symbiont into the host usually occurs by way of the root hair which usually becomes deformed, apparently owing to the production of auxin by the bacteria. The entry of the bacteria is apparently facilitated by the loosening of the cell walls in response to the auxin and by a digestion of the cell wall pectins by the enzyme polygalacturonase produced by host in response to the bacterial association. Under the light microscope the site of entry into the root hair can be identified as a small refractile spot.

After their entry into the hair, the bacteria become organized into a thread-like alignment and migrate directly toward the basal portion of the cell. The infection thread passes through the cells of the root cortex and is surrounded by a cellulose sheath. The matrix or zoogleal material in which the bacteria are embedded in the thread is of uniform density without visible structures and may be composed of bacterial mucilage. Each bacterium is embedded in a capsule in the infection thread; however, bacteria about to be released from the infection thread into the host cell have no detectable capsule.

The infection thread passes through or between host cortical cells until it enters the inner cortex. It has been generally accepted that the thread penetrated tetraploid cells which are dispersed in the host tissue as indicated in Fig. 1.3. However, more recent studies[377] indicate that entry is not necessarily into tetraploid cells and the polyploidy of the cortical cells may be induced by products secreted by the advancing infection thread. The bacteria are released into the cortical cell. At the point of release, the infection thread is surrounded by host cell plasmalemma but is devoid of cellulose wall and the bacteria are released into the host cell. It is not clear how this release occurs. One school[152, 224] of thought indicates that the bacteria as they are released from the infection thread are enclosed by the plant cell membrane which surrounds the infection thread in a type of pinocytosis. Others contend that the membrane which surrounds the bacteria arises as a result of *de novo* synthesis[135] or is derived from the endoplasmic reticulum of the host cell.[312] After being released from the infection thread, the bacteria divide. The plant membrane also divides so that each bacterium or a group of bacteria is surrounded by an envelope derived from the host. At the final stage of maturation the bacteria usually enlarge up to forty-fold and change form and undergo internal, structural and biochemical modification to produce the bacteroids (Fig. 1.4a, 1.4b).

Concurrent with the release of rhizobia from the infection threads, the invaded host cell and also several layers of non-invaded neighbouring cells undergo rapid cell division. Dissemination of the rhizobia is facilitated by the rapid mitotic division of the host cell. This increased cellular proliferation in the infected and adjacent cells results in the production of the characteristic nodules. The nodule can be divided into four distinct zones as follows: (a) A nodule cortex composed of undifferentiated non-infected parenchymatous cells; (b) A meristematic region which constitutes the growing point of the nodule and from which all the specialized tissue of the nodule originates; (c) The vascular system which functions to transport

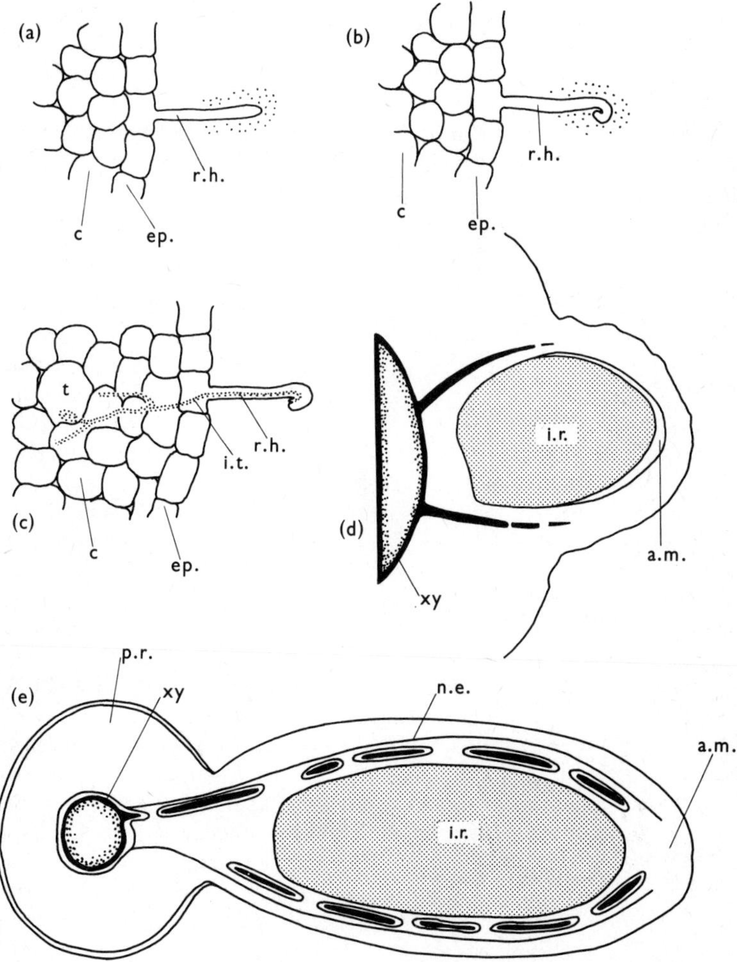

Fig. 1.3 Initiation and structure of pea nodules. (**a**) Rhizobia aggregate round root hairs. (**b**) Root hairs curl. (**c**) Rhizobia infect root hair and move through the root hair and inner cortex. This stimulates meristematic activity. (**d**) A central infected region and apical meristem become distinguished. (**e**) Longitudinal section through nodule showing central infected region, apical meristem and nodule endodermis. (a.m., apical meristem ; c., cortex ; ep., epidermis ; i.r., infected region ; i.t., infection thread ; n.e., nodule endodermis ; p.r., primary root ; r.h., root hair ; t., tetraploid cell ; xy, xylem.) From Stewart,[597] 1966, Fig. 2, p. 18. (Permission of Athlone Press.)

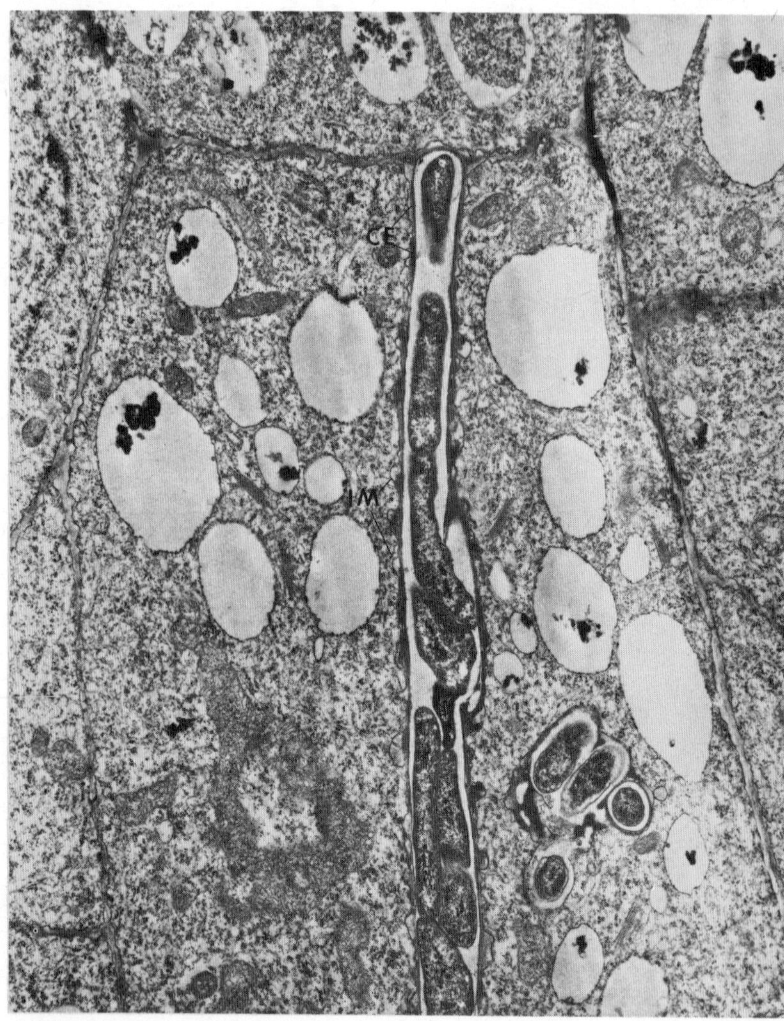

Fig. 1.4a Infection thread in soybean (*Glycine max*) root nodules (× 10 000). The thread is enclosed by cellulose (CE) and an infection thread membrane (IM). In soybean, unlike that in many other species, the infection thread is only transiently present and does not become filled with masses of zoogleal matrix. (From Goodchild and Bergersen,[224] Fig. 2, p. 207.)

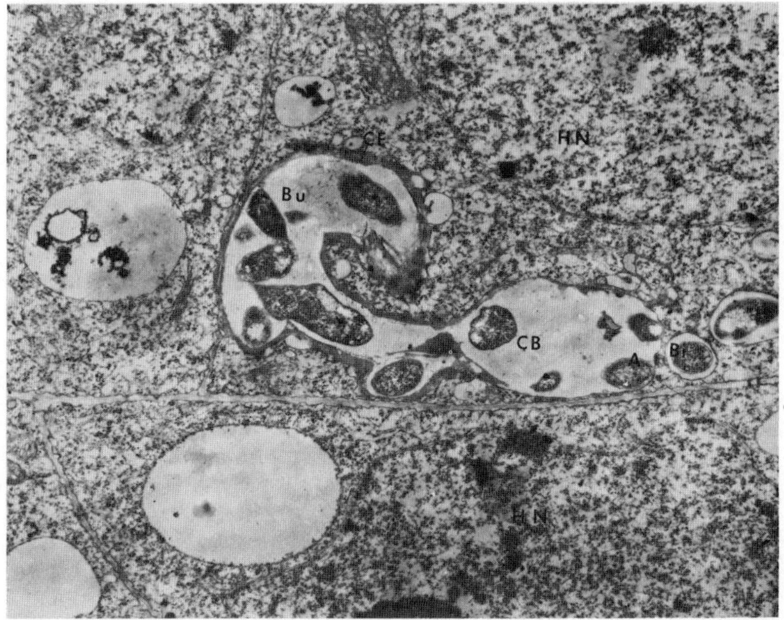

Fig. 1.4b Release of bacteria from an infection thread (× 8 000). A cellulose lined bulge (Bu) has developed and this has developed a cellulose-free bulge (CB). The bacteria within the latter appear to move into the cytoplasm by endocytosis after becoming applied to the inner surface of the membrane lining the bulge (A). HN, host nucleus ; Bi, bacterium with ingestion vacuole ; CE, infection thread cellulose. (From Goodchild and Bergersen,[224] Fig. 4, p. 208.)

plant nutrients to the nodule and in turn nitrogenous substances from the rhizobia to the host plant. This vascular tissue is continuous with that of the host plant and contains those components normally encountered in a root vascular system; (d) The central portion of the nodule is the bacteroidal zone. It is the site of the nitrogen fixation process and is composed principally of infected cells in which the rhizobia have undergone the characteristic transition to bacteroids. Associated with the production of bacteroids there is a production of the characteristic red pigment, leghaemoglobin. This pigment, which is according to current concepts synthesized in the host cell,[131] has been suggested to be located within the membrane envelope which surrounds the bacteroids.[48]

It is only following the establishment of the nodule with the commensurate differentiation of the bacteria into a bacteroid and the production of leghaemoglobin that nitrogen fixation is possible. Thus far, all attempts to induce nitrogen-fixing capabilities in free-living rhizobia have been unsuccessful.

In addition to the nitrogen fixation achieved by associated symbiotic rhizobia and legumes as indicated previously, an important contribution to nitrogen fixation is made by other symbionts associated with non-leguminous plants. Stewart[597] lists 10 genera of plants which possess nodules and have the capacity to grow on nitrogen-free media owing to their capacity to fix atmospheric nitrogen. Although many investigations have been carried out on these plants, particularly *Alnus*, *Ceanothus* and *Myrica*, the nature of the symbiont has not been determined and reports of the culture of the symbiont and subsequent re-inoculation have generally been unconfirmed. *In situ* studies using the electron microscope characterize the symbiont as an actinomycete. The endophyte is surrounded by a membrane of apparent host plant origin.[567] The nodules are pigmented; however, attempts to associate the pigmentation with haemoglobin have been unsuccessful.

Biochemistry of nitrogen fixation

Although historically the understanding of nitrogen fixation centred on studies of the root nodules of leguminous plants, current concepts of the biochemistry of the process have taken their lead from studies with the free-living nitrogen-fixing forms, notably *Azotobacter* and *Clostridium*. The increased utilization of the free-living forms is undoubtedly related to their relative ease of manipulation in contrast to the growth of plants to produce sufficient nodules required for experimentation.

In the 1940's with the availability of $^{15}N_2$ as a tracer for nitrogen fixation it was[87] demonstrated that fixation occurred in many bacteria provided that the assays were performed under anaerobic conditions. Additionally, the isotopic technique demonstrated that the product of nitrogen fixation is ammonia[457] and it was not possible to detect any other free intermediates in the reductive sequence. Original attempts to demonstrate N_2 fixation by cell-free extracts from the microorganisms *Azotobacter* and *Clostridium* were generally unconvincing and unrepeatable. However, in 1960 Carnahan *et al.*[97] were successful in demonstrating N_2 fixation by a cell-free preparation from *Clostridium pasteurianum*. The key to their success was that the assays were performed in the absence of oxygen with high levels of pyruvate which served, as was discovered later, to provide reductant and energy in the form of ATP. Unfortunately the discoveries with the *Clostridium* system were not immediately repeatable with extracts from *Azotobacter*. However, in 1965[85] cell-free extracts were prepared from *Azotobacter* which were capable of N_2 fixation. Adequate fixation was observed only in the presence of a low potential electron donor such as sodium hydrosulphite and a continuous supply of ATP. By utilizing this technique it has been possible to demonstrate N_2 fixation in cell-free preparations from other facultative anaerobes capable of N_2 fixation,[396] from heterocystic blue-green algae[256] and in cell-free preparations of bacteroids from soybean nodules.[340]

The capacity for nitrogen fixation in the cell-free extracts is attributed to the enzyme nitrogenase which can be fractionated by appropriate chromatography into two fractions, an Fe–Mo protein of molecular weight of 200 000 and an Fe protein of molecular weight 40 000 which is cold labile. The enzyme and the subunits from all sources examined are air-sensitive so that extensive purification involves anaerobic chromatography.

The functioning of ATP and the subunits in the reduction of N_2 by the nitrogenase is not elucidated; however current concepts of the process are illustrated in Fig. 1.5.

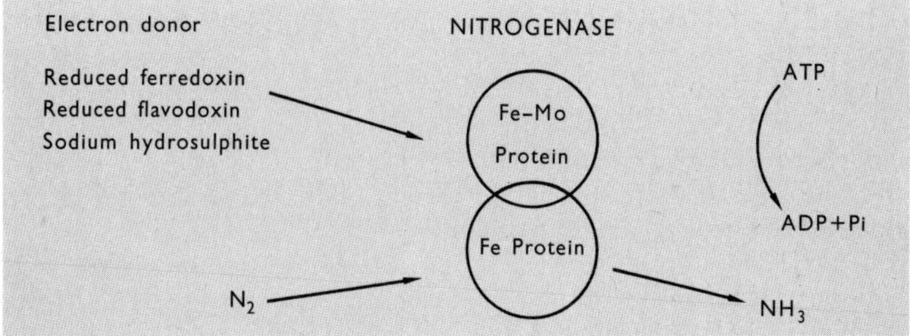

Fig. 1.5 Proposed scheme for biological nitrogen fixation by the enzyme nitrogenase. The reactions require ATP which is converted to ADP and orthophosphate; and a reducing reagent to function as electron donor.

The source of reductant utilized in nitrogen fixation is not completely resolved. In *Clostridium* it appears that the electrons are furnished from reduced ferredoxin[439] or flavodoxin generated in the phosphoroclastic reaction.[440]

1. Pyruvate $\xrightarrow[\text{Ferredoxin, Coenzyme A}]{\text{Thiamine pyrophosphate}}$ Acetyl CoA + Reduced ferredoxin + CO_2

2. Acetyl CoA + Phosphate $\rightleftharpoons$ Acetyl phosphate + CoA

3. Acetyl phosphate + ADP $\rightleftharpoons$ Acetic acid + ATP

The possibility that Acetyl CoA produced in (1) could generate ATP via reactions (2) and (3) was not initially recognized and whilst the role of reduced ferredoxin in nitrogen fixation was established there was initially some confusion concerning the involvement of ATP. This confusion was further confounded initially by the observation[97] that high levels of ATP inhibit nitrogenase activity. In current assays of nitrogenase it is customary to provide ATP by means of a generating system which provides a sustained supply at a low but effective concentration.

However, with the demonstration that nitrogen fixation by cell-free extracts of *Clostridium pasteurianum* could be inhibited by arsenate or glucose plus hexokinase[417] and that reductants such as dithionite or hydrogen could not function in the nitrogen fixation process in the absence of ATP, it became clear that the nucleotide was a necessary component of the nitrogen fixation reaction. The general consensus is that 4–5 molecules of ATP are required per each electron pair transferred.[133]

While the phosphoroclastic cleavage of pyruvate provides a mechanism for the generation of reductant, ferredoxin, in *Clostridium* the reaction does not occur universally and other mechanisms of the generation of reductant must be established.

It has been demonstrated that reduced ferredoxin can function as the electron donor for nitrogenase in the blue-green alga *Anabaena cylindrica*.[573] *In vitro* studies indicated that ferredoxin could be reduced by illuminating subcellular particles in the presence of ascorbate and dichlorophenol indophenol (DCPIP). In this system ascorbate can serve as a donor of electrons and they are transferred through the DCPIP into the photosynthetic electron chain and can ultimately reduce NADP by way of ferredoxin (Fig. 1.6). However it is not clear where the electrons for ferredoxin reduction are generated *in vivo*. In higher plants ferredoxin and ultimately NADP are reduced by electrons originally derived from water

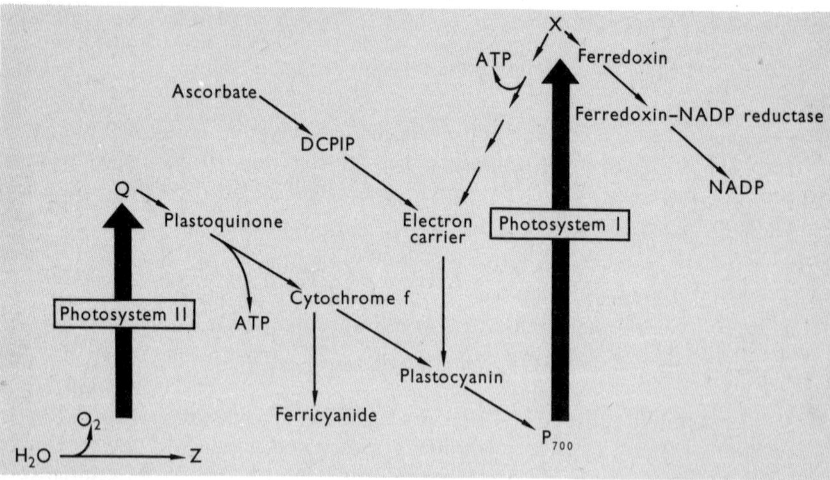

Fig. 1.6 Scheme of photosynthetic electron transport and photophosphorylation in higher plants.

and involves the activities of Photosystem II and Photosystem I (Fig. 1.6). Oxygen is evolved during reactions involving photosystem II. However investigations[624] have indicated that the heterocyst of *Anabaena* is deficient in pigments involved in photosystem II. This arrangement would prevent the evolution of oxygen which is inhibitory to nitrogenase activity and provide for a mechanism of generating required ATP by cyclic photophosphorylation; however it fails to generate the required reductant.

Ferredoxin can also be reduced by $NADPH^+$ by a reversal of NADP ferredoxin reductase.[611] Thus, if there is a means of producing $NADPH^+$ it is possible that the reduced nucleotide could indirectly serve as an electron donor for nitrogen fixation. In the non-photosynthetic aerobic bacterium *Azotobacter vinelandii* $NADPH^+$ appears to be generated by isocitrate dehydrogenase activity. The electrons are transferred through a specific azobacter ferredoxin to an azoflavodoxin which is the immediate electron donor for nitrogenase in this bacterium.[611]

It is possible to transfer the reducing power generated by illuminated chloroplasts or from $NADPH^+$ to nitrogen fixation in cell-free extracts[707,340] of bacteroids of *Rhizobium* in the presence of various cell constituents. However the characteristics of the cell constituents, functional in electron transfer, and their physiological role are unknown.

In symbiotic rhizobial systems it appears that the reductant may be derived by the oxidative metabolism of photosynthetic products produced by the host. It was suggested[339] that the metabolism of poly β-hydroxybutyrate, which accumulates in the bacteroids, could provide reductant. Although an NAD-dependent β-hydroxybutyrate dehydrogenase has been demonstrated in soybean nodules[706] there is as yet no satisfactory system to explain the possible transfer of electrons from $NADH^+$ to nitrogenase. More recent investigations indicate that poly β-hydroxybutyrate does not have a direct role as energy source for the maintenance of nitrogenase activity.

In addition to the requirement for a reductant in the conversion of nitrogen to ammonia in the nitrogen fixation step, there is also a need for a supply of ATP. As indicated previously in respect of *Clostridium*, this ATP can be generated during the metabolism of pyruvate to acetate. It has been suggested that in photosynthetic systems showing nitrogen fixation that ATP could be provided by photophosphorylation. In other organisms the ATP is apparently generated during the oxidation of respiratory substrates. Such a situation creates a paradox in which adequate oxygen is required to sustain the high rates of respiration to provide adequate ATP; at the same time, however, the nitrogenase is inhibited by oxygen. It appears that *in vivo* these conflicting situations are reconciled by various mechanisms. For *Azotobacter* it has been proposed[504] that the high respiratory activity for which this organism is renowned represents a physiological mechanism for preventing the nitrogen-fixing mechanism from damage by oxygen. The oxygen is consumed and scavenged during respiration. In the free-living blue-green algae it has been suggested[599] that the heterocyst wall prevents oxygen entry and allows nitrogen fixation to occur in such cells. It has been demonstrated[600] that nitrogenase activity can be induced in certain blue-green algae without heterocyst formation provided that the organism is maintained under strictly anaerobic conditions. Significantly in the mutualistic association with the fungi in lichens the blue-green algae again fix nitrogen without heterocyst formation.[425] It is proposed that respiratory activities of the symbiotic fungi reduce oxygen concentration sufficiently to prevent oxygen inactivation of nitrogen fixation.

In the symbiotic legume nodule it is proposed[48] that the leghaemoglobin surrounding the bacteroids serves a dual role, one of facilitating oxygen diffusion to the bacteroids and thus maintaining an adequate level of respiration to produce ATP required for nitrogen fixation. As a result of the ready formation of the oxygenated derivative of leghaemoglobin, it follows that the concentration of free oxygen in the vicinity of the bacteroid will be low and thus an alternate or ancillary role of leghaemoglobin is to scavenge oxygen from the nitrogen-fixing site and thereby prevent inhibition of nitrogenase.

Utilization and fate of fixation products

The availability of isotopically labelled $^{15}N_2$ provided a mechanism for tracing the path of nitrogen in the fixation process. The initial product was found to be NH_3. However, this ammonia rapidly becomes incorporated into organic combination in amino acids and amides. The carbon skeletons required for the formation of the amino acids are presumably derived from the respiratory metabolism of carbohydrates. The prime point of ammonia entry into organic combination was thought to be via condensation with α-ketoglutarate to form glutamic acid as occurs in higher plants (p. 26). However, recent evidence from various free-living bacteria,[611] extracts of *Rhizobium* bacteroids,[448] and blue-green algae[149] indicates the operation of an alternate pathway. Ammonia is initially converted to glutamine by the enzyme glutamine synthetase (p. 26). This glutamine is then deaminated in the presence of α-ketoglutarate by the recently identified[611] enzyme glutamate synthetase, with the resulting formation of glutamate.

$$NH_3 + \text{glutamate} + \text{ATP} \xrightarrow[\text{synthetase}]{\text{glutamine}} \text{glutamine} + \text{ADP} + \text{Pi}$$

$$\text{Glutamine} + \alpha\text{-ketoglutarate} + \text{NADPH}^+(\text{NADH}^+) \xrightarrow[\text{synthetase}]{\text{glutamate}} 2\ \text{glutamate} + \text{NADP(NAD)}$$

NITRIFICATION

Ammonia can be released into the soil either as a result of N_2 fixation by free-living microorganisms or from the decomposition of dead organisms. This ammonia, as discussed later, can be taken up directly by plants or it may be oxidized to nitrate and nitrite. This biological oxidation of ammonia to nitrite and nitrate is termed **nitrification** and is catalysed by bacteria and some fungi. The oxidation of ammonia is achieved by bacteria belonging to the Nitroso group of genera which convert ammonia from the -3 oxidation state of nitrogen to the $+3$ situation of nitrite. The oxidation is assumed to occur via 2 electron steps with the production of intermediates of -1 and $+1$ oxidation states.

On the evidence that intact *Nitrosomonas* cells would not oxidize ammonia under anaerobic conditions and that hydroxylamine and not nitrite accumulates when cells are treated with hydrazine,[269] it appears that the first stage of nitrification involves the oxidation of ammonia to hydroxylamine.

$$NH_3 + \tfrac{1}{2}O_2 \longrightarrow \underset{\text{Hydroxylamine}}{NH_2OH}$$

The mechanisms of this oxidation are not known at this time. It has not been possible to demonstrate ammonia oxidation by cell-free extracts; however, in view of the fact that ammonia oxidation by intact cells is inhibited by thiourea[357] and chelating agents, it is possible that a copper-containing oxidase may be operative.

Hydroxylamine to nitrite

Intact *Nitrosomonas* can readily oxidize hydroxylamine to nitrite. This oxidation can also be demonstrated in cell-free extracts of the organism provided that a suitable electron carrier such as cytochrome *c* is present.[459] This cell-free oxidation is stimulated by flavine mononucleotide (FMN) and inhibited by atabrine and cyanide and carbon monoxide. In addition, it can be demonstrated that the metabolism of hydroxylamine by intact cells is associated with changes in the oxido-reduction status of cytochrome *b*, *c*, and *a*.[10] On the basis of this evidence, it is considered that the metabolism of hydroxylamine involves a dehydrogenation with a transfer of electrons through a flavoprotein to the cytochrome electron transport chain.

$$NH_2OH \dashrightarrow \text{FLAVIN} \underset{\text{ATABRINE}}{\dashrightarrow} \text{cyt } b \dashrightarrow \text{cyt } c \dashrightarrow \text{cyt } a \underset{\substack{CN\\CO}}{\dashrightarrow} O_2$$

The product arising from the dehydrogenation of hydroxylamine has not been isolated or characterized, but it is believed to be nitroxyl (NOH) in which the nitrogen atom is in the +1 oxidation state. Under aerobic conditions, the nitroxyl can be converted to nitrite. According to Aleem *et al.*[9] this conversion is achieved in a semi-cyclic process involving the oxidation of a further intermediate nitrohydroxylamine, $NO_2.NHOH$. The oxidation of nitrohydroxylamine can be demonstrated using cell-free extracts. Oxidation occurs in the absence of added electron acceptors, and although cytochrome *c* is reduced during the oxidation of hydroxylamine the electron carrier is not reduced during the stoichiometric conversion of nitrohydroxylamine to nitrite. Although nitrohydroxylamine has not as yet been identified as an intermediate of hydroxylamine oxidation, the studies of Aleem[7] convincingly imply its functioning in the sequence outlined below.

$$NH_2OH + 2\text{ cyt } c\ Fe^{3+} \longrightarrow (NOH) + 2\text{ cyt } c\ Fe^{2+} + 2H^+$$
$$(NOH) + HNO_2 \longrightarrow NO_2.NHOH$$
$$NO_2.NHOH + \tfrac{1}{2}O_2 \longrightarrow 2HNO_2$$
$$2\text{ cyt } c\ Fe^{2+} + 2H^+ + \tfrac{1}{2}O_2 \longrightarrow 2\text{ cyt } c\ Fe^{3+} + H_2O$$

$$\text{Sum} \quad NH_2OH + O_2 \longrightarrow HNO_2 + H_2O$$

Nitrite to nitrate

The conversion of the nitrite, arising from the activities of the Nitroso group of genera, to nitrate involves the oxidation of nitrite. Bacteria belonging to the Nitro group of genera *Nitrobacter* and *Nitrocystis* are capable of this conversion.

In intact *Nitrobacter* treated with nitrite, cytochrome *c* and cytochrome oxidase-like components become reduced[358] suggesting that nitrite oxidation requires the participation of an electron carrier cytochrome system. Such an oxidative system can also be demonstrated[11] using cell-free cytochrome electron particles (nitrite oxidase) prepared from *Nitrobacter*, which can transfer electrons from nitrite to molecular oxygen according to the scheme.

$$NO_2 \longrightarrow \text{cyt } a_1 \longrightarrow \text{cyt } a_3 \longrightarrow O_2$$

Since nitrite oxidation is mediated by the electron transfer chain, the terminal acceptor oxygen must be reduced to water; however, clearly nitrite, the substrate which apparently stimulates electron flow, has no electrons to donate. Accordingly, it is suggested that the true substrate for nitrite oxidase is a hydrated form of nitrite and thus the oxidation of the nitrite molecule is achieved at the expense of an oxygen atom from water and not from atmospheric oxygen.

This feature was elegantly demonstrated[8] by the use of ^{18}oxygen and $H_2{}^{18}O$. When nitrite is oxidized in the presence of $^{18}O_2$ little of the isotope is recovered in nitrate. If the oxidation occurs in $H_2{}^{18}O$ the nitrate produced is extensively enriched in ^{18}O. Furthermore, the oxidation of nitrite to nitrate can occur in intact cells or cell-free extracts in the presence of an artificial electron acceptor in the absence of oxygen. Thus, it is convenient to depict the oxidation of nitrite as follows:

$$NO_2{}^- + H_2{}^{18}O \longrightarrow NO_2\text{—}H_2{}^{18}O$$

$$NO_2\text{—}H_2{}^{18}O + 2 \text{ cyt } a_1 \text{ } Fe^{3+} \longrightarrow N^{18}O_3 + 2 \text{ cyt } a_1 \text{ } Fe^{2+} + 2H^+$$

$$2 \text{ cyt } a_1 \text{ } Fe^{2+} + 2H^+ + \tfrac{1}{2}O_2 \longrightarrow 2 \text{ cyt } a_1 \text{ } Fe^{3+} + H_2O$$

Sum $NO_2 + \tfrac{1}{2}O_2 \longrightarrow NO_3 + \text{Energy}$

Chemosynthesis

The nitrifying bacteria are chemosynthetic autotrophs deriving the energy required for their biosynthetic synthesis from the oxidation of ammonia and nitrite. From the schemes outlined above, it is clear that energy derived from the oxidations could be conserved as the high energy bonds of ATP which could be generated as electrons flow through the cytochrome components of the electron transfer chain. The cytochrome electron transport system of *Nitrobacter* is somewhat peculiar in that during the oxidation of nitrite phosphorylation is achieved without the involvement of cytochrome *c*.

$$H_2O.NO_2 \longrightarrow \text{cyt } a_1 \longrightarrow \text{cyt } a_3 \xrightarrow{\text{ADP + Pi} \;\curvearrowright\; \text{ATP}} O_2$$

In *Nitrosomonas*, ATP is generated at conventional locations in the electron transfer chain between cyt *b* and cyt *c* and during the oxidation of reduced cyt *c* by cytochrome oxidase.

$$NH_2OH \longrightarrow \text{cyt } b \xrightarrow{\text{ADP + Pi} \;\curvearrowright\; \text{ATP}} \text{cyt } c \xrightarrow{\text{ADP + Pi} \;\curvearrowright\; \text{ATP}}$$

$$\text{cyt oxidase} \longrightarrow O_2$$

The fixation of carbon dioxide carried out by these autotrophs, in addition to ATP, requires reduced pyridine nucleotides. However, as depicted, the oxidation of ammonia by the Nitroso group or nitrite by the Nitro group does not result in the generation of reductant. Instead, it appears that reductant is generated by energy-linked reversed electron flow in which NAD reduction is achieved at the expense of ATP during the oxidation of ammonia or nitrite.

$$NO_2 \longrightarrow \text{cyt } a_1 \xrightarrow{\text{ATP}} \text{cyt } c \xrightarrow{\text{ATP}} \text{cyt } b \longrightarrow$$

$$\text{flavoprotein} \xrightarrow{\text{ATP}} \text{NAD}$$

Pathway of energy-linked transfer of electrons in *Nitrobacter*.

Clearly, the bioenergetics and chemosynthesis of *Nitrobacter* and *Nitrosomonas* pose extremely complex problems with regard to energy budgets; some of these are discussed at length by Gibbs and Schiff,[212] Aleem,[7] and Wallace and Nicholas.[664]

DENITRIFICATION

The nitrate produced as a result of the activities of the nitrifiers *Nitrosomonas* and *Nitrobacter* can be taken up by higher plants and further metabolized (see later). Additionally, however, the nitrate can be converted to nitrogen gas or nitrous oxide, or a mixture of these two gases in a process termed **denitrification**. The gases return to the atmosphere and thus denitrification represents a mechanism by which soil nitrogen is depleted.

Characteristically the denitrifying organisms are bacteria which can facultatively utilize nitrite or nitrate rather than oxygen as the terminal hydrogen acceptor in the energy-yielding electron transfer. Denitrification occurs most readily under anaerobic conditions. The process is inhibited by oxygen because this gas effectively competes with nitrite or nitrate as the terminal electron acceptor.

The first stage in denitrification involves the reduction of nitrate to nitrite. The enzyme involved in this reaction has been termed respiratory nitrate reductase and is, in contrast to the assimilatory nitrate reductase, a particulate enzyme. Particulate respiratory or dissimilatory nitrate reductase has been prepared from a variety of organisms and it can be demonstrated that the conversion of nitrate to nitrite is coupled with the generation of ATP.[490] The transfer of reductant in these nitrate reductases appears to be mediated by cytochromes and molybdenum.

Generalized scheme for respiratory nitrate reductase:

Substrate ⟶ reduced NAD (P) ⟶ FAD ⟶ cytochrome ⟶ cytochrome oxidase ⟶ O_2

cytochrome | Nitrate reductase ↓ NO_3^- ↓ NO_2^-

Nitrite reduction

The nitrite produced can be further metabolized to nitrogen or nitrous oxide by a variety of microorganisms. The sequence of intermediates and enzymes involved in these conversions are still incompletely resolved.

By utilizing whole cells and a suitable respiratory substrate, it can be demonstrated that under anaerobic conditions nitrite can be converted to nitrogen gas.[408] The production of nitrogen gas by the intact cells can be prevented by azide but there is a continued release of nitrous oxide.[408] These observations suggest that nitrous oxide may be a key intermediate in the conversion of nitrite to gaseous nitrogen. In some organisms, *Corynebacterium nephridii* for example, nitrite reduction appears to proceed only as far as nitrous oxide.[520] Nitrogen gas or nitrous oxide production is stimulated in many instances by the addition of hydroxylamine. This observation raises the possibility that the metabolism of nitrite by these denitrifying organisms involves hydroxylamine as an intermediate. However, current ideas favour the concept that hydroxylamine is not a functional intermediate in the conversion but can interact with as yet unidentified products to enhance gas production. The sequence of reactions involved has been proposed as follows:[407]

$$HNO_2 \xrightarrow[\text{Denitrifying enzyme}]{} NO \xrightarrow[\text{Nitric oxide reductase}]{} N_2O \xrightarrow[\text{Nitrous oxide reductase}]{} N_2$$

The initial metabolism of nitrite has been demonstrated in cell-free extracts and is mediated by an enzyme variously termed **denitrifying enzyme**[300] or **nitrite reductase.**[299] In certain species the enzyme contains cytochromes with two haem derivatives[458, 408] whereas in other species the enzyme is apparently a copper-containing protein.[299] The products of nitrite reduction by the cell-free extracts have been identified as nitric oxide.

The nitric oxide can be further metabolized to nitrous oxide by a particulate fraction under anaerobic conditions in the presence of lactate.[431] However, the electron donor component at this reduction step has not as yet been identified.

The subsequent conversion of nitrous oxide to nitrogen has not been achieved in cell-free systems. However,[408,431] the observed sensitivity of the reduction to azide in intact cells appears to demonstrate the functioning of metal ions in this step.

NITRATE METABOLISM

Nitrate taken up by the roots of the plant must be reduced to ammonia prior to its availability in the overall nitrogen economy of the plant. This reduction can occur in both the plant roots and leaves. Evidence to support the reduction of nitrate by roots is based upon the observation that in certain species nitrate is not present in the xylem exudate from cut stems even though the ion is present in the medium surrounding the roots. Additionally, excised roots can be maintained in culture with nitrate serving as the sole nitrogen source. In situations of adequate nitrate fertilization, nitrate can usually be detected in the aerial portions of plants and the greater levels of enzymes involved in nitrate metabolism in the leaf as opposed to root tissue suggests that the leaves may in many instances be the principal location of nitrate reduction. The reduction of nitrate to ammonia is achieved in two stages involving the participation of the enzymes nitrate reductase and nitrite reductase.

Nitrate reductase

The initial metabolism of nitrate to nitrite can be readily demonstrated *in vivo* by incubating plant tissue in the presence of nitrate in the dark under anaerobic conditions. Under such conditions nitrite accumulates in the tissue and the surrounding medium. Nitrite accumulation is enhanced by infiltrating the tissue with sugar phosphates[337] and is prevented by aeration.

The conversion of nitrate to nitrite is mediated by the enzyme nitrate reductase which in the presence of reduced pyridine nucleotide is capable of reducing nitrate to nitrite.

The enzyme catalysing this reaction was originally isolated from *Neurospora*.[453] Subsequently, a similar enzyme has been demonstrated in extracts from a variety of plant species. In the majority of instances nitrate reductase from higher plants shows a specific requirement for $NADH^+$ as electron donor.[41] In a few instances the reduction can be catalysed by $NADPH^+$; however in these situations it appears that the supplied electron donor $NADPH^+$ is converted to $NADH^+$ by a nucleotide phosphatase and in actuality it is $NADH^+$ which functions as electron donor.[678, 678a]

In addition to reduced pyridine nucleotides, reduced flavins $FMNH_2$ or $FADH_2$ can serve as electron donors in the reduction process. Nitrate reduction in crude plant extracts utilizing $NADH^+$ is often enhanced by the inclusion of FMN or FAD. Although preparations from higher plants have not been purified sufficiently to establish the fact unequivocally, it is generally considered that nitrate reductase is a flavoprotein.[42]

In 1939 Arnon and Stout[19] demonstrated the essentiality of molybdenum as a micro-nutrient for higher plants. Subsequently, it has been established that such molybdenum-deficient plants accumulate high concentrations of nitrate.[263] This accumulation of nitrate can be associated with a reduced capacity for nitrate reduction and a low nitrate reductase. Recent evidence has demonstrated that radio-active molybdate is incorporated into protein fractions possessing nitrate reductase activity and it appears that molybdenum is an essential component of the enzyme.[18, 465]

In addition to reduced pyridine nucleotides, a range of non-physiological chemicals such as reduced viologen dyes or dithionite can serve as electron donors in the reduction of nitrate. In purified extracts $NADPH^+$ can serve as an electron donor only in the presence of a nucleotide phosphatase or a flavoprotein NADP reductase and FMN or FAD. Purified nitrate reductase can reduce cytochrome *c* and possesses diaphorase activity (i.e. it can transfer electrons from $NADH^+$ to various synthetic dyes). The enzyme is currently considered to consist of two combined entities: a component which transfers electrons from the reduced pyridine nucleotide to the flavin-containing component, and a subunit transferring electrons from the flavin by way of molybdenum to nitrate.

The current concept of the enzyme nitrate reductase, the molecular weight of which has been estimated at 500 000 to 600 000, is depicted in Fig. 1.7.

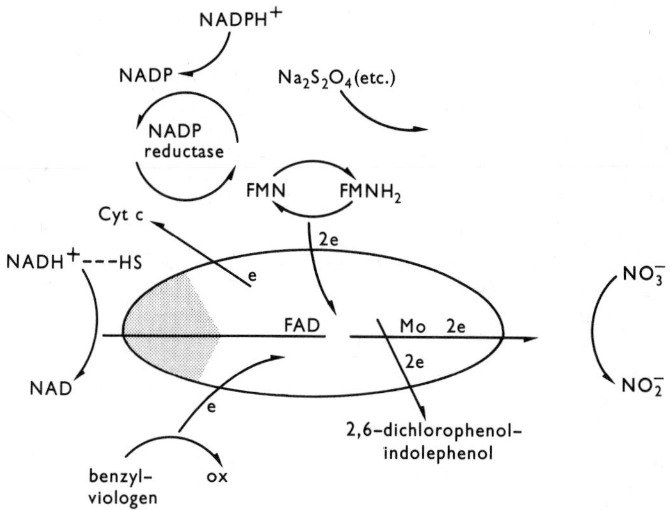

Fig. 1.7 An illustration of various donors of electrons to and acceptors of electrons from nitrate reductase and the proposed internal electron transport path. *In vivo*, $NADH^+$ is considered to be the electron donor for nitrate reduction. (Modified from Beevers and Hageman,[41] p. 498.)

All attempts to separate physically the two subunits in enzyme preparations from higher plants have failed. However, in molybdenum-deficient conditions or in the presence of tungstate, which competes with molybdenum, an inactive nitrate reductase is formed but the enzyme does possess diaphorase activity and is capable of reducing cytochrome *c*.[466,709] In *Neurospora* a series of mutants have been developed which appear to be deficient in one of the subunits and active nitrate reductase can be formed by appropriate complementation experiments.[452]

Nitrate utilization in plants is influenced by many environmental conditions. Particularly striking is the influence of illumination. At low light intensities nitrate accumulates in leaves of higher plants. It was originally speculated that this nitrate accumulation was due to a reduced photosynthetic production of reduced pyridine nucleotide under shaded conditions. However, the reductant for nitrate reduction is $NADH^+$ whereas the prime product of photosynthesis is $NADPH^+$ which is produced in the chloroplast and does not readily traverse the chloroplast membrane.[257] These observations tend to preclude the direct involvement of photosynthetically reduced pyridine nucleotide in nitrate reduction.

More convincingly, it appears that the $NADH^+$ is generated in the cytoplasm, the site of location of nitrate reductase. As a result of photosynthetic carbon dioxide fixation sugars and sugar phosphates are produced in the chloroplast. These intermediates can traverse the chloroplast membrane into the cytoplasm and be metabolized in glycolysis. Glyceraldehyde-3-phosphate produced in glycolysis or released from the chloroplast as a result of photosynthetic activity is oxidized to phosphoglyceric acid in the cytoplasm by the enzyme glyceraldehyde-3-phosphate dehydrogenase with the concomitant reduction of NAD to $NADH^+$. It is envisaged that the cytoplasmically generated $NADH^+$ furnishes the electrons for nitrate reduction to nitrite.[337] The proposed sequence is depicted in the scheme shown in Fig. 1.8.

This scheme is attractive in that it accounts for the intimate relationship between nitrate metabolism and photosynthesis.

Extensive studies with root tissues by Yemm and Willis[697,716] readily demonstrated the ability of non-green tissue to metabolize nitrate. More recent investigations with cell tissue culture also demonstrate the ability of non-green tissue to utilize nitrate.[709] In studies with roots it was indicated that nitrate utilization was associated with a changed respiratory metabolism. The utilization of reduced pyridine nucleotide produced during glycolysis for nitrate reduction as depicted in the above scheme adequately supports these observations and provides a mechanism for electron generation for the reduction of nitrate in non-green tissue as well as photosynthetic tissues.

In addition to influencing the supply of photosynthate from which reductant for nitrate reductase can be generated, illumination also directly affects the tissue level of the enzyme nitrate reductase.[42] As discussed later, nitrate reductase is an inducible enzyme and light can influence nitrate metabolism through effects on the synthesis of the enzyme nitrate reductase.

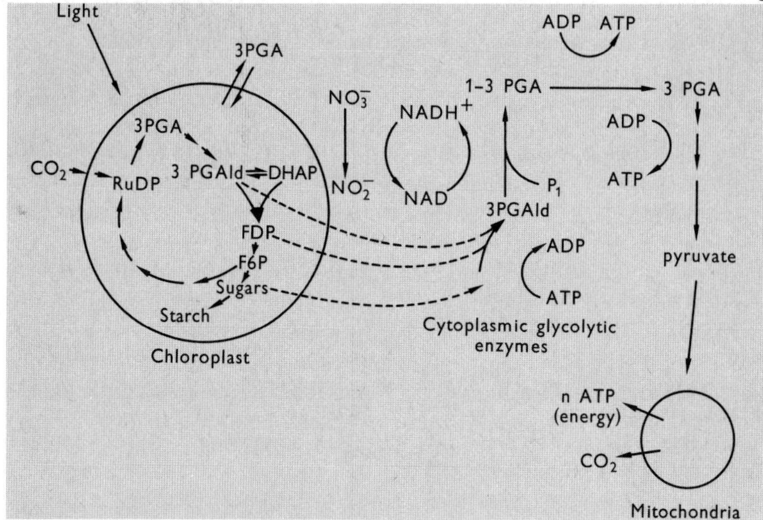

Fig. 1.8 Interrelationship between photosynthesis, glycolysis, and nitrate reduction. The $NADH^+$ generated would be used by other $NADH^+$-requiring cytoplasmic reactions as well as for nitrate reduction. (From Klepper, Flesher and Hageman,[337] Fig. 10, p. 589.)

Nitrite reductase

The nitrite produced by the activities of the enzyme nitrate reductase rarely accumulates in plant tissues under normal aerobic conditions. Usually the nitrite is converted to ammonia by the enzyme **nitrite reductase** which is present in considerably higher quantities than nitrate reductase.

The first convincing demonstration of the occurrence of nitrite reductase in extracts of higher plants was made by Hageman *et al.*[242] These workers were able to demonstrate that plant extracts, when fortified with reduced benzyl viologen as electron donor, were capable of stoichiometrically converting nitrite to ammonia. The crude enzyme preparations also possessed the capacity for hydroxylamine reduction and superficially it appeared that nitrate reduction could proceed according to the scheme originally proposed by Meyer and Schulze[421] in which hyponitrite and hydroxylamine functioned as intermediates.

The rate of reduction of hydroxylamine in the crude preparations was slower than that of nitrite; furthermore, hydroxylamine did not inhibit or retard nitrite reduction.

More recently, extensively purified preparations have been obtained which are active in nitrite reduction but are devoid of hydroxylamine reductase.[277] It is now believed that the reduction of nitrite to ammonia is catalysed by one protein nitrite reductase in which the iron porphyrin prosthetic group, siroheme, of the enzyme functions in the transfer of six electrons.[446a] There are no free intermediates.

Extensive studies have been made to determine the physiological electron donor for nitrite reductase. Preliminary investigations indicated that illuminated chloroplasts could catalyse the reduction of nitrite when fortified with an additional soluble enzyme component. Subsequently, it was shown that ferredoxin[390, 262] could replace reduced viologen dyes as electron donor for the enzymatic reduction and thus it appears that illuminated chloroplasts functioned in nitrite reduction by producing a source of reduced ferredoxin.

The discovery that ferredoxin, which could be reduced photochemically during non-cyclic photophosphorylation[494] or by $NADPH^+$ in the presence of ferredoxin NADP reductase,[313] could function as an electron donor provided a convenient and apparently satisfactory explanation for the frequently observed[651] acceleration of nitrite utilization in leaves by light. Moreover, subcellular fractionations indicate that nitrite reductase is located in the chloroplast.[524] These observations can be conveniently fitted into a scheme as follows:

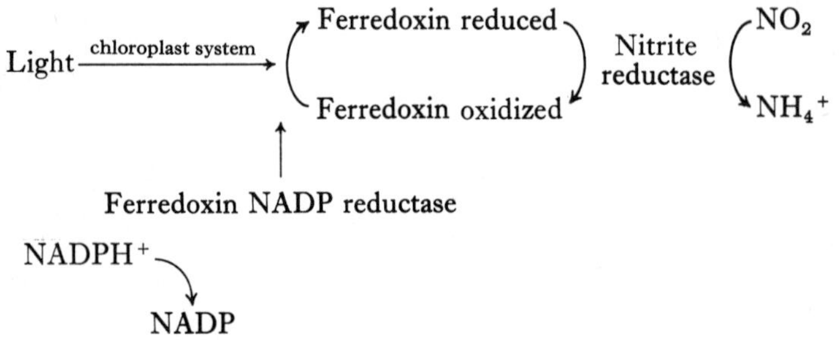

Light —chloroplast system→ Ferredoxin reduced / Ferredoxin oxidized — Nitrite reductase — NO_2 → NH_4^+

↑

Ferredoxin NADP reductase

$NADPH^+$ → NADP

Although this scheme has been readily accepted, there are still some aspects of nitrite metabolism which it does not accommodate.

Nitrite is readily metabolized by root tissues which are devoid of chloroplasts and which so far as can be determined lack ferredoxin.

Nitrite reductase has been found[132] to be associated with proplastids in root tissues and these organelles are enriched in enzymes of the pentose phosphate pathway which could produce $NADPH^+$ through the activities of glucose-6-phosphate dehydrogenase and 6-phosphogluconate dehydrogenase. However, $NADPH^+$ is unable to function directly as an electron donor for nitrite reduction. So far, no intermediate component mediating the possible transfer of electrons from $NADPH^+$ to nitrite reductase has been isolated from roots.

Nitrite utilization *in vivo* is prevented by anaerobiosis, a feature which allows for the *in vivo* assay of nitrate reductase. Upon aeration nitrite is rapidly utilized. These findings imply that nitrite metabolism is directly dependent upon aerobic metabolism and that nitrite does not function as a terminal electron acceptor for respiratory generated reductant. The aerobic metabolism of nitrite is prevented by uncouplers of oxidative phosphorylation and it has been speculated that the role of respiration is to provide high energy phosphate required for nitrite reduction. However, to date no energized intermediates of nitrite which might be required for its metabolism have been isolated and the ready reduction of free nitrite in cell-free systems tends to rule out the involvement of 'active' or 'energized' nitrite.

THE ENTRY OF AMMONIA INTO ORGANIC COMBINATION

Detailed investigations by Yemm and Willis in the 1950's[697, 716] indicated that the respiratory metabolism of roots of barley (*Hordeum vulgare*) was markedly influenced by nitrogen metabolism. Transfer of excised nitrogen-deficient roots to ammonium salts, or nitrate or nitrite resulted in a rapid increase in CO_2 evolution and increased oxygen uptake (Fig. 1.9).

Similar respiratory changes are observed when nitrogen-deficient *Chlorella* is transferred to medium enriched with ammonia, nitrate or

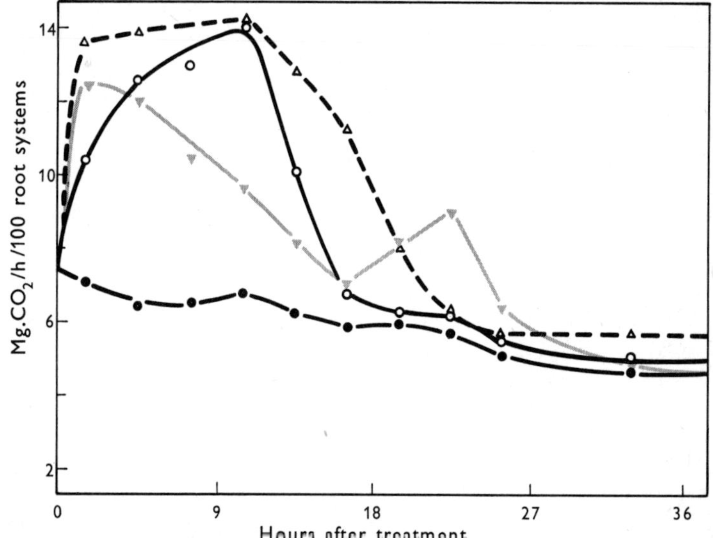

Fig. 1.9 The effect of different sources of nitrogen on carbon dioxide production by roots of 14-day-old barley seedlings. Sodium nitrate ○——○; sodium nitrite △– –△; ammonium dihydrogen phosphate ▼——▼; culture solution lacking nitrogen ●——●. From Willis and Yemm.[697]

nitrite.[617] In the studies of barley utilizing ^{15}N isotopic nitrogen[716] it was established that the high rates of respiration encountered during nitrogen nutrition were associated with a rapid synthesis of nitrogen compounds in the tissue. A massive accumulation of the amides glutamine and asparagine was accompanied by a rapid loss of tissue carbohydrate. The glutamine was labelled in both the amino and amide group; aspartic acid and asparagine, although labelled, had a much lower abundance of ^{15}N than glutamine and it was thus considered that they were not primary products of nitrogen assimilation.

The physiological observations can be adequately accounted for with the demonstration of enzymes capable of catalysing the appropriate syntheses. Until recently it was considered that the principal and possibly only modes of entry of ammonia into organic combination in higher plants were through glutamine and glutamate involving the activities of glutamine synthetase and glutamic dehydrogenase.

The incorporation of ammonia into glutamine[654] is mediated by the enzyme glutamine synthetase according to the scheme

$$\begin{array}{l} COOH \\ | \\ CHNH_2 \\ | \\ CH_2 \\ | \\ CH_2 \\ | \\ COOH \\ \text{Glutamate} \end{array} \begin{array}{c} \\ \\ \\ \\ + NH_3 + ATP \underset{\text{glutamine synthetase}}{\overset{Mg^{2+}}{\rightleftharpoons}} \\ \\ \\ \\ \\ \\ \end{array} \begin{array}{l} COOH \\ | \\ CHNH_2 \\ | \\ CH_2 + ADP + Pi \\ | \\ CH_2 \\ | \\ CONH_2 \\ \text{Glutamine} \end{array}$$

In the presence of glutamic dehydrogenase ammonia combines with α-ketoglutarate in a reductive amination to produce glutamate.

$$\begin{array}{l} COOH \\ | \\ CO \\ | \\ CH_2 \\ | \\ CH_2 \\ | \\ COOH \\ \alpha\text{-Ketoglutarate} \end{array} \begin{array}{c} \\ \\ \\ \\ + NH_3 + NADH^+ \underset{\text{glutamic dehydrogenase}}{\rightleftharpoons} \\ \\ \\ \\ \\ \\ \end{array} \begin{array}{l} COOH \\ | \\ CHNH_2 \\ | \\ CH_2 + NAD + H_2O \\ | \\ CH_2 \\ | \\ COOH \\ \text{Glutamate} \end{array}$$

The bulk of the enzyme is located in the mitochondria[524] and utilizes $NADH^+$ preferentially. Additionally, there have been reports of an $NADPH^+$ glutamic dehydrogenase[220, 394a] located in the chloroplasts and the enzyme has been suggested to function extensively in ammonia incorporation in leaves. However, the concentration of ammonia required for glutamate synthesis by the enzyme glutamic dehydrogenase is higher than that normally encountered in plant tissues. In fact the ammonia concentration required for optimal glutamic dehydrogenase activity is sufficient to uncouple photophosphorylation. These observations tend to preclude, but do not eliminate, glutamic dehydrogenase as a major route of entry of inorganic nitrogen into organic form.

Recent evidence suggests that the problem of high ammonia requirements is circumvented by the utilization of glutamine as the nitrogen donor in glutamate synthesis in a reaction catalysed by glutamate synthetase (glutamine oxaglutarate amino transferase). The enzyme from carrot (*Daucus carota*),[156a] sycamore (*Acer pseudoplatanus*) cell cultures and pea (*Pisum sativum*) roots[200a] is able to utilize $NADH^+$ or $NADPH^+$ as the reduced pyridine nucleotide and asparagine as well as glutamine can function as the amide donor. In contrast the enzyme characterized from chloroplasts requires reduced ferredoxin.[354a]

Thus an alternative and perhaps preferred method of entry of inorganic ammonia nitrogen into organic combination is through the combined activity of glutamine synthetase and glutamate synthetase. Such a system is comparable to the scheme described for the entry of ammonia into organic combination in nitrogen fixing organisms (p. 14).

Although there are other reports of the combination of ammonia with pyruvate and oxaloacetate to produce alanine and aspartate and of aspartate and ammonia to give asparagine, these findings are for the most part unconfirmed.

2

Amino Acids

PROTEIN AND NON-PROTEIN AMINO ACIDS: OCCURRENCE AND DISTRIBUTION

Extraction of plant tissues with aqueous ethanol results in the separation of the ethanol-soluble nitrogenous compounds from the ethanol-insoluble nitrogen. A large part of the soluble nitrogen fraction is made up of amino acids, whereas the ethanol-insoluble nitrogen arises to a large extent from protein.

The amino acids in the ethanol-soluble fraction can be separated from other components by appropriate treatment and subsequently identified by chromatography. This identification of amino acids present in the ethanolic extracts of plants was greatly facilitated by the development of paper chromatography and in the early 1950's this technique was utilized extensively by Steward's[593] and Fowden's[198] groups to characterize the soluble amino acids. Additionally it is possible to hydrolyse the alcohol-insoluble residue with hydrochloric acid to liberate protein amino acids. During acid hydrolysis certain amino acids, particularly tryptophan, are destroyed. If an analysis of the total amino acid complement of proteins is desired, an additional separate alkaline hydrolysis is performed. Such hydrolysates can be fractionated by chromatography and are usually found to contain twenty different amino acids. The amino acids commonly found in protein hydrolysates are the so-called protein amino acids and can be categorized as shown in Table 2.1.

When chromatographic analyses are performed on the ethanol-soluble fractions from plants several important observations are made. In addition to the twenty common amino acids encountered in protein hydrolysates

Table 2.1 The structures of amino acids present in most plant proteins and the aldo or keto acid analogues from which they are synthesized.

Monoamino monocarboxylic acids		
Glycine	$CH_2(NH_2).COOH$	Glyoxylic acid
Alanine	$CH_3.CH(NH_2).COOH$	Pyruvic acid
Valine	$CH_3.CH(CH_3).CH(NH_2).COOH$	α-Keto-isovaleric acid
Leucine	$CH_3.CH(CH_3).CH_2.CH(NH_2).COOH$	α-Keto-isocaproic acid
Isoleucine	$CH_3.CH_2.CH(CH_3).CH(NH_2).COOH$	α-Keto-β-methyl valeric acid
Hydroxy aliphatic acids		
Serine	$CH_2(OH).CH(NH_2).COOH$	Hydroxy pyruvic acid
Threonine	$CH_3.CH(OH).CH(NH_2).COOH$	α-Keto-β-hydroxy butyric acid
Dicarboxylic amino acids		
Aspartic acid	$HO_2C.CH_2.CH(NH_2).COOH$	Oxaloacetic acid
Glutamic acid	$HO_2C.CH_2.CH_2.CH(NH_2).COOH$	α-Ketoglutaric acid
Amides		
Asparagine	$H_2NOC.CH_2.CH(NH_2).COOH$	—
Glutamine	$H_2NOC.CH_2.CH_2.CH(NH_2).COOH$	—
Basic amino acids		
Lysine	$CH_2(NH_2).CH_2.CH_2.CH_2.CH(NH_2).COOH$	—
Arginine	HN, H_2N ⟩C.NH.$CH_2.CH_2.CH_2.CH(NH_2).COOH$	—
Histidine	N—CH, HC, C.$CH_2.CH(NH_2).COOH$, N H (imidazole ring)	Imidazoleacetol phosphate
Sulphur-containing acids		
Cysteine	$HS.CH_2.CH(NH_2).COOH$	—
Cystine	$S.CH_2.CH(NH_2).COOH$ | $S.CH_2.CH(NH_2).COOH$	— —
Methionine	$CH_3.S.CH_2.CH_2.CH(NH_2).COOH$	—
Imino acid		
Proline	H_2C—CH_2, H_2C, CH.COOH, N H (ring)	—
Aromatic amino acids		
Phenylalanine	(benzene ring) $CH_2.CH(NH_2).COOH$	Phenyl-pyruvic acid
Tyrosine	HO—(benzene ring)—$CH_2.CH(NH_2).COOH$	Hydroxy-phenyl-pyruvic acid
Tryptophan	(indole ring, N H) $CH_2.CH(NH_2).COOH$	—

there are other amino acids present. These have been collectively characterized as the **non-protein amino acids**. These non-protein amino acids exhibit a greater variety of chemical structure than their protein counterparts (Table 2.2) but they may also be divided into the groupings described

Table 2.2 Some non-protein aminoacids found in plants. After Fowden.[199]

Non-protein amino acid (corresponding protein amino acid and relationship in parentheses)	Structural formulae	Distribution
Acidic		
α-Aminoadipic acid (H)	$HO_2C.(CH_2)_3.CH(NH_2).COOH$	Many plants
γ-Hydroxyglutamic acid (S) (glutamic acid)	$HO_2C.CH(OH).CH_2.CH(NH_2).COOH$	Some Liliaceae and ferns
γ-Methyleneglutamic acid (S) (glutamic acid)	$HO_2C.C{=}(CH_2).CH_2.CH(NH_2).COOH$	Random, e.g. tulip and peanuts
Amides		
N-Ethylasparagine (S) (asparagine)	$(C_2H_5)HN.OC.CH_2.CH(NH_2).COOH$	Some Cucurbitaceae
Basic		
α,γ-Diaminobutyric acid (H) (lysine)	$H_2N.CH_2.CH_2.CH(NH_2).COOH$	Sporadic, some legume seeds and antibiotics
Homoarginine (H) (arginine)	$(HN{=})(H_2N)C.NH.(CH_2)_4.CH(NH_2).COOH$	Some *Lathyrus species*
Hydroxy		
Homoserine (H, I) (serine, threonine)	$HO.CH_2.CH_2.CH(NH_2).COOH$	Many plants
Aromatic		
m-Carboxyphenylalanine (S) (phenylalanine)	benzene ring with $CH_2.CH(NH_2).COOH$ and meta CO_2H	*Iris, Reseda*
Heterocyclic		
β-Pyrazol-1-ylalanine (I) (histidine)	pyrazole ring (HC=CH, HC=N, N) with $N.CH_2.CH(NH_2).COOH$	Some Cucurbitaceae; also as γ-glutamyl peptide
Sulphur-containing		
S-Methylcysteine (S) (cysteine)	$CH_3.S.CH_2.CH(NH_2).COOH$	Legumes, Cruciferae; also as γ-glutamyl peptide
Imino acids		
Azetidine-2-carboxylic acid (H) (proline)	four-membered ring: H_2C, $C(H_2)$, $CH.COOH$, $N(H)$	Many Liliaceae

(*Table continued on p. 31*)

Table 2.2 *continued*

Non-protein amino acid (corresponding protein amino acid and relationship in parentheses)	Structural formulae	Distribution
Imino acids		
Pipecolic acid (H) (proline)	H_2C–CH_2–CH_2 ring: H_2C, CH_2, H_2C, $CH.COOH$, N–H (six-membered ring)	Many plants, especially legume seeds
5-Hydroxypipecolic acid (H) (4-hydroxyproline)	six-membered ring: CH_2, $HO.HC$, CH_2, H_2C, $CH.COOH$, N–H	Dates, ferns, legumes
Cyclopropyl acids		
α-(Methylenecyclopropyl) glycine (A) (leucine)	$H_2C{=}C$—$CH.CH(NH_2).COOH$ (cyclopropane ring closed by CH_2)	*Litchi seed*
1-Aminocyclopropyl-1-carboxylic acid	H_2C—$C(NH_2)(COOH)$ (cyclopropane ring closed by CH_2)	Some unripe Rosaceae fruits

Types of relationship can be summarized as: H, homologue containing a different number of methylene groups in main chain; I, compounds exhibit positional isomerism of atoms or groups; S, compounds have substituent groups attached to related protein constituent; A, analogue possessing a molecular shape similar to protein constituent.

earlier. The membership in this group of amino acids is constantly being extended and detailed studies of their biosynthesis and relationship to protein amino acids are actively being conducted particularly by Fowden.[200] It is generally considered that the non-protein amino acids are secondary plant products and appear to play little direct role in plant growth and development. However, certain members of this group—for example, homoserine and canavanine—appear to function in nitrogen translocation and storage.

Another important observation concerning the ethanol-soluble nitrogen fraction is that the amino acid composition is extremely variable and bears little relationship to the amino acid composition of the protein fraction which tends to be relatively stable. The extensive studies of the Cornell group[591] indicated that the amino acid composition of the ethanol fraction is affected by plant variety, stage of growth, plant part and such environmental influences as nitrogen nutrition, photoperoid, and mineral nutrition (Tables 2.3, 2.4).

Table 2.3 The amino acid content of the protein fraction of normal and nutrient deficient turnip (*Brassica rapa*) leaves based on protein nitrogen: note the relative constancy of amino acid composition (values expressed as micrograms per milligram of nitrogen). Data from Thompson, Morris and Gering.[626]

Amino acid	Deficient nutrient			
	None	Nitrogen	Phosphorus	Sulphur
Aspartic acid	606	540*	621	616
Glutamic acid	612	606	631	634
Serine	201	220	210	219
Glycine	308	286	309	290
Threonine	262	265	277	260
Alanine	351	332	255	338
Methionine	110	121	116	121
Proline	282	262*	285	271
Valine	332	323	345	341
Tyrosine	136	197	198	212
Isoleucine	254	270	245	258
Leucine	466	450	493	475
Phenylalanine	261	269	306	284
Histidine	71	80	83	89†
Lysine	264	248	270	266
Arginine	300	273	308	294

* Significantly different from no deficiency at 5% level.
† Significantly different from no deficiency at 1% level.

The soluble amino acids arise as a result of biosynthesis or as a result of the hydrolysis of proteins; an arrested protein synthesis could equally result in an accumulation of soluble amino acids. However, the observation that the amino acid composition of the soluble fraction differs from that of the proteins indicates that the soluble amino acids are not necessarily either precursors of protein synthesis or direct products of protein degradation. Most recent thinking[51] implicates the existence of pools of amino acids, some of which are metabolically active and contribute directly to protein synthesis, whereas others located in storage pools are derived from both synthesis and protein breakdown. Thus, the composition of the soluble amino acids need have little relationship to the amino acids present in proteins. Additionally, the amino acid composition of the soluble fraction can fluctuate without grossly altering the amino acid composition of proteins.[626]

AMINO ACID BIOSYNTHESIS

Although reference to conventional metabolic charts indicates that the biosynthesis of amino acids is well defined it should be remembered that

Table 2.4 Amino acid content of the non-protein (ethanol-soluble) fraction of normal and nutrient deficient turnip (*Brassica rapa*) leaves. Note the great range of variability and influence of mineral nutrition on the various amino acids (micrograms per gram fresh weight) and the lack of relationship between amino acid composition of this fraction with the protein fraction in Table 2.3. Data from Thompson, Morris and Gering.[626]

Amino acid	Deficient nutrient			
	None	Nitrogen	Phosphorus	Sulphur
Aspartic acid	42.1	15.6*	226.0†	151.0†
Glutamic acid	49.6	51.7	228.0†	237.0†
Serine	41.9	30.9	142.0†	191.0†
Glycine	8.90	7.30	53.3†	134.4†
Asparagine	19.2	0.0	55.1	66.3
Threonine	22.1	12.3†	80.7*	112.0†
Alanine	71.7	49.8	200.0†	136.0†
Glutamine	39.0	18.4	1046.0†	1334.0†
Proline	34.0	11.1†	418.0†	520.0†
Valine	14.8	11.1	67.6†	35.0*
Tyrosine	4.62	5.26	25.4†	8.47
γ-Amino butyric acid	75.5	43.8	129.3	119.0
Pipecolic acid	2.42	16.3	12.9	0.13†
Methyl cysteine sulphoxide	73.1	66.0	234.0†	7.30*
Cysteine	9.26	17.2*	10.2	2.18*
Methionine	7.54	7.28	20.6*	9.68
Isoleucine	9.96	5.60	76.5†	17.9*
Leucine	7.30	5.50	17.1	10.9†
Phenylalanine	7.46	6.68	34.8†	20.7†
Histidine	7.08	4.10	39.3†	13.4
Lysine	13.4	7.52*	60.3†	16.9
Arginine	49.72	27.1	439.0*	66.4

* Significantly different from no deficiency at 5% level.
† Significantly different from no deficiency at 1 % level.

most of these schemes have been established from studies on microbial and animal systems. Very few detailed investigations have been conducted to describe the total enzymes and intermediates involved in the biosynthesis of specific amino acids in plants.

In general, studies of amino acid biosynthesis in plants have been conducted utilizing tracer techniques to establish relationships between possible precursors and products without identifying the intermediary enzymes and metabolites. An alternative approach to establish reaction sequences has been utilized productively by Dougall and Fulton[158] in which the influence of suspected intermediates on the transfer of label from labelled glucose to product has been followed. Such studies along with the more conventional labelled precursor-product studies have indicated that the biosynthesis of the amino acids follows to some extent the pathways generally established in bacterial and mammalian systems. Families of amino acids emerge which can be related through their origin from so-called head amino acids (Fig. 2.1).

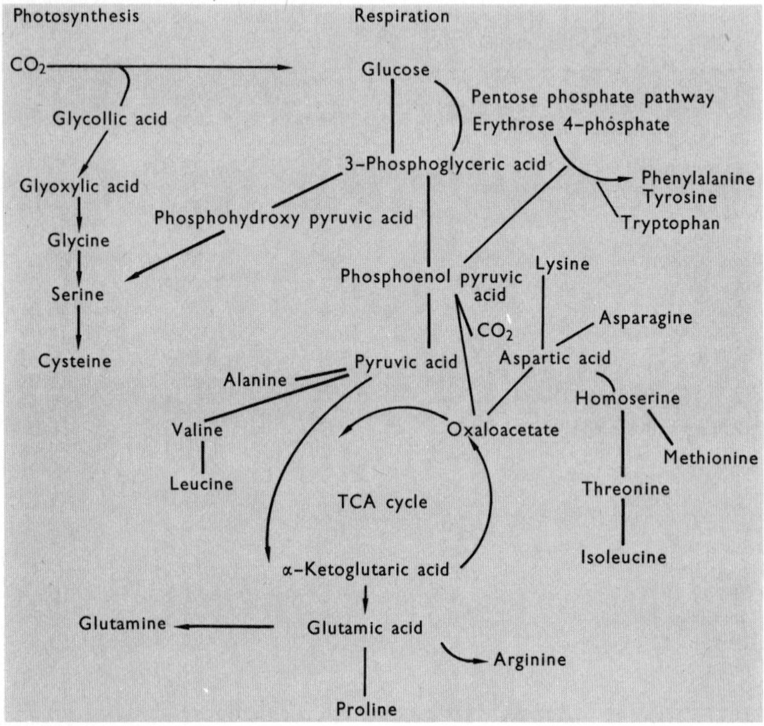

Fig. 2.1 Biosynthetic sequences in amino acid synthesis.

Transaminase

The 'head' amino acids arise to a great extent by a process of transamination in which the amino group of an amino acid is transferred to the keto group of a keto acid according to the general formula below.

$$\underset{\text{Amino acid}}{R_1\text{—}\overset{\overset{\Large H}{|}}{\underset{\underset{\Large NH_2}{|}}{C}}\text{—COOH}} + \underset{\text{Keto acid}}{R_2\text{—}\overset{\overset{\Large O}{\|}}{C}\text{—COOH}} \underset{\text{transaminase}}{\rightleftharpoons} R_1\text{—}\overset{\overset{\Large O}{\|}}{C}\text{—COOH} + R_2\text{—}\overset{\overset{\Large H}{|}}{\underset{\underset{\Large NH_2}{|}}{C}}\text{—COOH}$$

The reaction is catalysed by a transaminase or an amino transferase (the terms are synonymous). It can be demonstrated that plant extracts can catalyse the transfer of amino groups from amino acids to a range of keto acids[195] and there is an expanding literature characterizing amino transferases or transaminases in a range of plant species. There is some discussion as to whether there are different transaminases for each amino acid–keto acid combination.[195,196] Although it is possible to separate amino transferases demonstrating specificity for particular amino acids and keto acid combinations[518] it seems[196,177] that many amino acids can be transaminated to a range of keto acids by partially purified plant extracts and there may be only a few transaminases with multi-substrate specificity. Pyridoxal phosphate is an essential cofactor for the amino transferases of animal origin; however, there is no established requirement for this compound for transaminase activity in extracts and partially purified enzymes from plants.[330]

Given the capacity for amino group transfer, the biosynthesis of an amino acid in many instances hinges upon the capacity to form the appropriate keto acid precursor.

The keto acid analogues of many amino acids have been identified[591] in plants and thus the synthesis of these amino acids can proceed through the activities of amino transferases. While glutamate is frequently given a key role in these amino transferase reactions most amino acids can serve as amino donors to a variety of keto acids (Table 2.5).

Amino transferases are located in the cytoplasm, chloroplasts and microbodies. In photosynthetic tissues the activity of amino transferase located in the chloroplast may play a key role in the synthesis of amino acids. Labelling experiments demonstrate the rapid entry of carbon from carbon dioxide into the amino acids glutamic acid, aspartic acid, alanine, serine and glycine.[528] The generation of glutamic acid can proceed through the functioning of chloroplastic $NADPH^+$-dependent glutamic dehydrogenase,[220,394a] or glutamine synthetase and reduced ferredoxin-dependent glutamate synthetase.[354a] However, the synthesis of the amino acids in the chloroplasts requires the continued supply of appropriate keto acids. It is speculated that while α-ketoglutarate may be produced in the chloroplast pyruvic acid and oxaloacetate may be synthesized either in the cytoplasm or chloroplast from phosphoglyceraldehyde produced in the chloroplast in photosynthesis.[333,394a] These keto acids can undergo transamination in the chloroplast with the production of alanine and aspartate respectively as indicated in Fig. 2.2.

$$\text{Pyruvate} + \text{glutamate} = \text{alanine} + \alpha\text{-ketoglutarate.}$$
$$\text{Oxalacetate} + \text{glutamate} = \text{aspartate} + \alpha\text{-ketoglutarate.}$$

This requirement for keto acids, produced in the cytoplasm, for amino

Table 2.5 Rates of transamination reactions occurring in extracts of cotyledons from 8-day-old bushbean (*Phaseolus vulgaris*) seedlings when each of 22 amino acids and α-ketoglutarate or oxaloacetate, or pyruvate, or glyoxylate were provided as substrates.*

Amino acid substrate	Keto acid substrate			
	α-Keto-glutarate	Oxalo-acetate	Pyruvate	Gly-oxylate
Glycine	50	10	40	—
Alanine	530	30	—	300
Serine	0	0	50	120
Cysteine	310	190	230	10
Threonine	150	10	40	20
Methionine	40	50	30	25
Valine	70	20	30	10
Leucine	100	30	40	14
Isoleucine	120	30	20	10
Glutamic acid	—	800	750	365
Glutamine	—	150	300	200
Aspartic acid	670	—	130	25
Asparagine	130	—	70	20
Lysine	20	10	30	75
Arginine	40	40	50	60
Histidine	50	70	40	50
Tryptophan	50	35	20	5
Phenylalanine	80	50	30	10
Tyrosine	60	40	20	10
Proline	0	0	0	0
Hydroxyproline	0	0	0	0
Cystine	0	0	0	0

* Results are expressed in terms of nanomoles of the product amino acid per milligram of protein per hour of incubation.
Note transfer of amino group from one amino acid to range of keto acids.
(Data from Forest and Wightman,[195] Table 1, p. 540. Reproduced by permission of the National Research Council of Canada from the *Canadian Journal of Biochemistry*, (1972), **50**, pp. 538-562.)

acid biosynthesis in the chloroplast to some extent explains the paradox that while in intact photosynthetic tissues carbon rapidly enters amino acids very little amino acid synthesis occurs in isolated chloroplasts.[663] If carbon dioxide is supplied to isolated chloroplasts in the presence of appropriate keto acids then amino acid biosynthesis proceeds.[333]

GLYCINE AND SERINE

Synthesis of these amino acids could occur in the chloroplasts by way of amino transfer to the keto acids, glyoxylate and hydroxypyruvate. However, although enzymes involved in serine and glycine metabolism have been located in the chloroplast, the biosynthesis of these amino acids may involve more than a simple transaminase reaction and in all probability

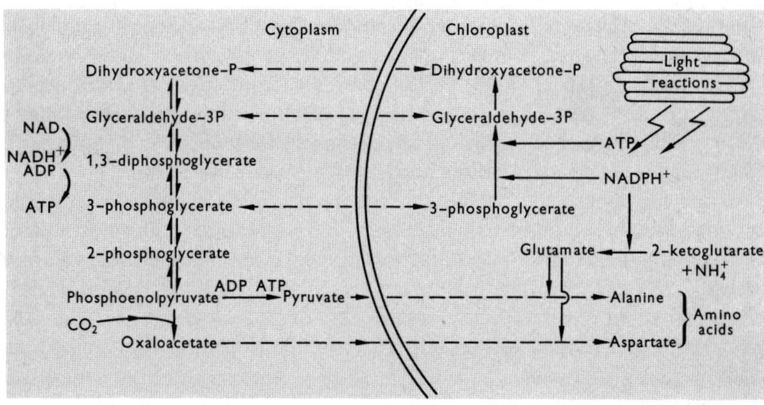

Fig. 2.2 Diagram showing the proposed extended shuttle system between the chloroplast. (From Kirk and Leech,[333] Fig. 1, p. 232.)

involves the cooperative function of chloroplasts, microbodies (peroxisomes) and mitochondria.[518]

Glycollate produced in the chloroplasts during photosynthesis is converted to glyoxylate in the microbodies. The glyoxylate can be converted to glycine through the activity of amino transferases located in the microbody.

$$\underset{\text{Glyoxylic acid}}{\begin{array}{l}COOH\\ |\\ CHO\end{array}} + \text{Glutamate} \longrightarrow \underset{\text{Glycine}}{\begin{array}{l}COOH\\ |\\ CH_2NH_2\end{array}} + \alpha\text{-Ketoglutarate}$$

The glycine can give rise to serine with the elimination of carbon dioxide and ammonia. This serine biosynthesis has been demonstrated in mitochondrial preparations.[334]

$$\underset{\text{Glycine}}{2\begin{array}{l}COOH\\ |\\ CH_2NH_2\end{array}} \rightleftharpoons \underset{\text{Serine}}{\begin{array}{l}COOH\\ |\\ CHNH_2\\ |\\ CH_2OH\end{array}} + NH_3 + CO_2$$

The carbon dioxide released in this reaction is considered to be that produced during the photorespiration demonstrated in many plant species.

The actual conversion involves several enzymatic steps requiring the decarboxylation and deamination of glycine with the production of a C_1-hydroxymethyl residue. The C_1 residue can combine with a further glycine molecule under the activity of the enzyme serine hydroxymethyl transferase.[125] This C_1 transfer and others to be encountered later require the participation of the cofactors derived from pteroyl-glutamic acid (folic acid) which have been demonstrated to occur in plants.

This mitochondrial biosynthesis of serine from glycine according to recent findings can be coupled to the generation of adenosine triphosphate, ATP.[52] This observation is of great significance since it provides an unsuspected energy-generating function for photorespiration.

While the above sequence allows for a mechanism for serine and glycine biosynthesis, other pathways are possible. The reduction of glyceric acid to hydroxypyruvate provides the keto acid which upon amination produces serine. This serine can in turn be converted to glycine by the elimination of the terminal hydroxymethyl group. The removal of the C_1 fragment from serine is mediated by serine hydroxymethyl transferase in conjunction with pteroyl-glutamate.

REACTIONS OF GLUTAMIC ACID

Glutamate serves as the precursor for the biosynthesis of the amino acids proline, hydroxyproline, ornithine, citrulline and arginine.

Investigations with labelled precursors and enzymatic studies[438] indicate that initially glutamic acid is converted to glutamic semialdehyde. The enzymatic conversion requires ATP, Mg^{2+} and $NADH^+$ and appears to involve the phosphorylation of the carboxyl group of glutamic acid before reduction:

$$\begin{array}{c} COOH \\ | \\ CH_2 \\ | \\ CH_2 \\ | \\ CH.NH_2 \\ | \\ COOH \end{array} \xrightarrow[\text{ATP} \quad \text{ADP}+\text{Pi}]{\text{NADH}^+ \quad \text{Mg}^{2+} \quad \text{NAD}} \begin{array}{c} CHO \\ | \\ CH_2 \\ | \\ CH_2 \\ | \\ CH.NH_2 \\ | \\ COOH \end{array}$$

Glutamic acid — Glutamic γ semialdehyde

Additionally glutamic acid may be acetylated and the *N*-acetyl-glutamic acid converted to *N*-acetyl-glutamic semialdehyde.[438]

$$\begin{array}{c} COOH \\ | \\ CH_2 \\ | \\ CH_2 \\ | \\ CHNH_2 \\ | \\ COOH \end{array} + \text{Acetyl CoA} \longrightarrow \begin{array}{c} COOH \\ | \\ CH_2 \\ | \\ CH_2 \\ | \\ CHNH.CO.CH_3 \\ | \\ COOH \end{array} + \text{CoA}$$

Glutamic acid — *N*-Acetyl-glutamic acid

Glutamic semialdehyde in solution exists in equilibrium with its cyclic form Δ′-pyrroline-5-carboxylic acid which can be reduced to proline by pyrroline carboxylate reductase[462] with $NADH^+$ providing the reductant.

$$\begin{array}{ccc} \begin{array}{ccc} H_2C & — & CH_2 \\ | & & | \\ HC & & CHCOOH \\ & \diagdown\!\!\diagdown N \diagup & \end{array} & \xrightarrow{NADH^+ \;\curvearrowright\; NAD} & \begin{array}{ccc} H_2C & — & CH_2 \\ | & & | \\ H_2C & & CHCOOH \\ & \diagdown N \diagup & \\ & H & \end{array} \end{array}$$

Δ′-pyrroline-5-carboxylic acid Proline

At this time it is not clear whether acetyl glutamic semialdehyde serves as a precursor for proline synthesis; however, the ready production of the acetyl derivatives indicate their possible participation in this pathway.[438]

Glutamic acid semialdehyde, or more usually its *N*-acetyl derivative, can be converted to ornithine in a transaminase reaction. Unlike previously discussed transaminases which involved keto groups, this transamination occurs at the aldehyde group and is mediated by the enzyme *N*-acetyl-L-ornithine δ-transaminase. The enzyme shows a specificity for glutamate as the donor of the amino group and is stimulated by pyridoxal phosphate.[157]

$$\begin{array}{lclcl} CHO & & COOH & & CH_2NH_2 & & COOH \\ | & & | & & | & & | \\ CH_2 & & CH_2 & & CH_2 & & CH_2 \\ | & & | & & | & & | \\ CH_2 & + & CH_2 & \rightleftharpoons & CH_2 & + & CH_2 \\ | & & | & & | & & | \\ CHNH.COCH_3 & & CHNH_2 & & CHNH.COCH_3 & & CO \\ | & & | & & | & & | \\ COOH & & COOH & & COOH & & COOH \end{array}$$

Acetyl glutamic semialdehyde Glutamate α-*N*-acetyl ornithine α-Ketoglutarate

Acetyl ornithine is converted to ornithine by an acylase cleaving enzyme.[438,589]

The amination of glutamic semialdehyde by the activity of ornithine transaminase[61] results directly in the production of ornithine. The relative efficiency of the unsubstituted amino acids in proline and ornithine biosynthesis has not been established in plant systems.

Although ornithine does not occur as a constituent amino acid of cellular proteins, its biosynthesis is required for the subsequent production of arginine via the intermediates citrulline and arginino-succinate.

The initial reaction of ornithine is with carbamyl phosphate $NH_2.COO.PO_3H_2$ which is synthesized from ammonia, ATP and bicarbonate. In many instances the amide nitrogen of glutamine or asparagine functions more effectively than ammonia.[480, 481, 482]

The condensation of carbamyl phosphate and ornithine catalysed by the enzyme ornithine transcarbamylase[335] results in the formation of citrulline.

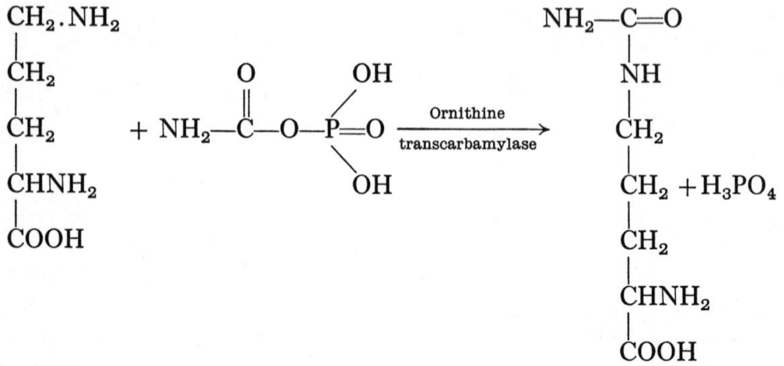

$$
\begin{array}{l}
CH_2.NH_2 \\
| \\
CH_2 \\
| \\
CH_2 \\
| \\
CHNH_2 \\
| \\
COOH
\end{array}
+ NH_2-\overset{\overset{O}{\|}}{C}-O-P(=O)(OH)_2
\xrightarrow[\text{transcarbamylase}]{\text{Ornithine}}
\begin{array}{l}
NH_2-C{=}O \\
| \\
NH \\
| \\
CH_2 \\
| \\
CH_2 \\
| \\
CH_2 \\
| \\
CHNH_2 \\
| \\
COOH
\end{array}
+ H_3PO_4
$$

Ornithine Carbamyl phosphate Citrulline

The production of arginine from citrulline involves the condensation of citrulline with aspartate and the production of the intermediate arginino-succinic acid.[560]

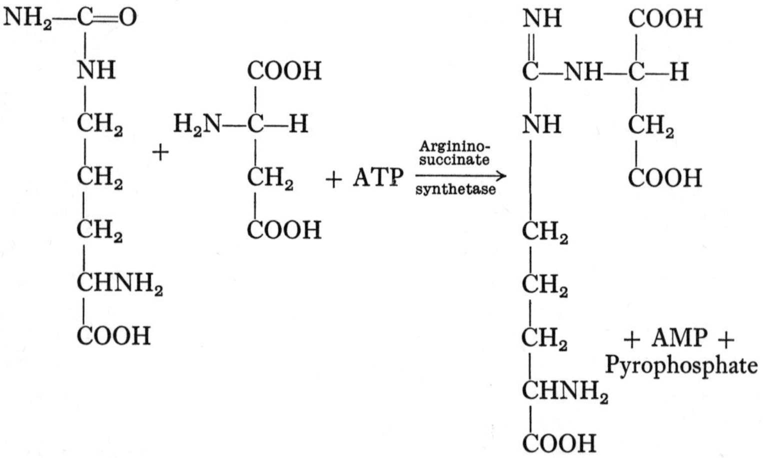

$$
\begin{array}{l}
NH_2-C{=}O \\
| \\
NH \\
| \\
CH_2 \\
| \\
CH_2 \\
| \\
CH_2 \\
| \\
CHNH_2 \\
| \\
COOH
\end{array}
+
\begin{array}{l}
COOH \\
| \\
H_2N-C-H \\
| \\
CH_2 \\
| \\
COOH
\end{array}
+ ATP
\xrightarrow[\text{synthetase}]{\text{Arginino-succinate}}
\begin{array}{l}
NH \qquad COOH \\
\| \qquad\quad | \\
C-NH-C-H \\
| \qquad\quad | \\
NH \qquad CH_2 \\
| \qquad\quad | \\
| \qquad COOH \\
CH_2 \\
| \\
CH_2 \\
| \\
CH_2 \\
| \\
CHNH_2 \\
| \\
COOH
\end{array}
+ AMP + \text{Pyrophosphate}
$$

Citrulline Aspartic acid Arginino-succinic acid

Arginino-succinic acid is converted to arginine and fumarate by the enzyme arginino-succinate lyase.[538] The operation of the enzyme arginino-succinate synthetase was originally only implied. Plant extracts had been shown to contain the lyase and were capable of converting

^{14}C-carbamyl citrulline + L-aspartate + ATP ⟶ ^{14}C-arginine + fumarate + AMP + pyrophosphate.

However, the synthetase has been separated from the lyase and it has the expected capacity of forming arginino-succinate from citrulline.[560]

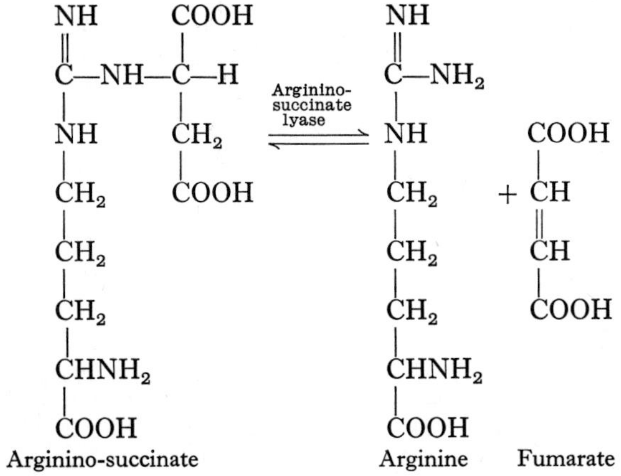

```
NH        COOH                      NH
‖         |                         ‖
C—NH—C—H                            C—NH2
|         |       Arginino-         |
|         |       succinate         |
NH        CH2     lyase   ⇌         NH          COOH
|         |                         |           |
CH2       COOH                      CH2     +   CH
|                                   |           ‖
CH2                                 CH2         CH
|                                   |           |
CH2                                 CH2         COOH
|                                   |
CHNH2                               CHNH2
|                                   |
COOH                                COOH
Arginino-succinate                  Arginine    Fumarate
```

As well as being a protein amino acid, arginine accumulates extensively in some plants as a soluble amino acid in which role it appears to function as a nitrogen storage compound and can undergo extensive metabolism and interconversions discussed later.

ASPARTIC ACID METABOLISM

As indicated (Table 2.5) aspartic acid may be synthesized by the amination of oxaloacetate. Glutamic-oxaloacetic transaminase was one of the earliest studied and most characterized plant transaminase. Partially purified preparations demonstrate substantial specificity for glutamate and oxaloacetate and are stimulated by pyridoxal phosphate.[177]

Studies with labelled aspartate demonstrate its ready conversion into asparagine. Webster and Varner[674] showed that extracts from lupin seedlings were able to catalyse the conversion of aspartate to asparagine in the presence of NH_3 (but not glutamine), Mg^{2+} and ATP. These workers stressed that the requirements for this enzymatic formation of asparagine were considerably different from the previously established conversion of glutamate to glutamine and indicated that the role of the enzyme in asparagine synthesis *in vivo* remained to be established. Subsequently a similar observation of aspartate to asparagine conversion in the presence of ATP, NH_3 and Mg^{2+} has been demonstrated in extracts from corn roots[473] and potatoes.[449] However, the role of such an enzyme in asparagine biosynthesis is still questioned.

The ^{15}N labelling data of Yemm and Willis[716] (see p. 26) implied that asparagine synthesis was a secondary event in barley following the initial incorporation of label into glutamate and glutamine. This observation can be accommodated by the findings that under appropriate conditions (presumably different from those of Webster and Varner) glutamine can serve as a more efficient source of nitrogen for asparagine synthesis than ammonia.[536, 607]

$$\text{Aspartate} + \text{Glutamine} + \text{ATP} \xrightarrow{\text{Asparagine synthetase}} \text{Asparagine} + \text{Glutamate} + \text{AMP} + \text{Pyrophosphate.}$$

Such a mechanism of biosynthesis seems to explain the rapid entry of labelled carbon from acetate[474] or carbon dioxide[629] into asparagine. Metabolism of acetate in the Krebs' tricarboxylic acid cycle produces oxaloacetate which upon transamination produces aspartate. Fixation of carbon dioxide by phospho-enol-pyruvate carboxylase also generates oxaloacetate which can be converted to aspartate. The conversion of the aspartate formed in this manner to asparagine could be achieved by asparagine synthetase. However other studies indicate that exogenously supplied aspartate is not an active precursor of asparagine in plant tissues.[607]

In addition to the possible biosynthesis of asparagine through the synthetase activity it has been found[54] that cyanide can serve as a precursor for the amide nitrogen of asparagine. This biosynthesis involves the reaction of cyanide with serine or cysteine to produce β-cyano-alanine[55] which is then degraded to asparagine.[98] There are several reports of the production of labelled β-cyano-alanine from labelled cyanide and the kinetics of the appearance of label in this compound and asparagine are consistent with its participation in asparagine biosynthesis as indicated below

$$\underset{\text{Serine}}{\text{HOOC-CH(NH}_2\text{)-CH}_2\text{-OH}} + \text{HCN} \xrightarrow[-\text{H}_2\text{O}]{} \underset{\beta\text{-cyano-alanine}}{\text{HOOC-CH(NH}_2\text{)-CH}_2\text{-C}\equiv\text{N}} \xrightarrow{+\text{H}_2\text{O}} \underset{\text{Asparagine}}{\text{HOOC-CH(NH}_2\text{)-CH}_2\text{-C(=O)-NH}_2}$$

Clearly the sustained functioning of this biosynthetic pathway is dependent upon the provision of cyanide. While cyanogenic glucosides (discussed later) could function in this capacity their restricted distribution in the plant kingdom eliminates them as candidates.[55] Labelled amino acids

α,β-dihydroxy acid dehydrase[317,345] produces α-keto-β-methylvalerate which is the keto acid precursor of isoleucine. The keto acid is converted to isoleucine by the activity of amino transferases.

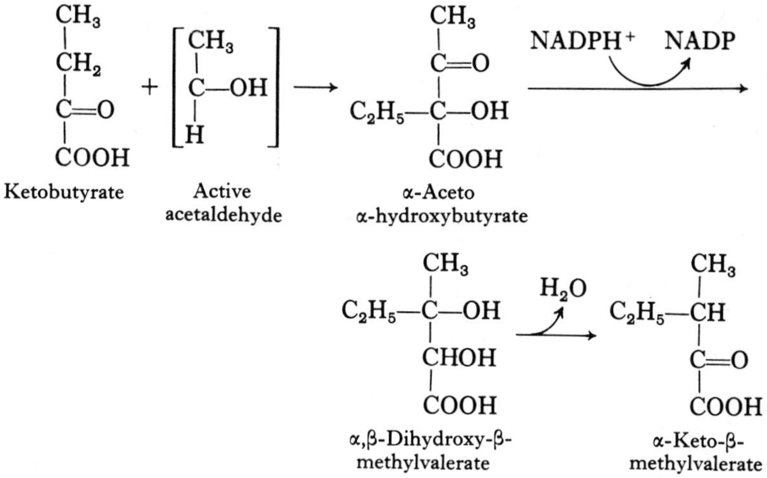

$$\begin{array}{c} CH_3 \\ | \\ CH_2 \\ | \\ C{=}O \\ | \\ COOH \end{array} + \left[\begin{array}{l} CH_3 \\ | \\ C{-}OH \\ | \\ H \end{array}\right] \longrightarrow \begin{array}{r} CH_3 \\ | \\ C{=}O \\ | \\ C_2H_5{-}C{-}OH \\ | \\ COOH \end{array} \xrightarrow{NADPH^+ \curvearrowright NADP}$$

Ketobutyrate — Active acetaldehyde — α-Aceto α-hydroxybutyrate

$$\begin{array}{r} CH_3 \\ | \\ C_2H_5{-}C{-}OH \\ | \\ CHOH \\ | \\ COOH \end{array} \xrightarrow{H_2O} \begin{array}{r} CH_3 \\ | \\ C_2H_5{-}CH \\ | \\ C{=}O \\ | \\ COOH \end{array}$$

α,β-Dihydroxy-β-methylvalerate — α-Keto-β-methylvalerate

SULPHUR-CONTAINING AMINO ACIDS: METHIONINE AND CYSTEINE

As well as serving as a precursor for threonine and hence isoleucine biosynthesis, homoserine has also been implicated in the biosynthesis of methionine. Dougall and Fulton[158] demonstrated the inhibition of incorporation of label from glucose to protein methionine by unlabelled homoserine, thus suggesting the operation of homoserine as an intermediate. In addition, experiments with rice callus tissue[202] have indicated a conversion of homoserine to methionine. This conversion appears to be reversible since the labelled carbon from methionine is incorporated into homoserine in pea seedlings which characteristically accumulate the latter amino acid.[230]

Metabolic events accounting for the homoserine to methionine conversion include an initial synthesis of cystathionine. Plant extracts are unable to utilize the free amino acid for this synthesis but in the presence of cysteine readily convert succinyl-homoserine or acetyl-homoserine to cystathionine.[216] However although acetyl-homoserine has been purified from peas[137] it appears to be of rather limited occurrence and on the basis of most recent evidence it is suggested[137] that *o*-phosphoryl homoserine is the physiological substrate for cystathionine synthase.

$$\underset{\text{O-Phosphoryl homoserine}}{CH_2O-\text{phosphate}-CH_2-CHNH_2-COOH} + \underset{\text{Cysteine}}{CH_2SH-CHNH_2-COOH} \xrightarrow[\text{synthase}]{\text{Cystathionine}} \underset{\text{Cystathionine}}{COOH-CHNH_2-CH_2-CH_2-S-CH_2-CHNH_2-COOH} + \text{phosphate}$$

Cystathionine can be hydrolysed in a β-elimination reaction to produce homocysteine, pyruvate and ammonia. Enzymes capable of homocysteine formation in this manner have been isolated from spinach (*Spinacia oleracea*).[217]

$$\underset{\text{Cystathionine}}{COOH-CHNH_2-CH_2-CH_2-S-CH_2-CHNH_2-COOH} \longrightarrow \underset{\text{Homocysteine}}{CH_2SH-CH_2-CHNH_2-COOH} + \underset{\text{Pyruvate}}{CH_3-C{=}O-COOH} + NH_3$$

Additionally spinach extracts are able to form homocysteine by direct sulphuration of *o*-acetyl-homoserine.[217]

Sulphide + *o*-acetyl-homoserine ⟶ homocysteine + acetate.

An apparently separate enzyme in spinach[217] and other plant extracts[625] catalyses the production of cysteine from sulphide and *o*-acetyl-serine.

Sulphide + *o*-acetyl-serine ⟶ cysteine + acetate.

The possibility of this mechanism serving in cysteine biosynthesis *in vivo* currently depends upon the successful demonstration of *o*-acetyl-serine in plants.

Methylation of homocysteine results in the formation of methionine. Many substances can function as methyl donors.[154]

S-methylmethionine + homocysteine ⟶ 2 methionine.
S-adenosyl L-methionine + homocysteine ⟶ methionine + S-adenosyl homocysteine.
S-methyl-pteroyl-glutamate + homocysteine ⟶ methionine + pteroyl-glutamate.

Of these three possibilities the last seems the most plausible[89, 154] since the other reactions would be dependent upon prior synthesis of methionine.

Methyl cysteine can also function as a potential methyl donor for methionine formation. However, according to Doney and Thompson[155] the methyl transfer is indirect and proceeds through the following sequence.

$$\text{Methyl cysteine} \longrightarrow \underset{\text{Methyl mercaptan}}{CH_3SH} + NH_3 + \text{pyruvate}$$

$$\underset{\text{Methyl mercaptan}}{CH_3SH} \longrightarrow \underset{\text{Formic acid}}{HCOOH} + SO_4^{2-}$$

The formic acid is reduced to formaldehyde and incorporated into the pteroyl-glutamate system for the synthesis of N^5 methyl-pteroyl-glutamate.

LYSINE

Knowledge of the mechanism of lysine biosynthesis is dependent upon labelling experiments which demonstrate that diamino-pimelic acid[188, 435a] is an effective precursor for lysine. The incorporation of label from glucose into protein lysine is also restricted by the addition of diamino-pimelic acid.[158] The terminal reaction of lysine biosynthesis from diamino-pimelic acid involves a decarboxylation of this compound and enzymes catalysing this reaction have been detected in plant extracts.[563] Collectively these data provide strong evidence for the biosynthesis of lysine from diamino-pimelic acid. In contrast α-amino-adipic acid which functions in an alternate pathway for lysine in fungi does not appear to function as a precursor in most higher plants.

Studies with bacterial systems indicate that diamino-pimelic acid is synthesized from aspartate by the following sequence.

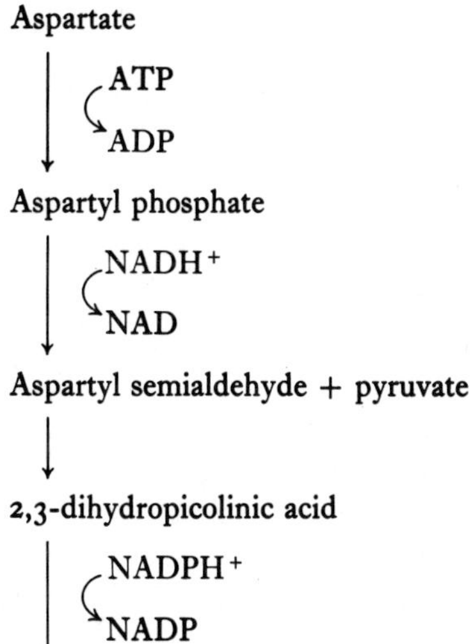

Δ′-piperidine-2,6-dicarboxylic acid
↓ Succinyl CoA → CoA
N-succinyl ε-keto L-α-amino-pimelic acid
↓ Transamination
N-succinyl α-ε-diamino-pimelic acid
↓ → succinate
L-α-ε-diamino-pimelic acid
↓
Meso α-ε-diamino-pimelic acid
↓ → CO_2
Lysine

To date few of the intermediates or enzymes involved in these conversions have been characterized in higher plants.

VALINE AND LEUCINE

These two amino acids derive their carbon skeletons from pyruvate. The initial event in their biosynthesis involves the condensation of pyruvate with an active acetaldehyde to produce acetolactate. This reaction is similar to that encountered previously in isoleucine biosynthesis except that pyruvate replaces α-ketobutyrate. The reaction in both instances is catalysed by the enzyme aceto hydroxyacid synthetase.[423] Acetolactate is converted to α,β-dihydroxyisovalerate by the enzyme reductoisomerase.[317] Again this type of reaction was previously encountered in the biosynthesis of isoleucine and is in all probability catalysed by the same enzyme. Activity of the dihydroxyacid dehydrase[317] converts α,β-dihydroxyisovalerate to α-keto-isovalerate. This dehydrase reaction is also common to the isoleucine pathway. Amination of α-keto-isovalerate produces valine; this reaction can be mediated by several transaminases as indicated by the capacity of amino transfer from valine to a range of keto acids.

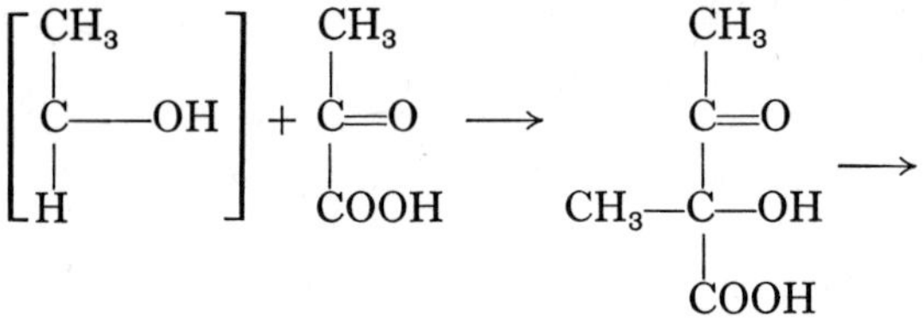

$$\left[\begin{array}{l} CH_3 \\ | \\ C\text{——}OH \\ | \\ H \end{array}\right] + \begin{array}{l} CH_3 \\ | \\ C{=}O \\ | \\ COOH \end{array} \longrightarrow \begin{array}{c} CH_3 \\ | \\ C{=}O \\ | \\ CH_3\text{—}C\text{—}OH \\ | \\ COOH \end{array} \longrightarrow$$

Active acetaldehyde | Pyruvic acid | Acetolactate

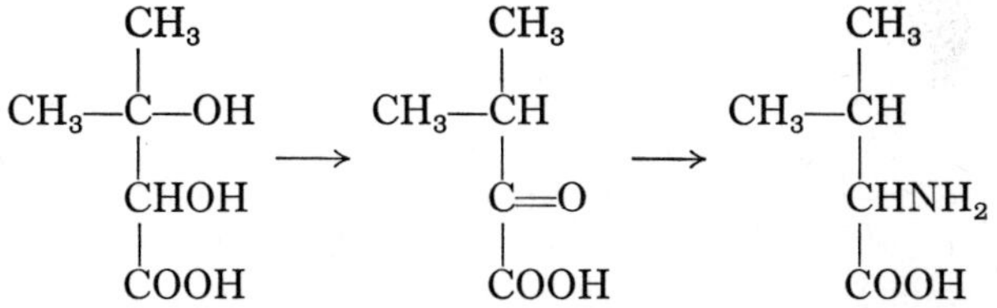

$$\begin{array}{c} CH_3 \\ | \\ CH_3\text{—}C\text{—}OH \\ | \\ CHOH \\ | \\ COOH \end{array} \longrightarrow \begin{array}{c} CH_3 \\ | \\ CH_3\text{—}CH \\ | \\ C{=}O \\ | \\ COOH \end{array} \longrightarrow \begin{array}{c} CH_3 \\ | \\ CH_3\text{—}CH \\ | \\ CHNH_2 \\ | \\ COOH \end{array}$$

α,β-Dihydroxy-isovalerate | α-Keto-isovalerate | Valine

Instead of being converted to valine, α-keto-isovalerate can condense with the methyl carbon atom of acetate (from acetyl CoA) to produce β-carboxy-β-hydroxy-isocaproic acid which undergoes structural rearrangement to form α-hydroxy-β-carboxy-isocaproic acid. This sequence of reaction is analogous to those involved in the formation of citric acid and its conversion to isocitrate in the Krebs' tricarboxylic acid cycle; however, a different series of enzymes is involved in each instance. Oxidative decarboxylation of α-hydroxy-β-carboxy-isocaproic acid leads to the formation of α-keto-isocaproic acid which undergoes an amino transferase reaction to produce leucine. The condensing enzyme, α-hydroxy-β-carboxy-isocaproic dehydrogenase, and amino transferase required in the reaction sequence have been characterized in extracts of maize.[470] The dehydrogenase is NAD-dependent and alanine appears to be the preferred amino acid in the transamination.

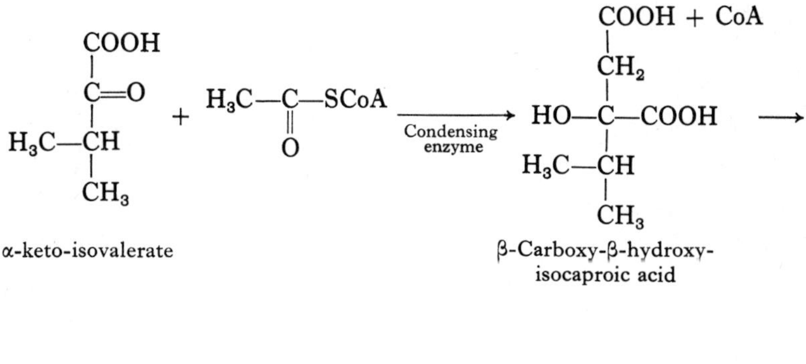

$$\begin{array}{c} COOH \\ | \\ C{=}O \\ | \\ H_3C\text{—}CH \\ | \\ CH_3 \end{array} + \begin{array}{c} H_3C\text{—}C\text{—}SCoA \\ \| \\ O \end{array} \xrightarrow[\text{Condensing enzyme}]{} \begin{array}{c} COOH + CoA \\ | \\ CH_2 \\ | \\ HO\text{—}C\text{—}COOH \\ | \\ H_3C\text{—}CH \\ | \\ CH_3 \end{array} \longrightarrow$$

α-keto-isovalerate | β-Carboxy-β-hydroxy-isocaproic acid

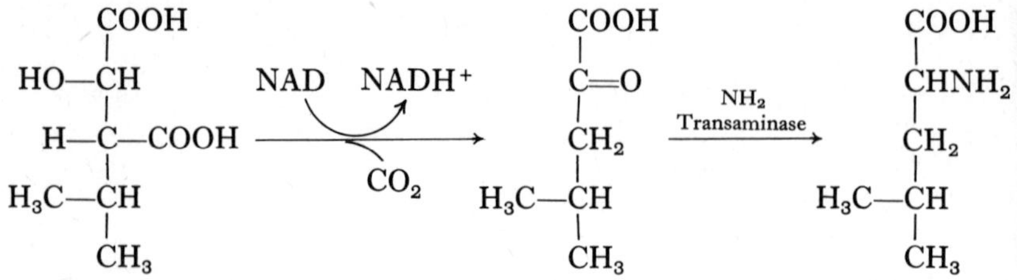

$$\underset{\text{α-Hydroxy-β-carboxy-isocaproic acid}}{HOOC{-}CH(OH){-}C(H)(COOH){-}CH(CH_3){-}CH_3} \xrightarrow[CO_2]{NAD \quad NADH^+} \underset{\text{α-Keto-isocaproic acid}}{HOOC{-}C(=O){-}CH_2{-}CH(CH_3){-}CH_3} \xrightarrow[\text{Transaminase}]{NH_2} \underset{\text{Leucine}}{HOOC{-}CHNH_2{-}CH_2{-}CH(CH_3){-}CH_3}$$

HISTIDINE

The pathway of histidine biosynthesis has not been established in plants. However, experiments with amino-triazole, an inhibitor of histidine biosynthesis in bacteria, and competitive labelling studies imply that the synthetic sequence is similar to that encountered in bacteria.

Plant cells treated with amino-triazole (which in bacterial systems inhibits the enzyme imidazole glycerol phosphate dehydratase involved in histidine synthesis) accumulate imidazole glycerol phosphate and its derivative imidazole glycerol.[140, 564] Addition of histidinol to rose cell cultures[158] eliminated the incorporation of label from glucose into histidine residues in proteins.

The biosynthesis of imidazole glycerol phosphate, according to bacterial studies, is initiated by the condensation of ATP and 5-phosphoribosyl pyrophosphate to yield phosphoribosyl adenosine triphosphate and pyrophosphate.

Phosphoribosyl ATP is converted to phosphoribosyl AMP which reacts with glutamine to form imidazole glycerol phosphate and 5-amino-ribosyl-4-imidazole carboxamide-5-phosphate which serves as a precursor in purine metabolism.

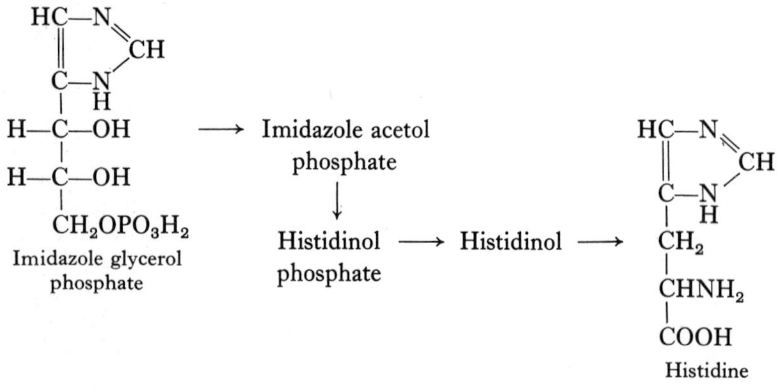

$$\underset{\text{Imidazole glycerol phosphate}}{\text{(imidazol-4-yl)}{-}C(H)(OH){-}C(H)(OH){-}CH_2OPO_3H_2} \longrightarrow \text{Imidazole acetol phosphate}$$

$$\downarrow$$

$$\text{Histidinol phosphate} \longrightarrow \text{Histidinol} \longrightarrow \underset{\text{Histidine}}{\text{(imidazol-4-yl)}{-}CH_2{-}CHNH_2{-}COOH}$$

(Imidazole ring as drawn: HC—N=CH—NH—C, with HC=C double bond.)

Imidazole glycerol phosphate is oxidized to the keto derivative, imidazole acetol phosphate, which can participate in amino transferase reactions with the production of histidinol phosphate. Dephosphorylation of histidinol phosphate produces histidinol which is then oxidized to histidine. The isolation of imidazole glycerol phosphate from plants coupled with the label competition experiments is consistent with the operation of this pathway; however, other components of the reaction sequence have yet to be characterized in higher plants.

THE AROMATIC AMINO ACIDS

There is now convincing evidence that in plants the amino acids tyrosine, phenylalanine and tryptophan are synthesized by the so-called 'shikimate pathway' originally characterized in microbial systems.

Labelling experiments indicate that shikimic acid is synthesized from phospho-enol-pyruvate and erythrose-4-phosphate through the intermediate 5-dehydroshikimic acid.[146]

The original condensation of phospho-enol-pyruvate and erythrose-4-phosphate is catalysed by the enzyme 3-deoxy-D-arabino-heptulosonic acid-7-phosphate synthase.[428]

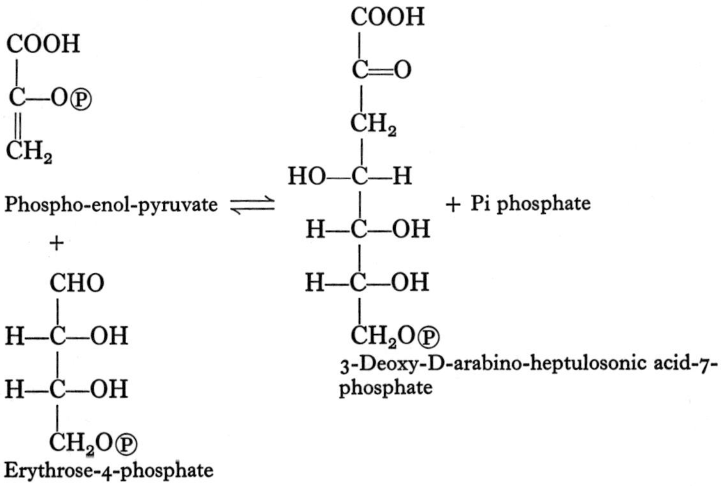

$$\begin{array}{l} COOH \\ | \\ C—O\text{Ⓟ} \\ \| \\ CH_2 \end{array}$$

Phospho-enol-pyruvate $\rightleftharpoons$

+

$$\begin{array}{c} CHO \\ | \\ H—C—OH \\ | \\ H—C—OH \\ | \\ CH_2O\text{Ⓟ} \end{array}$$

Erythrose-4-phosphate

$$\begin{array}{c} COOH \\ | \\ C=O \\ | \\ CH_2 \\ | \\ HO—C—H \\ | \\ H—C—OH \\ | \\ H—C—OH \\ | \\ CH_2O\text{Ⓟ} \end{array}$$

+ Pi phosphate

3-Deoxy-D-arabino-heptulosonic acid-7-phosphate

The 3-deoxy-D-arabino-heptulosonic acid is converted[451] to the cyclic compound 5-dehydroquinic acid which is transformed to 5-dehydroshikimic acid through the activity of 5-dehydroquinate dehydratase.[26, 27]

Activity of the enzyme shikimate dehydrogenase[545] converts 5-dehydroshikimate to shikimic acid.

5-Dehydroquinic acid —(5-DQ dehydratase, $-H_2O$)→ 5-Dehydroshikimic acid —(Shikimate dehydrogenase; NADPH$^+$ → NADP)→ Shikimic acid

The intermediates 5-dehydroquinic acid, 5-dehydroshikimic acid and shikimic acid have all been demonstrated in plant tissues.[25] Labelled shikimic acid is readily converted to phenylalanine, tyrosine and tryptophan. Few of the intermediates or enzymes involved in these conversions have been isolated from higher plants; however, on the basis of labelling data and the observation that plant extracts[429] can convert shikimate to phenylalanine and tyrosine in the presence of ATP, Mg^{2+}, glutamate and phospho-enol-pyruvate, it appears that the synthesis of the aromatic acids from shikimate proceeds in the following sequence originally established in microbial systems.

Shikimic acid + ATP ⟶ 5-phosphoshikimic acid + ADP.

Attempts to demonstrate this enzyme in plants have so far been unsuccessful.

5-Phosphoshikimate + phospho-enol-pyruvate ⟶ 3-enol-pyruvyl-shikimate-5-phosphate

3-Enol-pyruvyl-shikimate-5-phosphate ⟶ chorismic acid

The conversion of chorismic acid to prephenic acid has been demonstrated in plant extracts[127] and the reaction is catalysed by chorismate mutase. The reaction involves a transfer of the enol-pyruvyl moiety from the C_3 to the C_1 position.

HOOC— —OH
O
C
CH_2 COOH

Chorismic acid

COOH
—OH
CH_2—C—COOH
O

Prephenic acid

Activity of prephenic dehydrase[204] results in the simultaneous elimination of water and decarboxylation of prephenic acid with the formation of phenyl pyruvic acid which by appropriate amino transferase activity is converted to phenylalanine. Alternatively prephenic acid can undergo a reaction catalysed by an NADP-dependent prephenate dehydrogenase;[205] the resulting oxidative decarboxylation leads to the formation of hydroxy phenyl pyruvic acid which in amino transfer reactions is converted to tyrosine.

COOH
—OH
CH_2—C—COOH
O

Prephenic acid

Dehydrase →

CH_2—C—COOH
O

Phenyl pyruvic acid → Phenylalanine

Dehydrogenase →

—OH
CH_2—C—COOH
O

Hydroxyphenyl pyruvic acid → Tyrosine

In addition to functioning as a precursor for tyrosine and phenylalanine biosynthesis, chorismic acid is also utilized in tryptophan synthesis. The original step in the synthesis of tryptophan from chorismate involves the formation of anthranilic acid. This formation is achieved by the reaction of chorismate with glutamine in a reaction catalysed by the enzyme anthranilate synthetase.[46, 689]

The anthranilic acid is subsequently converted to *N*-5-phosphoribosyl anthranilate by a condensation with phosphoribosyl pyrophosphate; enzymes catalysing this reaction have recently been demonstrated in extracts from cells of *Daucus carota.*[690]

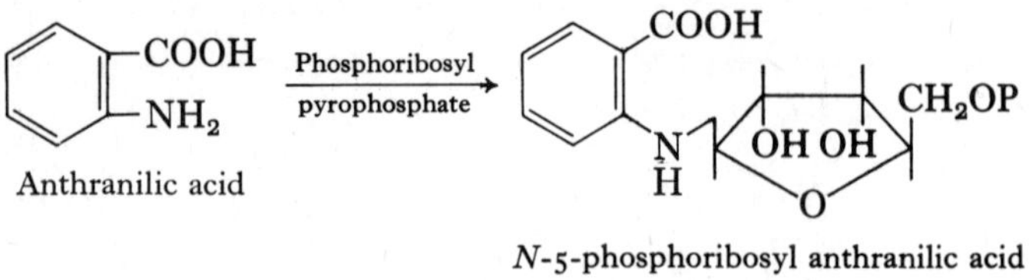

N-phosphoribosyl anthranilic acid is converted by the enzyme P-ribosyl anthranilate isomerase[690,691] to 1-(*o*-carboxyphenyl amino)-1-deoxyribulose-5-phosphate. This compound following a decarboxylation and dehydration catalysed by indole-3-glycerol phosphate synthetase[691] produces indole-3-glycerol phosphate. Indole-3-glycerol phosphate can be converted to tryptophan by plant extracts in the presence of pyridoxal phosphate and serine. However, at this stage it is not clear if this involves a direct exchange of the glycerol phosphate and serine residues or whether the indole glycerol phosphate is first converted to indole which then undergoes reaction with serine to form tryptophan. This tryptophan biosynthesis is catalysed by the enzyme tryptophan synthetase[145, 689, 691] which is usually assayed by the reactivity of free indole with serine; however such an assay does not preclude the direct exchange of serine and glycerol phosphate in an *in vivo* situation.

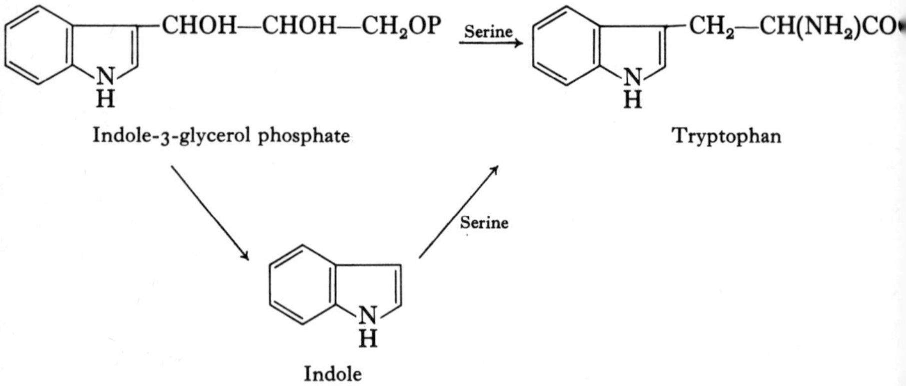

INTERACTION OF AMINO ACID BIOSYNTHESIS WITH RESPIRATION AND PHOTOSYNTHESIS

The utilization of organic acids to provide the carbon skeleton for amino acid biosynthesis explains the previously described (p. 25) change in respiratory metabolism when nitrogen-deficient plants are supplied with ammonia or nitrate. The organic acids are generated as a result of respiratory metabolism; depletion of these organic acids during amino acid biosynthesis allows for an enhanced flow of respiratory substrates into glycolysis and the TCA cycle producing the observed enhanced respiration.

Glycolytic activity provides the phosphoglyceric acid and pyruvic acid which can serve as precursors for the amino acids shown in Fig. 2.1. Metabolism of pyruvate in the mitochondria generates oxaloacetate and α-ketoglutarate which can be transformed into the 'head' amino acids aspartate and glutamate respectively. However, this depletion of carbon skeletons from the Krebs' TCA cycle for the biogenesis of amino acids presents a potential hazard to respiratory metabolism. The sustained oxidation of pyruvate in the Krebs' cycle requires a continued regeneration of oxaloacetate; this regeneration could be restricted if carbon skeletons are used in amino acid synthesis. To circumvent this depletion additional oxaloacetate can be generated through the activities of phospho-enolpyruvate carboxylase which generates oxaloacetate from phospho-enolpyruvate and carbon dioxide. The oxaloacetate can condense with acetate derived from pyruvate allowing for sustained utilization of carbohydrate and continued operation of the TCA cycle.

The TCA cycle is functioning in two capacities: on the one hand metabolism of pyruvate results in energy conservation and production of ATP; at the same time the cycle provides a source of keto acids for amino acid biosynthesis. It has been suggested[189] on the basis of labelling experiments that these somewhat conflicting functions may be separated with one class of mitochondria serving in amino acid biosynthesis and another class primarily generating ATP (Fig. 2.3).

In photosynthetic tissues there is also competition for carbon residues.

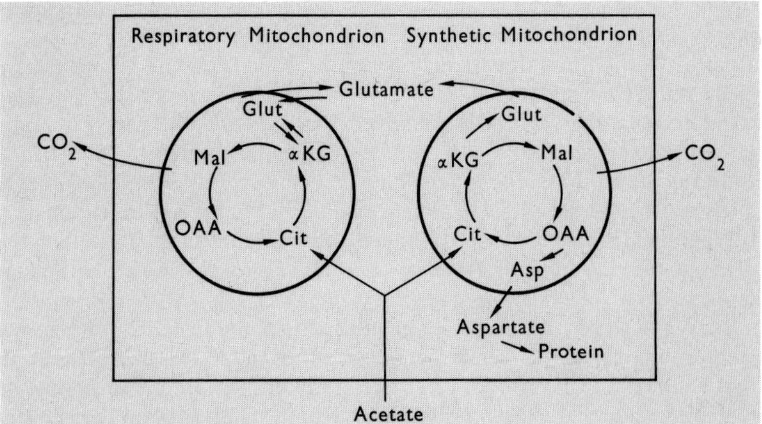

Fig. 2.3 Model proposing two populations of mitochondria in higher plant cells. (From Fletcher,[189] Fig. 3, p. 467.)

The initial photosynthetic product can be utilized in carbohydrate synthesis or alternatively it can be diverted into the glycolytic sequence to provide carbon skeletons for amino acid biosynthesis. Provision of ammonia to nitrogen-depleted *Chlorella* cells results in a reduction of carbohydrate synthesis and enhanced utilization of carbon for amino acid biosynthesis.[318] Part of this enhanced flow of carbon to amino acid biosynthesis may be attributed to an enhanced activity of the enzyme pyruvic kinase. This enzyme is stimulated by monovalent ions; if ammonia functions in this capacity there will be an accelerated traffic of carbon to pyruvate which can then be converted into other keto acid precursors of amino acids.

CONTROL OF AMINO ACID BIOSYNTHESIS

From the biosynthetic pathways described it is apparent that a particular precursor can function in the synthesis of a number of amino acids. Thus for example glutamate functions as a precursor for proline and arginine biosynthesis; however these amino acids accumulate at different rates suggesting that the diversion of glutamate to proline or arginine is regulated and not a random phenomenon.

Various mechanisms have been proposed to account for the differential synthesis of different products from a common precursor. In the case of proline and arginine it has been suggested, although no experimental evidence is available from higher plants, that the relative contribution of glutamate to either amino acid is dependent upon the degree of acetylation of glutamate. It is postulated that the substituted glutamate could serve as the preferential substrate for arginine biosynthesis whereas the free acid is utilized for proline synthesis.

The regulation of the biosynthesis of amino acids derived from aspartate is equally complex and factors determining the relative diversion of aspartate carbon into threonine, methionine, isoleucine and lysine remain to be resolved. From the view of controlling lysine synthesis it could be suggested that regulation could be exerted at the level of aspartyl kinase.[83] However, control of this enzyme by lysine could also regulate aspartate conversion into threonine, methionine and isoleucine.

During the biosynthesis of alanine, valine and leucine there is potential competition for carbon derived from pyruvate; again factors controlling the diversion of carbon from pyruvate into any of these amino acids remains to be resolved.

It has, however, been demonstrated that in some instances the flow of carbon into a particular amino acid is a function of the concentration of that amino acid. Oaks[471, 472] has successfully utilized feeding experiments to demonstrate that exogenously supplied amino acids prevent the entry of carbon from glucose or acetate into specific amino acids (Table 2.6).

This type of feedback regulation in which the end product prevents entry

Table 2.6 Effect of exogenous amino acids on the incorporation of acetate-2-^{14}C in amino acids of maize root tips.

Amino acids	(mg/2 ml)	c.p.m./20 tips Sucrose	Sucrose and amino acids
Serine	0.347	2 755	2 830
Glycine	0.168	2 629	2 348
Threonine	0.198	13 760	4 319
Alanine	0.476	29 560	31 860
Valine	0.33	5 725	2 846
Leucine–Isoleucine	1.009 and 0.237	35 560	9 369
Lysine	0.088	7 090	2 140
Arginine	0.15	11 745	9 110
Proline	0.621	25 990	3 980
γ-Aminobutyric	—	48 100	48 400
Glutamic	—	32 924	34 165
Aspartic	—	5 875	5 482

(Excised root tips were pretreated for 3 hours in a solution of salts containing 1% sucrose and a mixture of amino acids. At this time the roots were washed with water and fresh salts, sucrose and amino acids were added together with acetate-2-^{14}C for an additional 2 hours. (Data recalculated from Oaks,[472] Table II, p. 151.)

of metabolites, i.e. carbon, into a particular metabolic sequence can be attributed to two causes: **end product inhibition** and **end product repression.**

In the case of end product inhibition the end product of a particular metabolic pathway inhibits the activity of enzymes involved in initial reactions of that pathway.

In end product repression a particular end product causes a termination of the synthesis of enzymes required for initial reactions of a metabolic sequence.

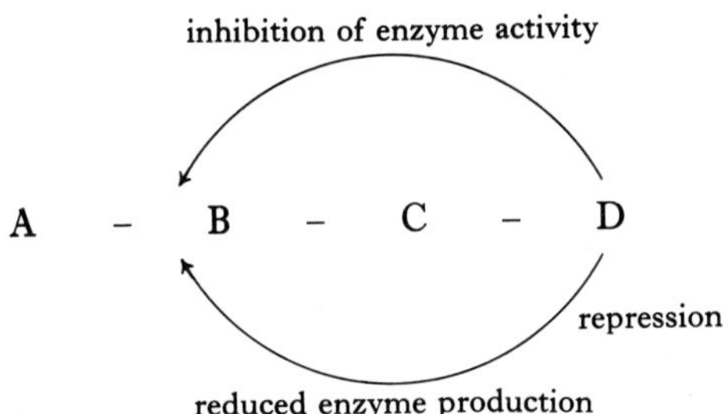

In higher plants the evidence to date, although limited, indicates that end product inhibition is the principal mechanism of regulating amino acid metabolism. This type of control, if suspected, can be tested for by studying the influence of the end product D on the intermediary enzyme reactions

Table 2.7 Enzymes involved in amino acid biosynthesis in plants which have been shown to be subject to end product inhibition.

Enzyme	Inhibitor
Acetohydroxy acid synthetase	Isoleucine[423] Leucine Valine
Homoserine dehydrogenase	Threonine[82, 1a]
Threonine deaminase	Isoleucine[156]
Isopropyl malate synthetase	Leucine[470]
Anthranilate synthetase	Tryptophan[46, 690]
Asparto kinase	Lysine[83, 109, 704] Threonine[1a]

A – B, B – C, etc. Using this approach, it has been shown that a variety of amino acids function in feed-back inhibition of enzymes involved in their synthesis (Table 2.7).

That this type of feedback and product regulation is a normal aspect of metabolism and not an artifact that is induced by exogenous amino acids is implicated by the experiments of Fletcher and Beevers.[190] In these studies application of the protein synthesis inhibitor, cycloheximide, inhibited protein synthesis and also markedly curtailed the incorporation of precursors into the amino acids arginine, lysine, isoleucine, valine and proline. The rapid cessation of synthesis of these amino acids was interpreted as being due to a transient accumulation of the amino acids when their incorporation into protein was prevented. The accumulated amino acids by the process of end product inhibition prevented entry of labelled carbon into metabolic sequences leading to their synthesis.

3

Amino Acid Metabolism

In addition to their function as precursors in protein synthesis amino acids serve as intermediates in the synthesis of other cellular constituents and are also capable of undergoing extensive interconversion.

The capacity for interconversion is most readily demonstrated at the time of seed germination (Table 3.1.) During germination the reserve proteins provide a carbon and nitrogen source for the synthesis of amino acids and cellular proteins in the growing seedling. These proteins of the growing seedling have a different amino acid composition from that of the reserve proteins so that there appears to be extensive interconversion of amino acids. Such changes are evident from the analyses by Folkes and Yemm,[194] who measured the change in content of specific amino acids of barley (*Hordeum vulgare*) seedlings over a ten-day germination period in the absence of added nitrogen.

Table 3.1 Changes in total content of individual amino acids during germination. Results are expressed as mg N/100 seedlings. (From Folkes and Yemm.[194])

	Total content on day 0	Change 0–10 days
Aspartic acid	2.91	+0.26
Glutamic acid	7.99	−3.86
Proline	4.88	−2.67
Lysine	2.82	+0.95
Histidine	2.05	−0.06
Arginine	6.75	+0.42

Similar interconversions occur during the process of seed development and maturation. In this case the reserve proteins being synthesized have a different amino acid composition from that of the cellular proteins or of the free amino acids in the xylem stream arriving at the site of seed maturation.

Although these situations indicate the possibility of amino acid interconversion, very few studies have been directed at elucidating the mechanisms involved. Foreseeably the transformations involve a deamination or transamination of a particular amino acid with the resulting keto acid derivative being degraded into components which can be utilized in the Krebs' tricarboxylic acid cycle. The carbon from the tricarboxylic acid cycle can then be reincorporated into other amino acids. Evidence that the carbon skeletons of amino acids can be metabolized in this manner is provided by experiments in which labelled amino acids are supplied to a germinating seedling (Table 3.2).

Table 3.2 Metabolism and interconversion of leucine U†-^{14}C and aspartate U-^{14}C in pea (*Pisum sativum*) cotyledons and castor bean (*Ricinus communis*) endosperm during germination. Results are recalculated and expressed as % of ^{14}C recovered in the named component.

Component	Pea cotyledon		Castor bean endosperm	
	Leucine[45]	Aspartate[353]	Leucine[595]	Aspartate[595]
Sugars	1	1	12	30
Organic acids	11	19	3	30
Soluble amino acids	26*	42*	84	30
Carbon dioxide	12	30	2	6
Insoluble residue	50	7	N.D.	N.D.

N.D.—Not determined.
* The predominant radioactive soluble amino acids in the pea cotyledon are respectively homoserine, glutamate and aspartate.
† U—Uniformly labelled.

In germinating peas (*Pisum sativum*) labelled carbon from supplied leucine is evolved as carbon dioxide,[45, 353] enters organic acids and is incorporated into other amino acids. In contrast labelled carbon from exogenously supplied phenylalanine is not extensively incorporated into other amino acids.[467] This disparity may be related to the relative proximity in metabolic terms of a particular amino acid to the tricarboxylic acid cycle. Those amino acids which are closely linked to the tricarboxylic acid cycle can apparently readily provide carbon skeletons which can undergo modifications during excursions through the cycle and can thus become incorporated into other amino acids.

In some instances the carbon skeletons of amino acids derived from the reserve proteins may be converted into carbohydrate during germination.[595]

This gluconeogenic function of amino acids, although extensive in animals, appears to be of restricted occurrence in plants. Significantly it occurs, as so far determined, in those species which show a lipid to sugar transformation during germination and is probably associated with the capacity of these plants to convert oxaloacetate to phospho-enol-pyruvate which can be subsequently metabolized to glucose by a reversal of glycolysis.

THE UREA CYCLE

The reserve proteins of seeds are frequently enriched in specific amino acids, particularly arginine which seems to function as a nitrogen storage compound. Additionally soluble amino acids appear to function in this capacity in certain instances. In some legumes the non-protein amino acid canavanine accumulates in large amounts in the developing seed. During seed germination the amino acid is depleted commensurate with the development of the growing embryo suggesting that this amino acid functions as a reserve which is metabolized and utilized during germination. In some woody perennials soluble arginine appears to function as a storage reserve.

Arginine and canavanine are related chemically and their metabolism during germination and breaking of dormancy in woody perennials has been the subject of extensive recent studies. Incubation of tissue with arginine demonstrates the capacity of cotyledonary[311] or bud tissue[169] to convert arginine to ornithine and related compounds such as citrulline, glutamic acid, Δ'-pyrroline-5-carboxylic acid, proline and γ-amino butyric acid. These compounds can all be derived from ornithine so that it appears that the initial metabolism of arginine occurs through its cleavage to ornithine and some other compound. When arginine, labelled in the guanido carbon, is incubated with cotyledonary slices or expanding bud tissue, labelled urea is formed. Such findings are consistent with the initial metabolism of arginine by the enzyme arginase[311] to form ornithine and urea (see Fig. 3.1). The enzyme arginase can hydrolyse canavanine to canaline and urea.[537, 688]

The production of urea in animal systems also occurs through the activities of arginase and it has been suggested that the production of urea in this fashion provides a convenient mechanism for the removal of ammonia in a cyclic series of reactions proposed originally by Krebs and Henseleit[344] in 1932.

Although the enzymes for such a cyclic metabolism occur in plant systems it appears that the cycle rarely operates.[311] Rather it seems that the products of arginase activity are degraded in other fashions. The observations that glutamate, Δ'-pyrroline-5-carboxylic acid, proline and γ-amino butyric acid become labelled from exogenously supplied arginine suggest that ornithine produced by arginase activity is converted to the intermediate glutamic

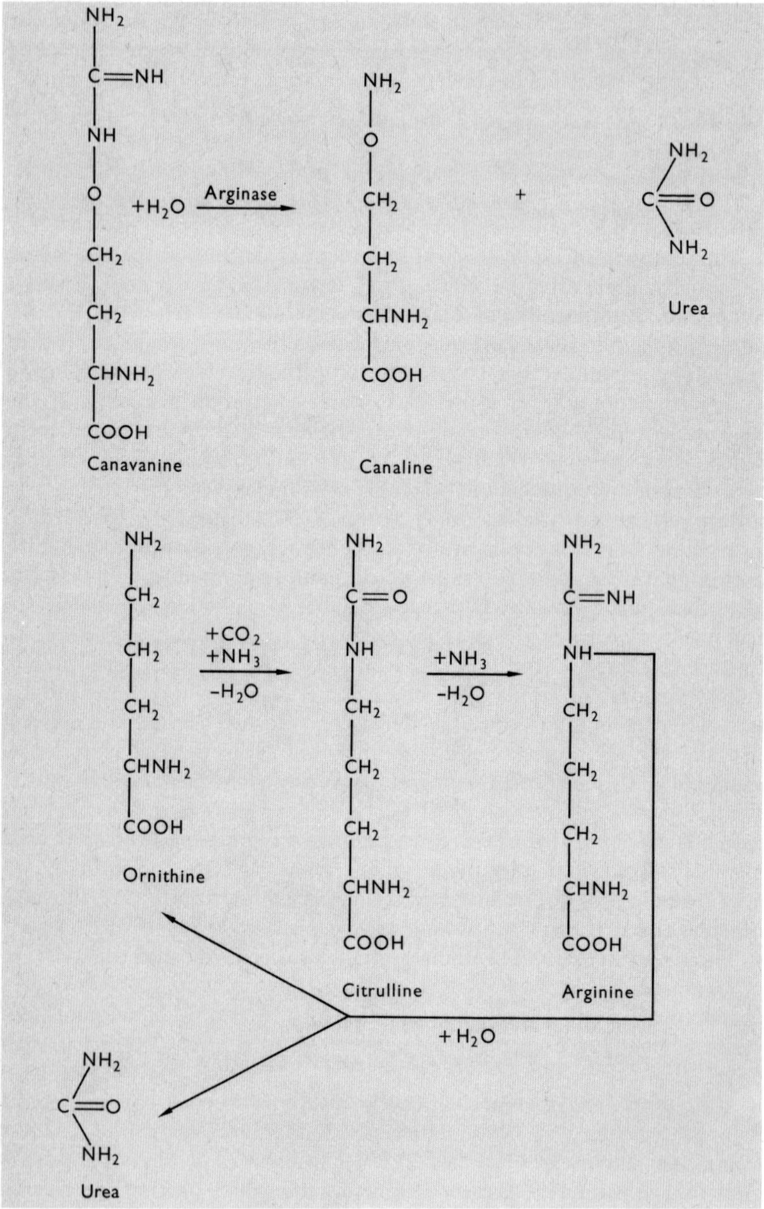

NH_2
|
$C{=}NH$
|
NH
|
O $+H_2O$ $\xrightarrow{\text{Arginase}}$
|
CH_2
|
CH_2
|
$CHNH_2$
|
$COOH$
Canavanine

NH_2
|
O
|
CH_2 +
|
CH_2
|
$CHNH_2$
|
$COOH$
Canaline

NH_2
$C{=}O$
NH_2
Urea

NH_2
|
CH_2
|
CH_2 $\xrightarrow[-H_2O]{+CO_2 \; +NH_3}$
|
CH_2
|
$CHNH_2$
|
$COOH$
Ornithine

NH_2
|
$C{=}O$
|
NH $\xrightarrow[-H_2O]{+NH_3}$
|
CH_2
|
CH_2
|
CH_2
|
$CHNH_2$
|
$COOH$
Citrulline

NH_2
|
$C{=}NH$
|
NH
|
CH_2
|
CH_2
|
CH_2
|
$CHNH_2$
|
$COOH$
Arginine

$+H_2O$

NH_2
$C{=}O$
NH_2
Urea

Fig. 3.1 A modified urea cycle.

acid semialdehyde which can then give rise to other products (see Chapter 2, p. 38).

Urea produced as a result of arginase activity in higher plants may be degraded by the enzyme urease to ammonia and carbon dioxide.

$$\begin{matrix} NH_2 \\ / \\ C{=}O \\ \backslash \\ NH_2 \end{matrix} + H_2O \xrightarrow{\text{Urease}} CO_2 + 2NH_3$$

The ammonia may be utilized in the formation of glutamate or glutamine and subsequently incorporated into other amino acids by transaminase activity. Such a sequence demands the functioning of the enzyme urease. This enzyme has been extensively characterized from plants and appropriately was originally crystallized in copious amounts from Jack Bean (*Canavalia ensiformis*) cotyledons which are enriched in the amino acid canavanine.[613]

DECARBOXYLATION OF AMINO ACIDS

A frequent component of the non-protein amino acids encountered in plant extracts is γ-amino butyric acid. Although the physiological role of this amino acid is not completely understood, in addition to its widespread occurrence it has been shown to be produced extensively in germinating seeds[283] and in leaf tissues under conditions of water stress and/or anaerobiosis.[608] The consensus indicates that γ-amino butyric acid is produced by decarboxylation of glutamic acid. This decarboxylation is catalysed by the enzyme glutamic acid decarboxylase.[36]

$$\begin{matrix} COOH \\ | \\ CHNH_2 \\ | \\ CH_2 \\ | \\ CH_2 \\ | \\ COOH \\ \text{Glutamic acid} \end{matrix} \longrightarrow \begin{matrix} CO_2 \\ + \\ CH_2NH_2 \\ | \\ CH_2 \\ | \\ CH_2 \\ | \\ COOH \\ \gamma\text{-Amino butyric acid} \end{matrix} \longrightarrow \begin{matrix} CHO \\ | \\ CH_2 \\ | \\ CH_2 \\ | \\ COOH \\ \text{Succinic semialdehyde} \end{matrix} \longrightarrow \begin{matrix} COOH \\ | \\ CH_2 \\ | \\ CH_2 \\ | \\ COOH \\ \text{Succinic acid} \end{matrix}$$

Labelling studies[153, 283, 000] indicate that the γ-amino group can be incorporated into other amino acids, presumably through an amino transfer reaction. The succinic semialdehyde produced by the deamination can subsequently be oxidized by an NAD dependent oxidoreductase[609] to succinic acid which can enter the TCA cycle.[153, 283]

Although less extensively studied a similar decarboxylation of aspartic acid could result in the formation of β-alanine which is occasionally encountered as a soluble non-protein amino acid in plant extracts.

$$\underset{\text{Aspartic acid}}{\begin{array}{c} COOH \\ | \\ CH_2 \\ | \\ CHNH_2 \\ | \\ COOH \end{array}} \longrightarrow CO_2 + \underset{\beta\text{-Alanine}}{\begin{array}{c} COOH \\ | \\ CH_2 \\ | \\ CH_2NH_2 \end{array}}$$

β-Alanine is of additional importance because of its function as a component of coenzyme A (see p. 103).

Amine synthesis

Although occurring in low concentrations, amines are widely distributed in plants. The amines can be divided into simple aliphatic primary amines, diamines, aromatic amines and amino sugars.

PRIMARY AMINES The aliphatic primary amines can occur free and in this situation they are present in relatively high concentrations in flowers. This is a feature of many species with inconspicuous green or white flowers and in many instances amine production seems to serve as an insect attracting mechanism which facilitates pollination.[575] As well as occurring free, amines or amine derivatives such as ethanolamine or choline are important constituents of phospholipids.

The primary aliphatic amines may be formed by the decarboxylation of the appropriate amino acid (Table 3.3) or transamination of the precursor aldehyde. However, as Smith[575] indicates, the biosynthetic mechanism for the production of many amines remains to be established.

$$\text{e.g.}\quad \underset{\text{Amino acid}}{R_1\text{—}\underset{\displaystyle NH_2}{\underset{|}{CH}}\text{—}COOH} \xrightarrow{\text{Decarboxylation}} \underset{\text{Amine}}{R_1\text{—}\underset{\displaystyle NH_2}{\underset{|}{CH}}\text{—}H} + CO_2$$

$$\underset{\text{Aldehyde}}{R\text{—}C\begin{array}{l} \diagup H \\ \diagdown\!\!\diagdown O \end{array}} + \underset{\text{Amino acid}}{R_1\text{—}\underset{\displaystyle NH_2}{\underset{|}{CH}}\text{—}COOH} \xrightarrow{\text{Aldehyde transaminase}} \underset{\text{Amine}}{R\text{—}\underset{\displaystyle NH_2}{\underset{|}{CH}}\text{—}H} + \underset{\alpha\text{-Keto acid}}{R_1\text{—}\underset{\displaystyle O}{\underset{\|}{C}}\text{—}COOH}$$

Table 3.3 Amino acid precursors of some primary aliphatic amines, diamines, and aromatic amines.

Amino acid	Amine	Formula
Serine	Ethanolamine	$CH_2(OH)-CH(NH_2)-H$
Valine	Butylamine	$(H_3C)_2CH-CH(NH_2)-H$
Leucine	Amylamine	$(H_3C)_2CH-CH_2-CH(NH_2)-H$
Methionine	Methyl mercaptopropyl amine	$CH_2(S-CH_3)-CH_2-CH(NH_2)-H$
Lysine	Cadaverine	$CH_2(NH_2)-(CH_2)_3-CH(NH_2)-H$
Ornithine	Putrescine	$CH_2(NH_2)-(CH_2)_2-CH(NH_2)-H$
Arginine	Agmatine	$CH_2(NH-C(=NH)-NH_2)-(CH_2)_2-CH(NH_2)-H$
Phenylalanine	Phenylethylamine	$C_6H_5-CH_2-CH(NH_2)-H$
Tyrosine	Tyramine	$HO-C_6H_4-CH_2-CH(NH_2)-H$
Tryptophan	Tryptamine	indol-3-yl$-CH_2-CH(NH_2)-H$ (indole ring with N–H)

Ethanolamine apparently functions as a precursor in choline biosynthesis; the conversion involves progressive methylation of the amine nitrogen with methionine serving as the methyl donor.

$$\underset{\text{Ethanolamine}}{CH_2(OH)-CH(NH_2)-H} + CH_3 \longrightarrow \underset{\text{Choline}}{CH_2(OH)-CH_2-\overset{+}{N}(CH_3)_3}$$

Diamines

In bacterial systems decarboxylation of the amino acids arginine and ornithine leads to the formation of the amines agmatine and putrescine respectively. These amines also occur in plants and while ornithine can be converted to putrescine the process is slow and it thus appears that ornithine decarboxylation is not the major mechanism of formation of this amine. More convincingly it appears that putrescine is derived from agmatine, the decarboxylation product of arginine according to the scheme below.[574]

$$
\begin{array}{lclclcl}
NH_2 & & NH_2 & & NH_2 & & \\
| & & | & & | & & \\
C{=}NH & & C{=}NH & & C{=}O & & NH_2 \\
| & & | & & | & & | \\
NH & \xrightarrow{(1)} & NH & \longrightarrow & NH & \xrightarrow{(2)} & (CH_2)_3 \\
| & & | & & | & & | \\
(CH_2)_3 & & (CH_2)_3 & & (CH_2)_3 & & | \\
| & & | & & | & & | \\
CHNH_2 & & CH_2NH_2 & & CH_2NH_2 & & CH_2NH_2 \\
| & & & & & & \\
COOH & & & & & & \\
\text{Arginine} & & \text{Agmatine} & & N\text{-carbamyl putrescine} & & \text{Putrescine}
\end{array}
$$

(1) = Arginine decarboxylase
(2) = *N*-carbamyl putrescine amido-hydrolase

The physiological and metabolic role of these amines remains to be established; however, agmatine and to a greater extent putrescine accumulate in leaves of potassium-deficient plants. Normal plants of barley (*Hordeum vulgare*) contain approximately 30 μg/g fresh weight of agmatine and about 20 μg/g fresh weight of putrescine; in deficient leaves levels of 200 μg/g fresh weight agmatine and 1 mg/g fresh weight putrescine have been reported.[574] It is postulated that amine production provides a mechanism by which the plant maintains a constant internal pH. Under conditions of potassium deficiency internal cellular pH may be lowered, triggering the production of amines. Consistent with this concept are the observations of Smith[574] that the level of the enzymes arginine decarboxylase and *N*-carbamyl putrescine amido-hydrolase is increased in potassium-deficient plants or in plants treated with solutions of dilute acid.

Aromatic amines

Decarboxylation of the aromatic amino acids phenylalanine and tyrosine produces the phenolic amines phenylethylamine and tyramine, important precursors in the biosynthesis of other aromatic amines and alkaloids discussed later (p. 76).

Amino sugars

Although the occurrence of amino sugars was recognized early in animal and microbial systems, their occurrence in higher plants had not been widely demonstrated until the 1960's. Since that time it has become apparent that these compounds are fairly widespread, being present in some cell wall proteins,[351] and in reserve protein hydrolysates[511] as well as occurring in the cytoplasmic proteins of some species.[529, 530] Additionally, derivatives of amino sugars have been detected in combination with nucleotides.[411]

Little work has been done on amino sugar biosynthesis. However, it has been found that glucosamine and its derivatives are synthesized from fructose-6-phosphate and glutamine by the action of the enzyme L-glutamine D-fructose-6-phosphate amido-transferase.[656]

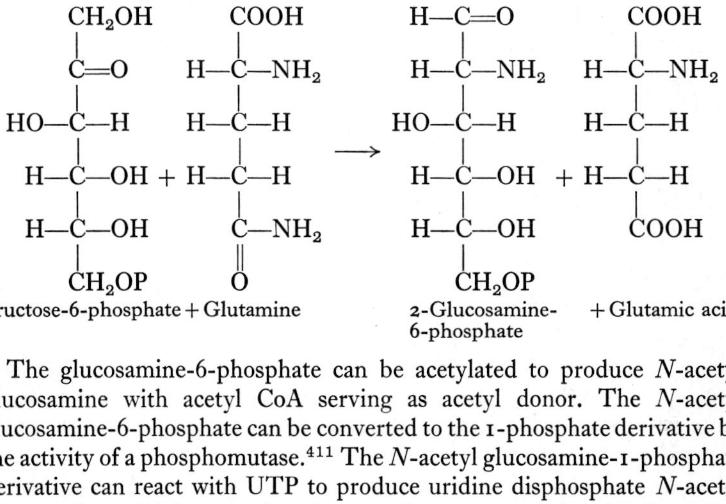

$$
\begin{array}{c}
CH_2OH \\ | \\ C{=}O \\ | \\ HO{-}C{-}H \\ | \\ H{-}C{-}OH \\ | \\ H{-}C{-}OH \\ | \\ CH_2OP
\end{array}
+
\begin{array}{c}
COOH \\ | \\ H{-}C{-}NH_2 \\ | \\ H{-}C{-}H \\ | \\ H{-}C{-}H \\ | \\ C{-}NH_2 \\ \| \\ O
\end{array}
\longrightarrow
\begin{array}{c}
H{-}C{=}O \\ | \\ H{-}C{-}NH_2 \\ | \\ HO{-}C{-}H \\ | \\ H{-}C{-}OH \\ | \\ H{-}C{-}OH \\ | \\ CH_2OP
\end{array}
+
\begin{array}{c}
COOH \\ | \\ H{-}C{-}NH_2 \\ | \\ H{-}C{-}H \\ | \\ H{-}C{-}H \\ | \\ COOH
\end{array}
$$

Fructose-6-phosphate + Glutamine 2-Glucosamine-6-phosphate + Glutamic acid

The glucosamine-6-phosphate can be acetylated to produce *N*-acetyl glucosamine with acetyl CoA serving as acetyl donor. The *N*-acetyl glucosamine-6-phosphate can be converted to the 1-phosphate derivative by the activity of a phosphomutase.[411] The *N*-acetyl glucosamine-1-phosphate derivative can react with UTP to produce uridine disphosphate *N*-acetyl glucosamine which may function as the precursor for the formation of polymeric products or be converted to other amino sugar derivatives.[411]

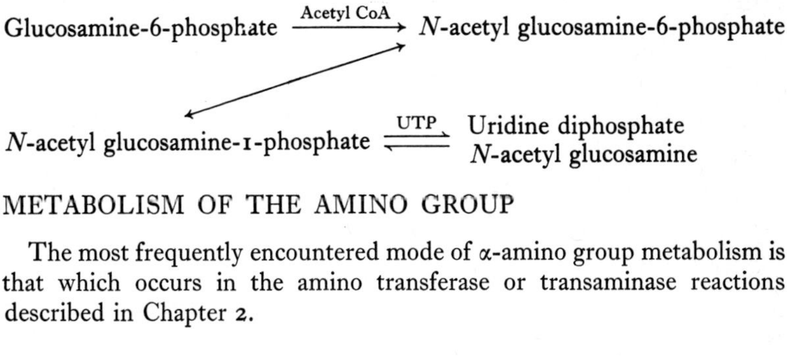

$$\text{Glucosamine-6-phosphate} \xrightarrow{\text{Acetyl CoA}} N\text{-acetyl glucosamine-6-phosphate}$$

$$\updownarrow$$

$$N\text{-acetyl glucosamine-1-phosphate} \overset{\text{UTP}}{\rightleftharpoons} \text{Uridine diphosphate } N\text{-acetyl glucosamine}$$

METABOLISM OF THE AMINO GROUP

The most frequently encountered mode of α-amino group metabolism is that which occurs in the amino transferase or transaminase reactions described in Chapter 2.

Amine oxidase

Plants contain enzymes capable of converting amines to the corresponding aldehydes according to the general equation:

$$R.CH_2.NH_2 + O_2 + H_2O \rightleftharpoons R.CHO + NH_3 + H_2O_2.$$

The subsequent metabolism of the generated hydrogen peroxide by catalase produces the overall equation:

$$R.CH_2.NH_2 + \tfrac{1}{2}O_2 \rightleftharpoons R.CHO + NH_3$$

The amine oxidases can be divided into the monomine oxidases and the diamine oxidases and while the monoamine oxidases have been reported in plants they have been little studied. In contrast Mann and coworkers[265] have performed detailed investigations on diamine oxidases. The partially purified enzyme is a copper-containing oxidase and is most active against diamines and less rapidly attacks aliphatic monoamines and lysine.[265]

Such oxidases could potentially also convert amino acids to corresponding aldehydes; however diamine acids such as ornithine and lysine are oxidized but slowly. Thus the participation of this enzyme in amino group metabolism is questionable. More significantly it has been suggested that the enzyme may be involved in indolyl acetic acid formation (p. 85) and in the formation of heterocyclic derivatives involved in alkaloid biogenesis (p. 73).

Ammonia lyases

The amino group can be metabolized in a non-oxidative manner by the activities of the so-called ammonia lyases. The most fully characterized enzyme of this type is phenylalanine ammonia lyase (PAL). Also described is a tyrosine ammonia lyase; however, the most recent evidence indicates that the non-oxidative deamination of both tyrosine and phenylalanine is achieved by the same moiety.[248]

$$C_6H_5-CH_2-CHNH_2-COOH \underset{}{\overset{PAL}{\rightleftharpoons}} C_6H_5-CH{=}CH-COOH + NH_3$$

L-phenylalanine → Trans cinnamic acid

$$HO-C_6H_4-CH_2-CHNH_2-COOH \underset{}{\overset{PAL}{\rightleftharpoons}} HO-C_6H_4-CH{=}CH-COOH + NH_3$$

Tyrosine → Trans *p*-coumaric acid

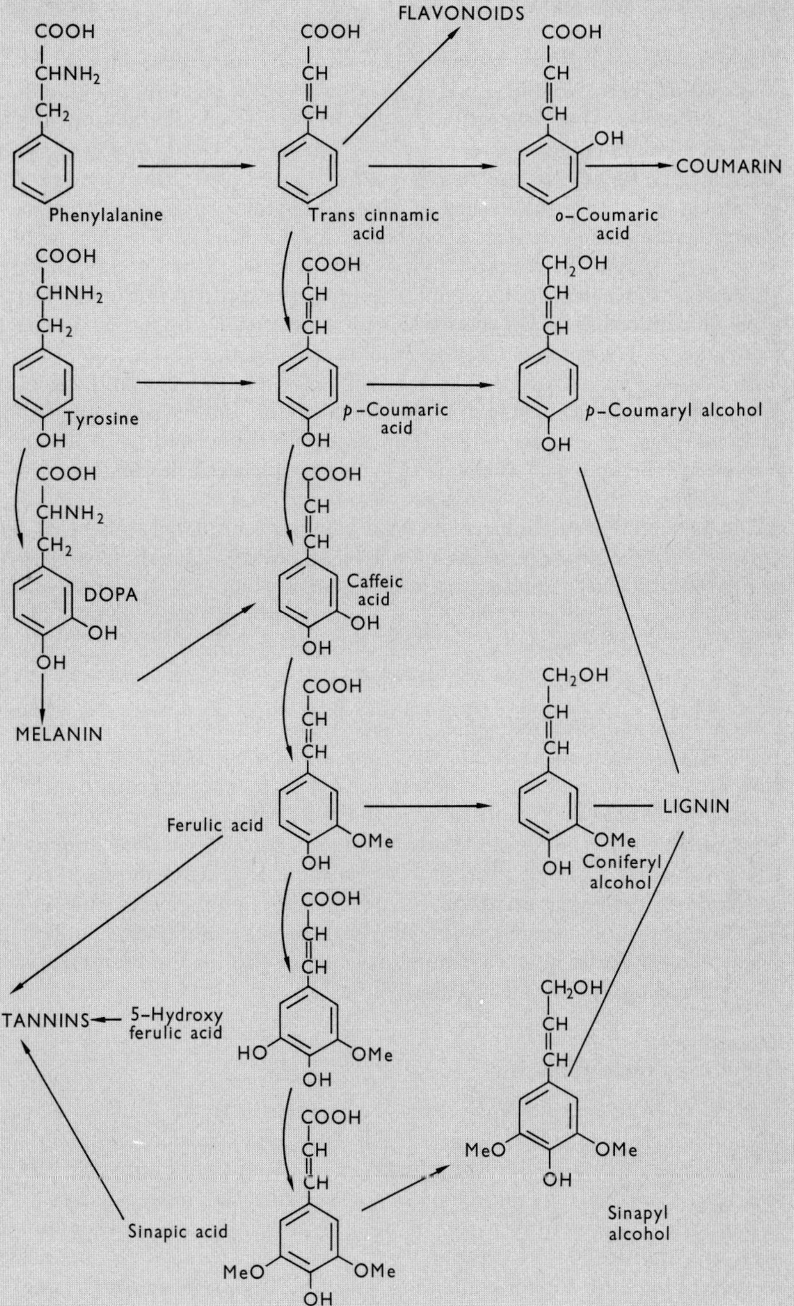

Fig. 3.2 Some secondary products from cinnamic acid and *p*-coumaric acid derived from phenylalanine and tyrosine.

The tissue level of the enzyme PAL is in many instances increased by illumination; however the nature of the photoreceptor pigment stimulating the enhanced enzyme production has not been resolved. The level of the enzyme in sweet potato (*Ipomoea batatas*) roots is increased by wounding. This facility to change the tissue level of the enzyme has provided a system in which to study the synthesis of a specific enzyme. Although inhibitor studies suggest that the increased enzyme level is associated with *de novo* synthesis, there is still considerable conflict in this area.[571] The enzyme is of great metabolic significance since the products of the reaction trans cinnamic acid and *p*-coumaric acid are extremely important in the biosynthesis of many secondary metabolites (Fig. 3.2).

In addition *p*-coumaric acid serves as the precursor for the synthesis of *p*-hydroxybenzoic acid which is involved in the biosynthesis of ubiquinone,[628] a coenzyme which functions in mitochondrial electron transport. The keto acid analogue of tyrosine which can be produced by transaminase activity is *p*-hydroxy phenyl pyruvic acid. This compound can be metabolized to homogenistic acid which is considered to be a precursor of the aromatic nucleus of plastoquinone[628] which functions in photosynthetic electron transport in chloroplasts (Fig. 3.3).

HYDROXYLATION OF AMINO ACIDS

In animal and some bacterial systems hydroxylation of phenylalanine plays an important role in the interconversion of this amino acid to tyrosine. However the biosynthesis of tyrosine in this manner appears to be of limited occurrence in higher plants. In instances where the phenylalanine–tyrosine conversion has been recorded it appears that the initial metabolism has involved the deamination of phenylalanine to phenyl pyruvic acid which upon hydroxylation forms *p*-hydroxy phenyl pyruvic acid, the keto acid precursor of tyrosine. According to Leete *et al.*[362] some limited direct hydroxylation of phenylanine may be involved in the biosynthesis of some alkaloids from phenylalanine.

Dihydroxy-phenylalanine

A frequently encountered soluble non-proteinaceous amino acid encountered in some plant extracts is dihydroxy-phenylalanine or DOPA. This amino acid is formed by the hydroxylation of tyrosine. The hydroxylation may be mediated by a specific hydroxylase; however, most evidence[509] favours the conversion by means of the enzyme tyrosinase or phenolase. This enzyme belongs to a general group of little characterized phenol oxidases which are copper-containing enzymes. In addition to catalysing

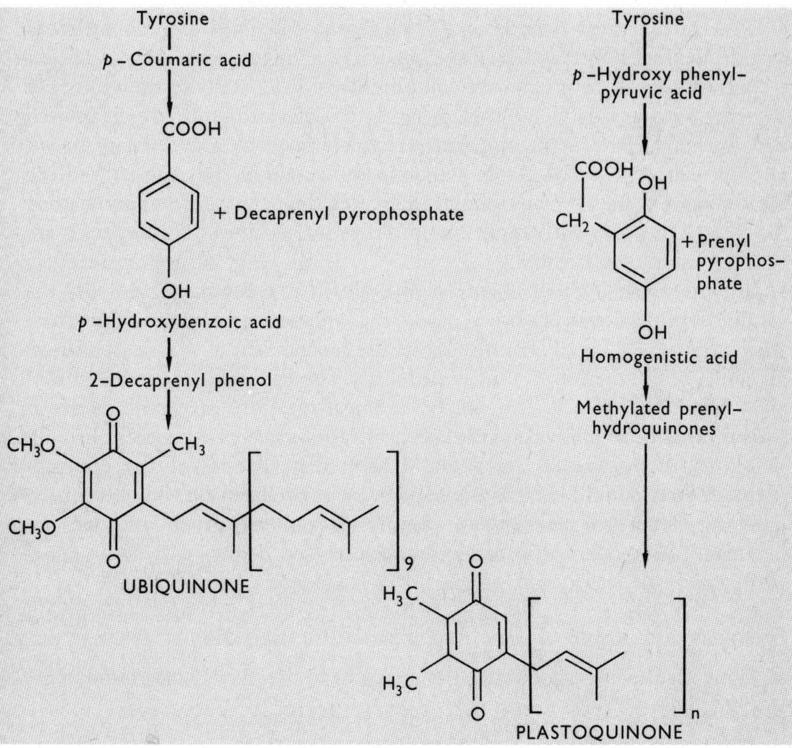

Fig. 3.3 Proposed mechanisms for biosynthesis of ubiquinone and plastoquinone with tyrosine providing the aromatic residue.

DOPA formation the same or similar enzymes can form quinone derivatives of DOPA and bring about the coupling of phenols with the hydroxyl group at the ortho position. Metabolism of DOPA in this manner gives rise to the melanin pigments which are complex polymers.

HO–C_6H_4–CH_2–CH(H_2N)–COOH (Tyrosine) —Tyrosinase→ (HO)$_2$$C_6H_3$–$CH_2$–CH($H_2N$)–COOH (Dopa)

Tyrosine | Dopa

In addition to serving as a precursor in melanin biosynthesis, DOPA or its decarboxylation product DOPAMINE function extensively in alkaloid biosynthesis described later.

ALKALOID BIOSYNTHESIS

An important function of amino acids is to provide the carbon skeleton and the nitrogenous components of alkaloids. Alkaloids are characterized as nitrogenous bases which occur naturally in plants. They form an extremely heterogeneous group of compounds which usually contain nitrogen as part of a heterocyclic system. A particular alkaloid is usually of limited distribution in the plant kingdom being confined to a specific genus or closely related group of plants. Much is known about the chemistry of the alkaloids. A conservative estimate indicates the occurrence of some 4000 different alkaloids, with more being described as more plant species are assayed for alkaloid content.

In spite of their great chemical diversity it is customary to divide the alkaloids into categories based upon the organic nitrogenous base from which they are derived. Knowledge of the mechanism of synthesis of most alkaloids is fragmentary. This is primarily due to the fact that the rate of alkaloid biosynthesis is slow, many chemical reactions are involved and it is thus difficult to follow incorporation of radioactive constituents through many of the proposed metabolic intermediates. Few of the enzymes involved in alkaloid biosynthesis have been isolated but on the basis of the labelling structures and known chemical reactions various biosynthetic sequences have been proposed and are described below.

Biosynthesis of pyrrolidine and piperidine alkaloids

Many alkaloids can be related chemically to the cyclic compounds

PYRROLIDINE ALKALOIDS These are considered to be synthesized primarily from ornithine. Evidence for the concept is derived from degradation of alkaloids from plants which had been incubated in the presence of 2-^{14}C ornithine. The radioactivity in the intact alkaloid could be accounted for in terms of the activity of individual carbon atoms at sites predicted on the basis of synthesis from ornithine.

Activity of 2-^{14}C ornithine for example is incorporated into the pyrrolidine moiety of the alkaloids nicotine, tropine, retronecine and hyoscyamine in a manner consistent with a derivation from ornithine; however, the mode of incorporation differs in each case.

In the case of nicotine the carbons 2 and 5 of the pyrrolidine nucleus are equally labelled suggesting that this pathway involves a symmetrical intermediate.[366] In contrast the activity in tropine derived from 2-^{14}C ornithine is confined to a single carbon atom and thus a symmetrical intermediate cannot be involved.[360] These labelling experiments indicate the operation of different pathways for the biosynthesis of pyrroline precursor of the alkaloids. The labelling distribution can be accounted for by the schemes outlines below (Fig. 3.4).

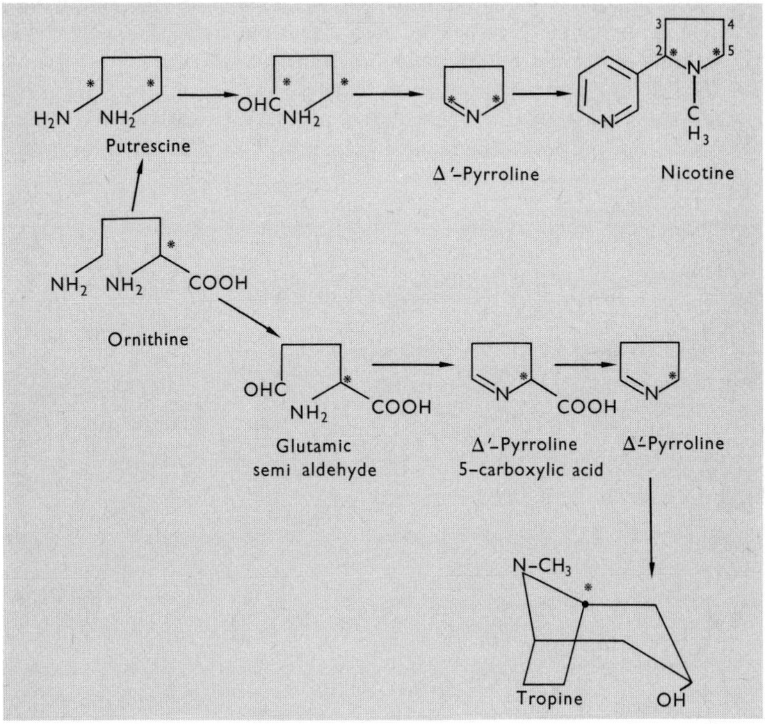

Fig. 3.4 Proposed mechanisms for ornithine incorporation into pyrrolidine alkaloids.

In the case of nicotine biosynthesis ornithine is decarboxylated to yield putrescine which may undergo transamination and ring closure to yield Δ′-pyrroline. Oxidation of putrescine by diamine oxidase could also yield the amino aldehyde intermediate which is spontaneously converted to Δ′-pyrroline. This reaction has in fact been demonstrated by Mann's group.[265] Δ′-Pyrroline produced from ornithine in this manner is symmetrically labelled and would account for the uniform labelling of carbons 2 and 5 in the pyrroline nucleus of nicotine. *In vivo* it appears that the actual conversion involves the methylated derivative[361] of putrescine and appropriate enzymes, ornithine decarboxylase, putrescine methyltransferase and *N*-methylputrescine oxidase, have been characterized in extracts from tobacco roots; conditions which enhanced nicotine biosynthesis increased the level of enzymes involved in the alkaloid biosynthesis.[433]

In the case of tropine biosynthesis ornithine is apparently metabolized by transamination to glutamic semialdehyde which undergoes ring closure to yield Δ′-pyrroline-5-carboxylic acid which undergoes decarboxylation to yield Δ′-pyrroline appropriately asymmetrically labelled to account for the labelling pattern observed in tropine.

PIPERIDINE ALKALOIDS These derive part of their carbon skeleton and nitrogen from the amino acid lysine (Fig. 3.5). Use of labelled lysine and

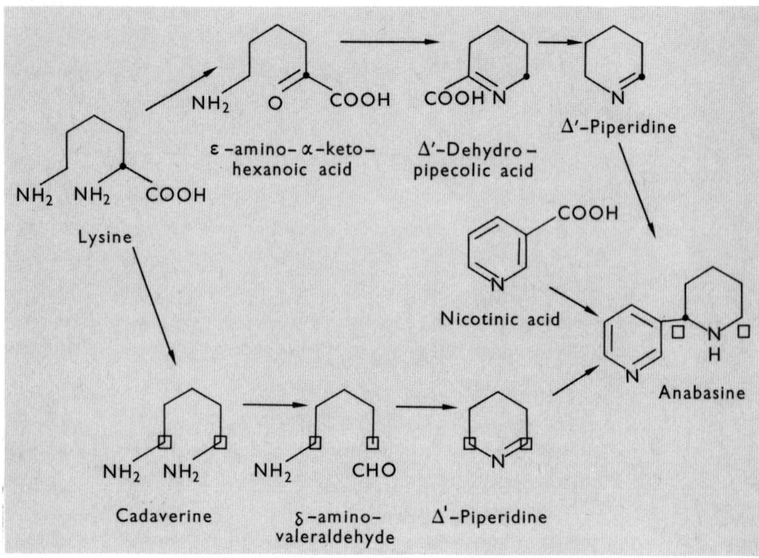

Fig. 3.5 Suggested alternate routes for the biogenesis of the piperidine alkaloid anabasine from lysine.

cadaverine, the decarboxylation product of lysine, indicates that there are two possible routes for the biosynthesis of the Δ′-piperidine precursor.[359] In one pathway lysine is deaminated to ε-amino-α-ketohexanoic acid which upon dehydration produces Δ′-dehydropipecolic acid, the carboxylated derivative of Δ′-piperidine. In an alternative pathway lysine can be decarboxylated to its amine derivative cadaverine which can by transamination produce δ-amino-valeraldehyde in a manner similar to the conversion of ornithine to glutamic semialdehyde. Ring closure produces Δ′-piperidine. The cadaverine pathway has the potential of producing Δ′-piperidine symmetrically labelled at carbons 2 and 6.

The conversion of cadaverine to piperidine again could be mediated by the enzyme diamine oxidase as indicated by Mann and coworkers.[265] The enzyme lysine decarboxylase has been characterized,[254] and the product cadaverine has recently been demonstrated in extracts from higher plants.[13]

Biosynthesis of pyrroline and piperidine from ornithine and lysine respectively requires the loss of a nitrogen residue. Attempts have been made to determine which of the nitrogen atoms of the precursor amino acid is preserved. The results from such experiments are difficult to interpret owing to the possibility of transamination, etc., of the applied ^{15}N labelled precursor. However it appears that in the biogenesis of anabasine from lysine the alpha amino nitrogen is lost in the formation of the piperidine precursor.[364] Likewise the alpha amino nitrogen is lost during the formation of the pyrrolidine ring of nicotine.[363]

NICOTINIC ACID BIOSYNTHESIS The pyridine residues condensed with Δ'-pyrroline derivatives in nicotine or the Δ'-piperidine derivative in anabasine originate from nicotinic acid. Labelling studies indicate that nicotinic acid is synthesized from glyceraldehyde phosphate and aspartic acid with quinolinic acid as an intermediate.

CH_2OⓅ–CHOH–CH_2OH + HOOC–CH_2–CH(NH_2)–COOH ⟶ [pyridine ring with COOH at 2 and 3 positions]

Glyceraldehyde phosphate + Aspartic acid Quinolinic acid

Although decarboxylation of quinolinic acid can potentially give rise to nicotinic acid it appears that the interconversion is more indirect and involves the operation of the pyridine nucleotide cycle (Fig. 3.6). Quinolinic acid condenses with 5-phosphoribosyl-1-pyrophosphate with an elimination of carbon dioxide to form nicotinic acid mononucleotide. Nicotinic acid mononucleotide can be either changed directly to nicotinic acid with the elimination of ribose-5-phosphate or it is converted to nicotinic acid adenine dinucleotide by condensation with ATP. The **nicotinic** acid adenine dinucleotide may be converted to **nicotinamide** adenine dinucleotide (NAD) by a transamidation of the amide nitrogen from glutamine. This NAD can be hydrolysed to nicotinamide which can then be deaminated to nicotinic acid. Evidence for the operation of such a cycle has been provided in wheat leaves[222] and castor beans.[666]

Prior to its condensation into the alkaloids nicotinic acid must be decarboxylated; enzymes capable of catalysing this reaction have recently been demonstrated in tobacco roots.[101]

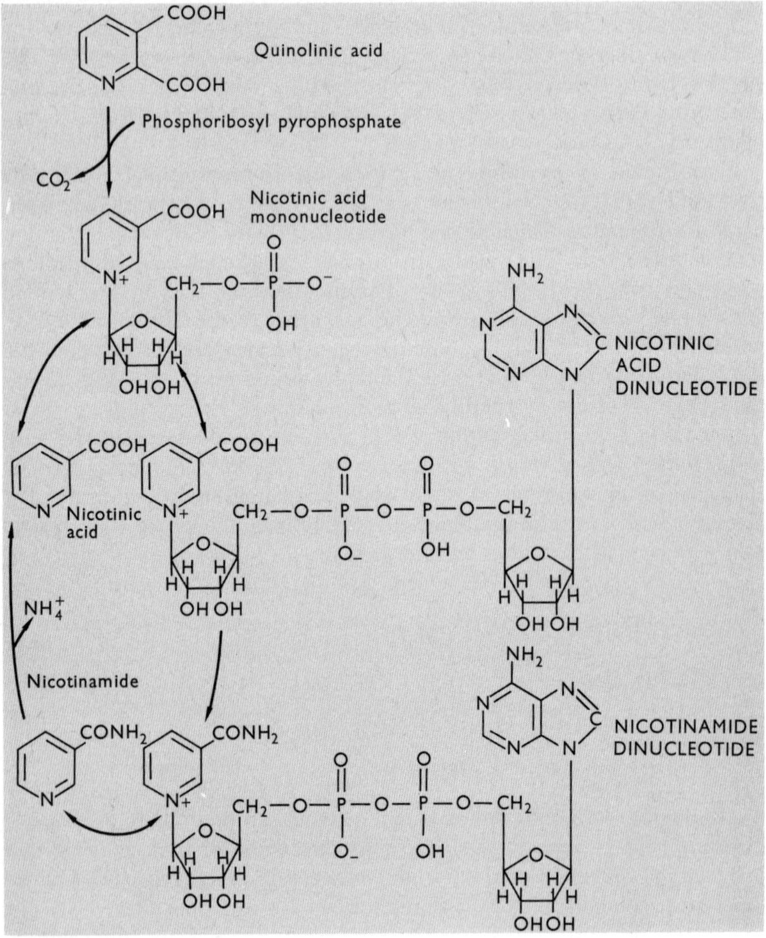

Fig. 3.6 The Nicotinic Acid Cycle. Modified from Waller *et al.*[666]

Alkaloids containing phenylethylamine and isoquinoline residues

Tyrosine and phenylalanine can serve as precursors of a large group of alkaloids characterized by the phenylethylamine or isoquinoline ring structures.

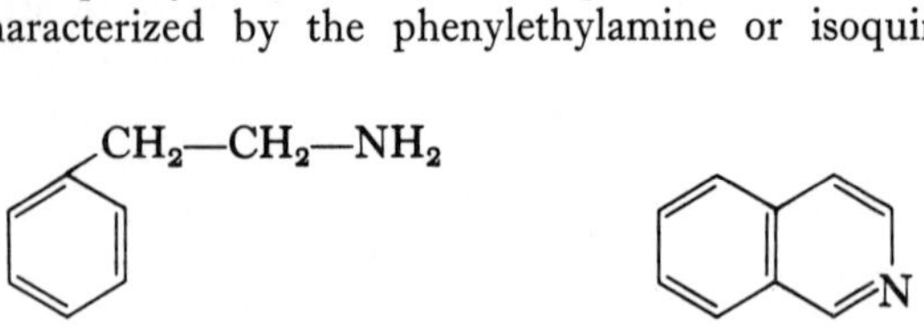

Phenylethylamine

Isoquinoline

Enzymes capable of decarboxylating tyrosine to tyramine[272] have been detected in barley plants and on this basis and labelling evidence it appears that tyrosine can serve as the precursor of the alkaloid hordenine derived from barley.[365] The methylation of tyramine to produce the intermediate methyltyramine and ultimately hordenine is believed to involve the amino acid methionine which frequently functions as a methyl donor. Similarly decarboxylation and methylation of DOPA give rise to the alkaloid mescaline[5] (Fig. 3.7).

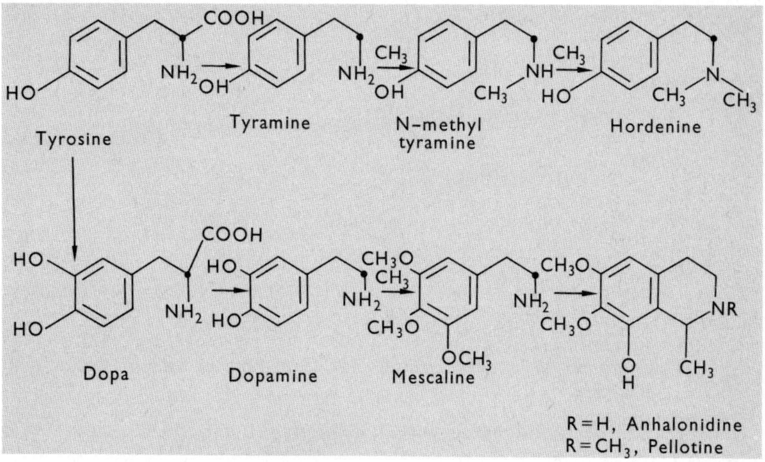

Fig. 3.7 Proposed biosynthetic pathways of some phenylethylamine alkaloids.

The phenylethylamine derivatives can undergo condensation with aldehydes to produce isoquinoline derivatives in the frequently encountered Mannich reaction. In this reaction an aldehyde reacts with an amine and an actual or potential carbanion.

$$-\overset{|}{\underset{|}{C}}{}^{-} + R-\overset{H}{\overset{|}{C}}=O + NHR_1R_2 + H^+ \longrightarrow$$

$$-\overset{|}{\underset{|}{C}}-\underset{\underset{R}{|}}{CH}-N\begin{matrix} R_1 \\ R_2 \end{matrix} + H_2O$$

The ortho and para carbon atoms of phenolic systems exhibit carbanion characteristics so that, for example, the synthesis of the relatively simple isoquinoline alkaloid pellotine can be considered as a condensation of hydroxylated tyramine with acetaldehyde followed by methylation.

The biochemical intermediates and the enzymes involved in such reactions have not been characterized but the labelling of pellotine from specifically labelled tyrosine is consistent with this mode of synthesis[33] (Fig. 3.8).

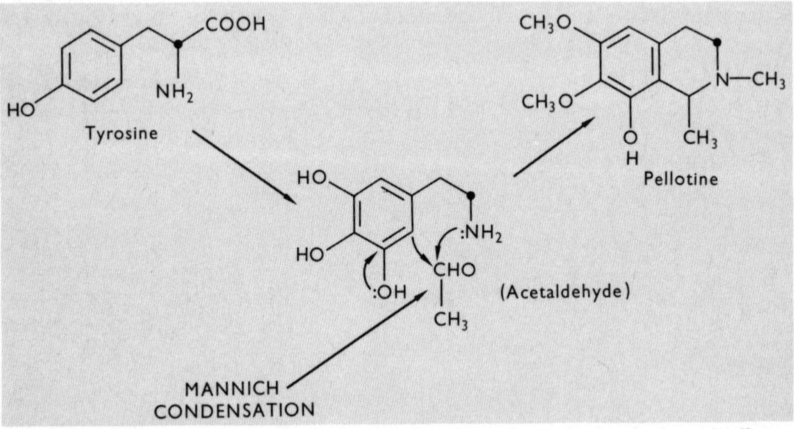

Fig. 3.8 Proposed biosynthetic routes for formation of simple isoquinoline alkaloids.

If the aldehyde condensing with the phenylethylamine derivative is an aromatic aldehyde then more complex polycyclic alkaloids are produced. Thus when a tyramine derivative reacts with a C_6–C_1 unit, benzyl isoquinoline alkaloids are produced (Fig. 3.9).

Condensation of the tyramine derivative dopamine (3,4-dihydroxyphenylethylamine) with a C_6–C_2 unit as indicated in the figure gives rise to norlaudanosoline, a precursor of many alkaloids of the isoquinoline morphine type.[31] Consistently C_6–C_2 units involved in Mannich condensations are derived from tyrosine whereas C_6–C_1 and C_6–C_3 units involved in condensations are derived from phenylalanine.[581]

In addition to the reaction of aromatic residues in the Mannich type condensation other cyclic structures can be formed by the intramolecular oxidative coupling between two phenol nuclei.[32] Such a coupling between adjacent phenolic residues results in increasing heterocyclic complexity and has been used to account for the transformation of methyl norbelladine into a series of related alkaloids (Fig. 3.10).

Indole and quinoline alkaloids

The amino acid tryptophan functions as a precursor for the biosynthesis of the indole and quinoline alkaloids[278, 393] (Fig. 3.11). Again

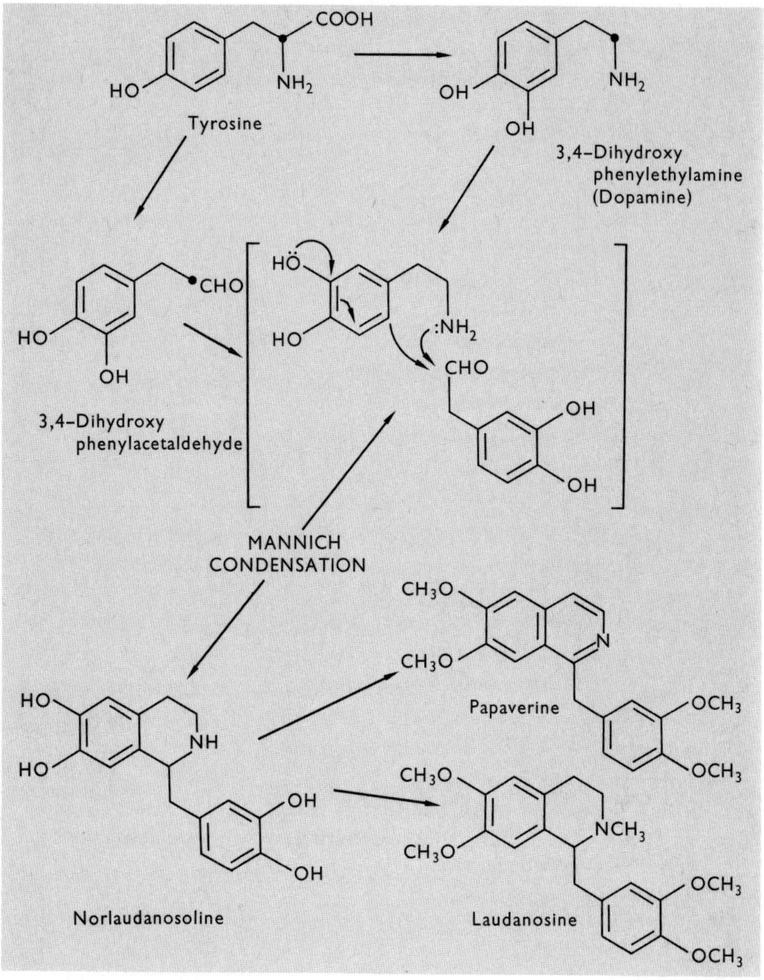

Fig. 3.9 Suggested route of biosynthesis of some benzyl isoquinoline alkaloids.

the evidence for such precursor function is primarily based on labelling studies in which specifically labelled tryptophan is incorporated into specific alkaloids.

The pattern of incorporation of tryptophan is consistent with its conversion to tryptamine by a decarboxylation reaction. However feeding experiments with labelled tryptamine indicated that this material was not incorporated into the alkaloids tested. Thus decarboxylation of the

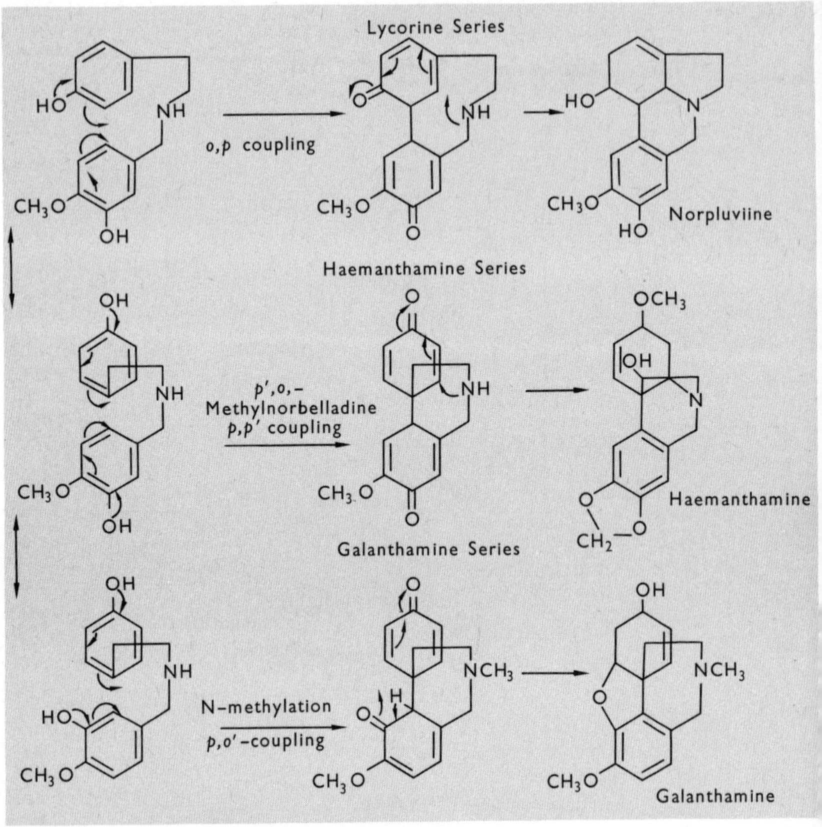

Fig. 3.10 Interconversion of methylnorbelladine involving oxidative coupling of adjacent phenolic nuclei.

tryptophan residue may occur after its condensation with other components.[581]

Frequently in this group of alkaloids additional carbon residues are introduced by condensation of the tryptophan with derivatives of mevalonic acid.

BIOSYNTHESIS OF CYANOGENIC GLYCOSIDES AND MUSTARD OILS

Many plants possess the capacity to produce HCN. In most instances it appears that the cyanogenesis is related to the capacity to produce certain cyanohydrin aglycone components which condense with a sugar

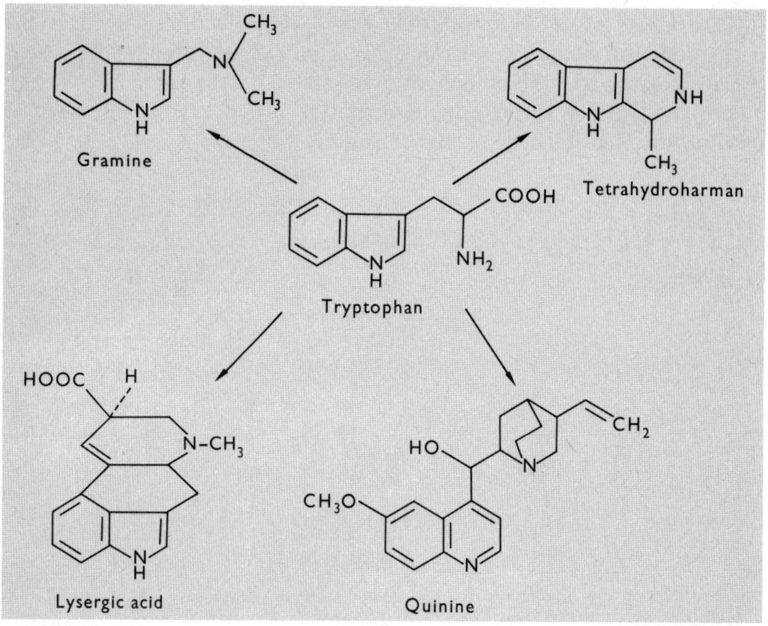

Fig. 3.11 Relationship of tryptophan to some of indole and quinoline alkaloids.

residue to produce so-called cyanogenic glycosides (Table 3.4). Some of these glycosides can occur in appreciable quantities in certain instances; dhurrin for example can reach a concentration of 3–5% of the dry weight of sorghum seedlings.

Table 3.4 Cyanogenic glycosides of known structure (from Conn and Butler[124]).

Glycoside	Sugar	Aglycone	Occurrence
Linamarin	D-Glucose	α-Hydroxyisobutyronitrile (acetone cyanohydrin)	*Linum usitatissimum, Phaseolus lunatus, Trifolium repens, Lotus* sp.
Lotaustralin	D-Glucose	α-Hydroxy-α-methyl butyronitrile (methyl ethyl ketone cyanohydrin)	(See linamarin)
Acacipetalin	D-Glucose	β-Dimethyl-α-hydroxyacrylonitrile	*Acacia* sp. (South African)
Prunasin	D-Glucose	D-Mandelonitrile	*Prunus* sp., many Rosaceae, *Eucalyptus* sp.
Sambunigrin	D-Glucose	L-Mandelonitrile	*Sambucus nigra, Acacia* sp. (Australian)
Amygdalin	Gentiobiose	D-Mandelonitrile	*Prunus* sp.
Vicianin	Vicianose	D-Mandelonitrile	*Vicia angustifolia* and other *Vicia*
Dhurrin	D-Glucose	L-*p*-Hydroxymandelonitrile	*Sorghum* sp.
Taxiphyllin	D-Glucose	D-*p*-Hydroxymandelonitrile	*Taxus* sp.
Zierin	D-Glucose	*m*-Hydroxymandelonitrile	*Zieria laevigata*
Gynocardin	D-Glucose	Gynocardinonitrile	*Gynocardia odorata, Pangium edule*

The release of HCN from these cyanogenic glycosides involves the participation of two enzymes. An initial glycosidase[90] cleaves the cyanogenic glycoside into the cyanohydrin aglycone and the component sugar. The cyanohydrin is then further degraded by a hydroxynitrile lyase[554] to produce the free aldehyde or ketone and HCN.

$$\text{HO}-\text{C}_6\text{H}_4-\text{CH}(\text{O}-\text{C}_6\text{H}_{11}\text{O}_5)-\text{C}\equiv\text{N} \xrightarrow{\beta\text{-Glucosidase}} \text{HO}-\text{C}_6\text{H}_4-\text{CH}(\text{OH})-\text{C}\equiv\text{N} + \text{C}_6\text{H}_{12}\text{O}_6 \xrightarrow{\text{Hydroxynitrile lyase}}$$

Dhurrin — *p*-Hydroxy mandelonitrile + glucose

$$\text{HO}-\text{C}_6\text{H}_4-\text{CH}{=}\text{O} + \text{HCN}$$

p-Hydroxy benzaldehyde

During the elucidation of this degradative sequence it was suggested that the biosynthesis of the cyanogenic glycosides might involve a simple reversal of the reactions. However in studies in which labelled HCN was fed to plants no evidence was found for its incorporation into the cyanogenic glycosides. It was independently demonstrated by Gander[206] and Conn and Akazawa[123] that tyrosine or its precursor shikimic acid functioned as the precursor of the cyanohydrin component of dhurrin. As a result of these and later investigations it is generally agreed that the

$$(\text{H}_3\text{C})_2\text{CH}-\text{CH}(\text{NH}_2)-\text{COOH} \longrightarrow (\text{H}_3\text{C})_2\text{CH}-\text{CH}{=}\text{NOH} \longrightarrow (\text{H}_3\text{C})_2\text{CH}-\text{C}\equiv\text{N} \longrightarrow$$

Valine — Isobutyraldoxime — Isobutyronitrile

$$(\text{H}_3\text{C})_2\text{C}(\text{OH})-\text{C}\equiv\text{N} \xrightarrow{+\text{Glucose}} (\text{H}_3\text{C})_2\text{C}(\text{O}-\text{Glc})-\text{C}\equiv\text{N}$$

Hydroxy isobutyronitrile — Linamarin

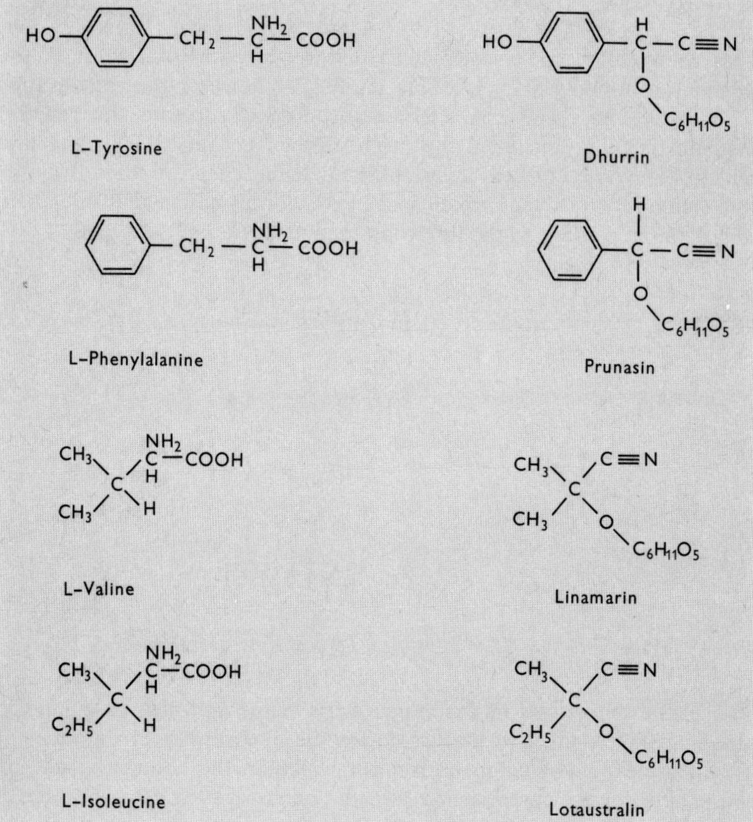

Fig. 3.12 The amino acid precursors of four cyanogenic glucosides.

various amino acids serve as the precursors of the cyanohydrin aglycones (Fig. 3.12).[124, 395]

The nitrile (CN) nitrogen is derived from the α-amino nitrogen of the appropriate amino acid. Thus the synthesis of the cyanogenic glucoside involves a reaction sequence in which the carboxyl carbon of the amino acid is lost, the alpha carbon is oxidized to the level of a nitrile and the beta carbon acquires a hydroxyl group that is subsequently glycosylated to yield the cyanogenic glucoside.

Although the evidence is still fragmentary it appears that the amino acid is converted to an oxime;[124, 395] dehydration of the oxime group yields a nitrile, which is subsequently converted to a hydroxynitrile. This hydroxynitrile or cyanohydrin derivative condenses with the appropriate sugar moiety, in a reaction apparently involving the nucleotide sugar derivative.

This proposed reaction sequence involving oximes as precursors of the aglycone is supported by labelling studies;[619] however the enzymes involved have not been characterized. Additionally before the reaction sequence is fully elucidated the mechanism of conversion of the amino acid to the oxime needs to be established. It has been proposed that this conversion proceeds via formation of an *N*-hydroxy amino acid and a keto acid oxime according to the following sequence:[395]

$$(R_1)(R_2)C(H)\text{—}CH(NH_2)\text{—}COOH \xrightarrow{O} (R_1)(R_2)C(H)\text{—}CH(NHOH)\text{—}COOH \xrightarrow{-2H}$$

Amino acid — *N*-hydroxy amino acid

$$(R_1)(R_2)C(H)\text{—}C(=NOH)\text{—}COOH \xrightarrow{-CO_2} (R_1)(R_2)C(H)\text{—}CH{=}NOH$$

Ketoxime — Aldoxime

The oxime derivatives of the amino acids valine and phenylalanine can also function as effective precursors for the biosynthesis of the mustard oil glucosides. These compounds occur widely in the Cruciferae and are responsible for the characteristic pungent odour of such leaves as those of nasturtium (*Tropaeolum majus*) and Brussels sprouts (*Brassica oleracea*). The close relationship of the mustard oils to the oximes is apparent from the structural formulas below.

$$(H_3C)_2CH\text{—}C(\text{S—Glucose}){=}NOSO_3$$

Isopropyl glucosinolate

$$C_6H_5\text{—}CH_2\text{—}C(\text{S—Glucose}){=}NOSO_3$$

Benzyl glucosinolate (glucotropaeolin)

The C and the N of the aromatic aglycones are derived from phenylalanine.[649] The derivative phenylacetaldoxime is also an effective precursor suggesting that metabolism of phenylalanine to the aglycone probably proceeds through the aldoxime derivative.[618] Tryptophan can also function as a precursor of the aglycone in the mustard oil glucobrassicin,[350] a suspected intermediate in the synthesis of indolyl acetic acid in some species (see later).

BIOSYNTHESIS OF GROWTH REGULATORS INDOLYL ACETIC ACID AND ETHYLENE

Indolyl acetic acid

The close structural resemblance between the amino acid tryptophan and the auxin indolyl acetic acid (IAA) are immediately apparent (Fig. 3.13). Thimann[623] was the first to suggest that the growth regulator was synthesized from tryptophan. The possible role of tryptophan in indolyl acetic acid biosynthesis has been extensively investigated and there is general but not universal agreement that tryptophan is the precursor of IAA. There is still confusion concerning the reaction sequences by which the amino acid is converted to IAA and it appears that the interconversion can occur by several pathways and in some species the conversion can occur by more than one route. Additionally epiphytic bacteria are capable of metabolizing tryptophan and can give confusing results in studies aimed at elucidating biosynthetic pathways in higher plants.[378] In an extensive survey, Gibson *et al.*[213] indicate that five possible pathways of the synthesis of IAA from tryptophan have been established (Fig. 3.13).

The operation of pathways 4 and 5 appears to be restricted to the Cruciferae and allied families. The detection of the intermediates in reaction sequences 1, 2, and 3 in extracts from other plants indicates that IAA biosynthesis could proceed by any of the outlined mechanisms. The functioning of these pathways is confirmed by labelling experiments and the demonstration of some of the necessary enzymes in plant extracts. Although feeding experiments demonstrate the conversion of indolyl lactic acid to indolyl acetic acid, the direct decarboxylation of indolyl lactic acid to the suspected intermediate tryptophol has not been demonstrated in plant extracts. It appears that indolyl lactic acid conversion to IAA probably proceeds through indolyl pyruvic acid.[213]

Enzymatic decarboxylation of tryptophan to tryptamine has been demonstrated[501] and the tryptamine can be converted by an amine oxidase[119] to indolyl acetaldehyde. The conversion of indolyl acetaldehyde to indolyl acetic acid in pathways 1 and 2 has been achieved in partially purified[692] extracts and is apparently mediated by an NAD dependent aldehyde dehydrogenase.

Given the possible alternative pathways for the generation of indolyl acetic acid from tryptophan via tryptamine or indolyl pyruvic acid, the possibility of controlling the metabolism of tryptophan by the alternative routes arise. However no mechanisms of control have been established. The capacity to produce IAA is generally considered to be a function of apical meristems; the biochemical restrictions which limit auxin production in other parts of the plants have not been elucidated.

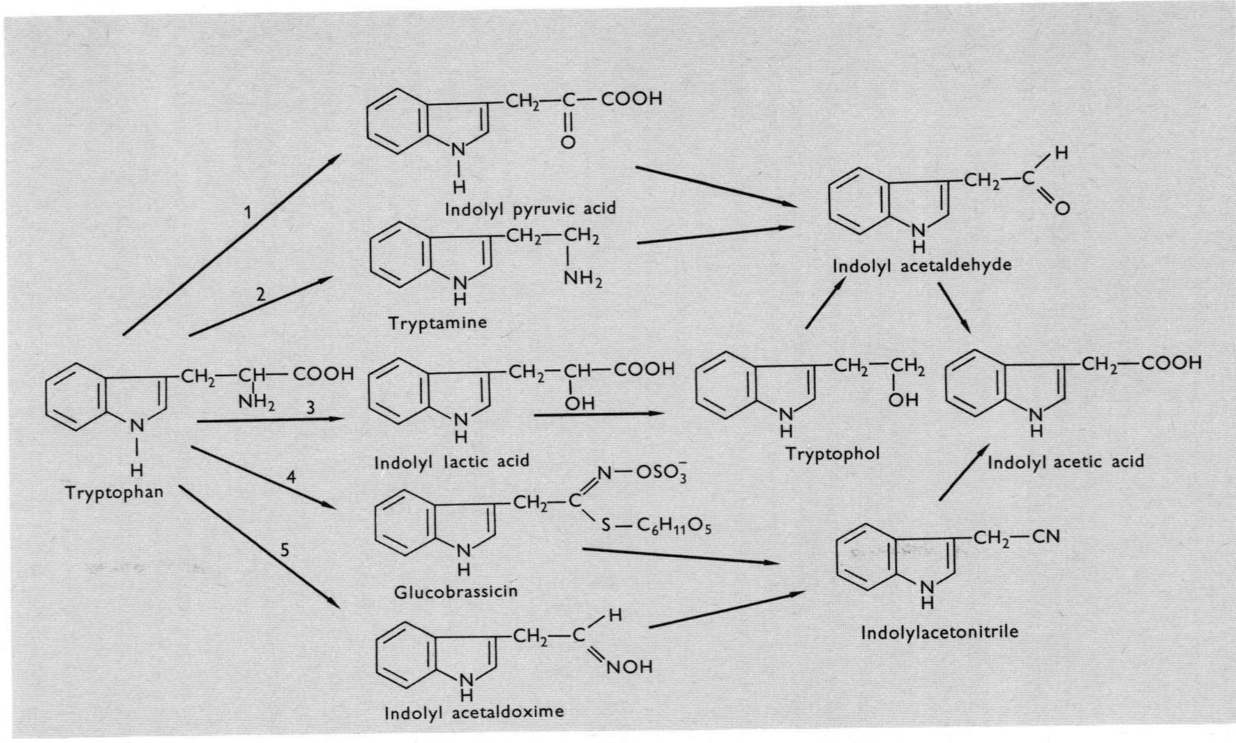

Fig. 3.13 Proposed pathways for the synthesis of IAA from tryptophan.

Ethylene

Many of the growth and developmental effects mediated by the auxin IAA whose biosynthesis has just been described can also be regulated by the gas ethylene.[2] Interest in this growth regulator was originally restricted to its possible role in fruit ripening; however interest in the functioning of ethylene in other regulatory processes has exploded in the last decade. These increased investigations into ethylene and its effects can be directly related to improved technology, primarily the development of gas–liquid chromatography which has facilitated the detection of the small quantities of the gas which are responsible for inducing effects on plant growth and development. Associated with an increased understanding of the extensive involvement of ethylene in growth and development there have been detailed investigations into its mode of biogenesis. It is now well established, following the original discovery of Lieberman and Mapson,[379] that the amino acid methionine is the precursor of the gas.

To date the mechanism by which methionine is converted to ethylene *in vivo* has not been satisfactorily elucidated. Model *in vitro* systems have been described in which peroxidases and phenols serve to produce free radical intermediates which react with methional to produce ethylene.[712] However methionine did not function in this enzymatic system. It has been speculated that the original metabolism of methionine may involve a transamination to produce α-keto-γ-methyl-mercaptobutyric acid which can then be transformed to methional and ethylene by the peroxidase system. However feeding of labelled intermediates has given conflicting results and methionine is usually a better precursor than the proposed intermediates.[1]

Methionine		α-Keto-γ-methyl mercaptobutyrate		Methional		
1. COOH		COOH				1. CO_2, NH_3
2. $CHNH_2$		C=O		CHO		2. H—COOH
3. CH_2	⟶	CH_2	⟶	CH_2	⟶	3. CH_2 ‖
4. CH_2		CH_2		CH_2		4. CH_2
S		S		S		
5. CH_3		CH_3		CH_3		5. CH_3—S

Consistently feeding experiments indicate that the ethylene is derived from carbons 3 and 4 of the methionine precursor.

The realization of the interplay between the auxin indolyl acetic acid and ethylene in many growth responses raises the possibility that their biogenesis might be interrelated. However attempts to demonstrate changes in enzyme levels which might account for a changed production of one growth regulator by application of the other have so far been inconclusive.[1]

The sustained production of ethylene requires a continued supply of methionine. In most tissues this could provide a potential problem because methionine is not an abundant amino acid. In fruit tissue it appears that methionine levels can to a great extent be sustained by a re-utilization of the CH_3–S units in further methionine biosynthesis according to the scheme outlined in Fig. 3.14.

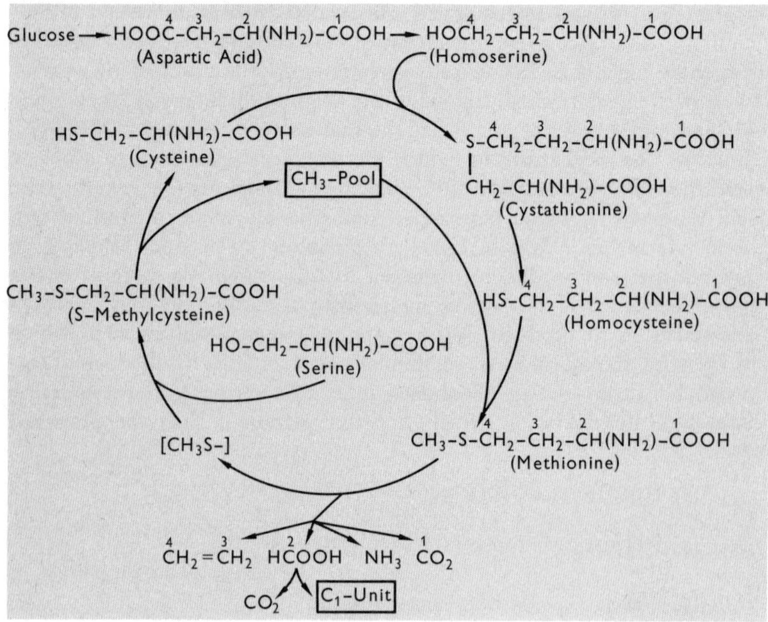

Fig. 3.14 Methionine cycle in relation to ethylene biosynthesis. (From Baur and Yang,[34] 1972, *Phytochem.*, **11**, 3207.)

PORPHYRIN BIOSYNTHESIS

Porphyrins can be considered as cyclic tetrapyrroles whose constituent pyrroles are joined through their α-carbon atoms by unsaturated methene bridges. The α-carbons are bound to the nitrogen atom and are fully substituted but the β-carbon may be substituted with a variety of side chains giving rise to a wide range of compounds. In addition to variation in the substitution on the eight pyrrole β-carbons, the porphyrins can differ from one another by the presence or absence of metal atoms held at the centre of the tetrapyrrole by bonding to the pyrrole nitrogen atoms. Among the physiologically important metalloporphyrins are the chlorophylls in which magnesium is the metal atom and the cytochromes, catalases and peroxidases in which iron is the metal substituent (Fig. 3.15).

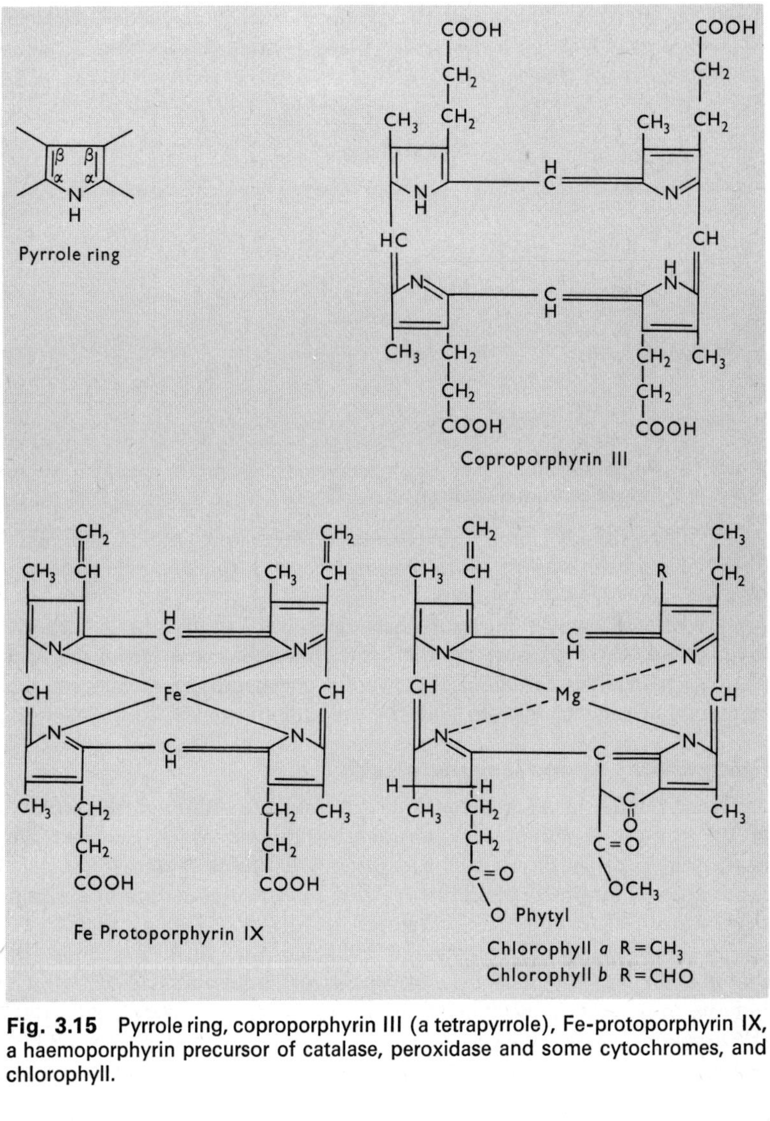

Pyrrole ring

Coproporphyrin III

Fe Protoporphyrin IX

Chlorophyll *a* R=CH_3
Chlorophyll *b* R=CHO

Fig. 3.15 Pyrrole ring, coproporphyrin III (a tetrapyrrole), Fe-protoporphyrin IX, a haemoporphyrin precursor of catalase, peroxidase and some cytochromes, and chlorophyll.

Pyrroles also occur in non-cyclic configurations as open chain tetrapyrroles; such structures are characteristic of the bile pigments. Again the constitutent pyrroles are joined to one another by carbon bridges between adjacent α-carbon atoms. Different substituents can occur at the eight β-carbon atoms of the pyrrole residues; in addition different bile pigments may have different residues attached to the outside α-carbon atoms of the linear tetrapyrrole. The prosthetic group of the photoreactive pigment phytochrome which mediates many plant photoresponses is a straight chain tetrapyrrole (Fig. 3.16).

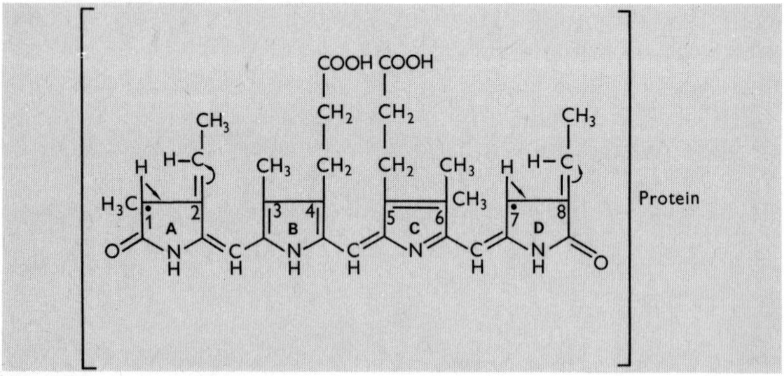

Fig. 3.16 Proposed structure for the phytochrome bile pigment prosthetic group (modified from Siegelman *et al.*).[565]

The biosynthesis of prophyrins has been extensively studied in animal, bacterial and plant systems, with the emphasis in plants being devoted to a study of chlorophyll biosynthesis.

Current concepts[516] indicate that chlorophyll biosynthesis can be divided into three phases: (a) the synthesis of δ-amino laevulinic acid (ALA), (b) the formation of the pyrrole porphobilinogen, and (c) the formation of the tetrapyrrole from the porphobilinogen.

Synthesis of δ-amino laevulinic acid

The formation of ALA is considered to take place through the activity of the enzyme δ-amino laevulinic acid synthetase which catalyses the condensation of succinyl Co-A and glycine. α-Amino β-keto-adipic acid is a suspected intermediate in this reaction but undergoes decarboxylation to produce ALA. To date, attempts to detect ALA synthetase in plants have failed. Circumstantial evidence for ALA formation in this manner was provided by experiments of Harel and Klein[249] which indicate that ALA accumulation in corn (*Zea mays*) and bean (*Phaseolus vulgaris*) leaves is enhanced by succinate and glycine. ALA accumulation in this tissue was owing to the fact that its conversion to chlorophyll was prevented by the presence of added laevulinic acid. However, more recent evidence indicates that the glycine label is not incorporated into the ALA as predicted by the formula below[35] and it is suggested that succinate and glycine may be metabolized to α-keto glutaraldehyde which can then be transaminated to δ-amino laevulinic acid.[258]

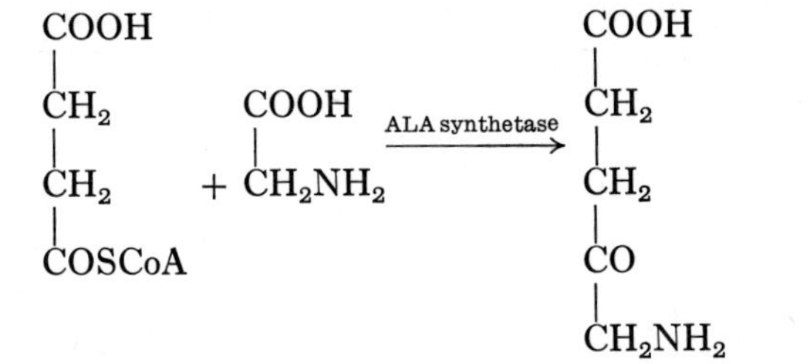

$$\begin{array}{l} COOH \\ | \\ CH_2 \\ | \\ CH_2 \\ | \\ COSCoA \end{array} + \begin{array}{l} COOH \\ | \\ CH_2NH_2 \end{array} \xrightarrow{\text{ALA synthetase}} \begin{array}{l} COOH \\ | \\ CH_2 \\ | \\ CH_2 \\ | \\ CO \\ | \\ CH_2NH_2 \end{array}$$

Succinyl Co-A Glycine δ-Amino laevulinic acid

Synthesis of porphobilinogen

The subsequent metabolism of ALA involves the condensation of two molecules of ALA to produce the monopyrrole porphobilinogen. This reaction is catalysed by the enzyme ALA dehydrase (porphobilinogen synthetase) which has been detected and partially purified from plants.[450,601] The enzyme is inhibited by laevulinic acid.

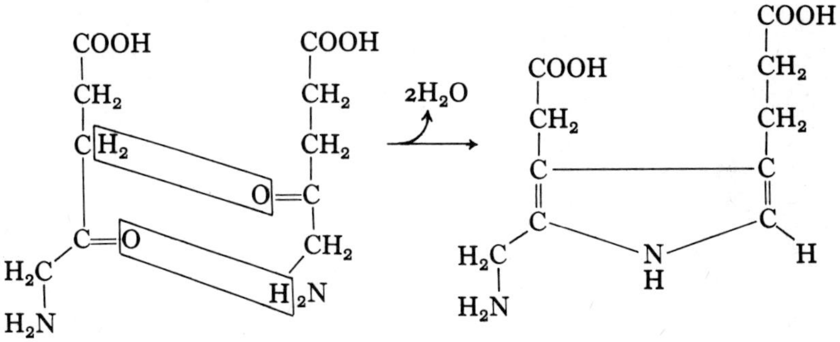

COOH, CH₂, CH₂, C=O, H₂C, H₂N (ALA) + COOH, CH₂, CH₂, O=C, CH₂, H₂N (ALA) → (−2H₂O) porphobilinogen

$$\text{ALA} + \text{ALA} \xrightarrow{-2H_2O} \text{Porphobilinogen}$$

ALA + ALA Porphobilinogen

Formation of uroporphyrinogen III

The formation of the tetrapyrrole from the porphobilinogen is complex and the mechanisms are not completely resolved. The expected product formed from the linear condensation of four porphobilinogen residues through the amino methyl substituent followed by cyclization is uroporphyrinogen I. However this derivative is of infrequent occurrence. The most frequently encountered porphyrins are derived from uroporphyrinogen III in which the orientation of the D ring appears to be switched.

Formation of this uroporphyrinogen III appears on the basis of evidence from spinach (*Spinacia oleracea*) and *Chlorella*[56] to require the participation of two enzyme moieties called urogen I synthetase and urogen III cosynthetase. In the presence of urogen I synthetase the product is uroporphyrinogen I. Urogen III cosynthetase is unable to convert porphobilinogen into a tetrapyrrole; however, in the presence of urogen I synthetase and urogen III cosynthetase uroporphyrinogen III is produced. It is believed[56] that the substrate for urogen III cosynthetase may be a tripyrrole intermediate with biosynthesis of the tetrapyrrole proceeding as indicated in Fig. 3.17.

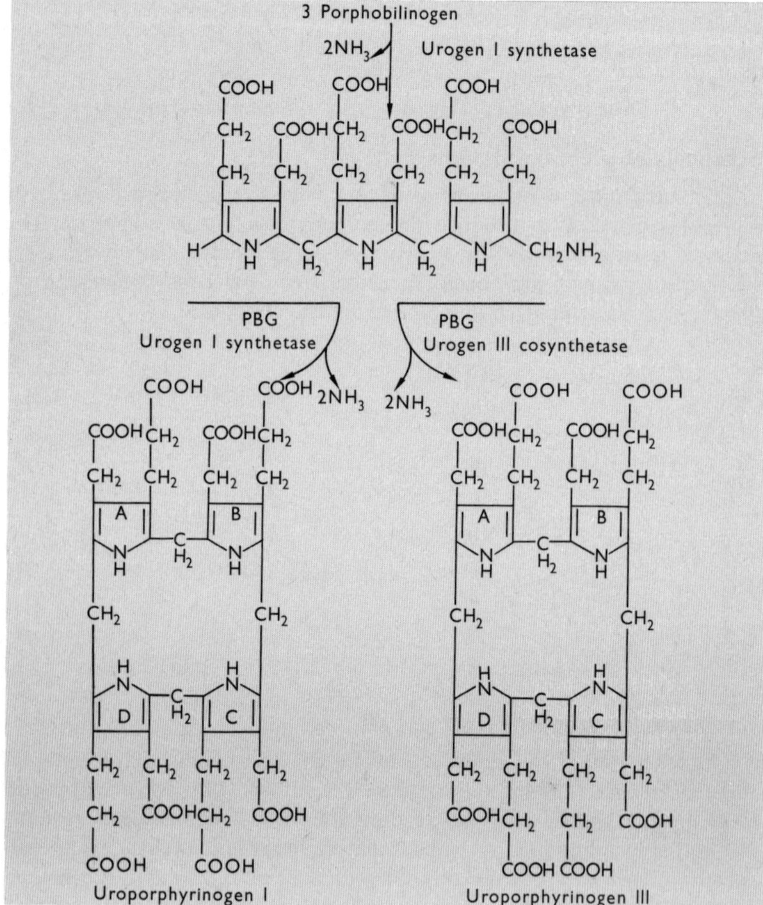

Fig. 3.17 Proposed sequence of uroporphyrinogen III synthesis involving urogen I synthetase in association with urogen III cosynthetase.

The fate of the ammonia liberated in the above condensation has not been established.

Protoporphyrin synthesis

Decarboxylation of the acetic acid substituents of the α-carbon residues of the tetrapyrrole uroporphyrinogen III by the enzyme uroporphyrinogen decarboxylase[57] produces coproporphyrinogen (coprogen) III.

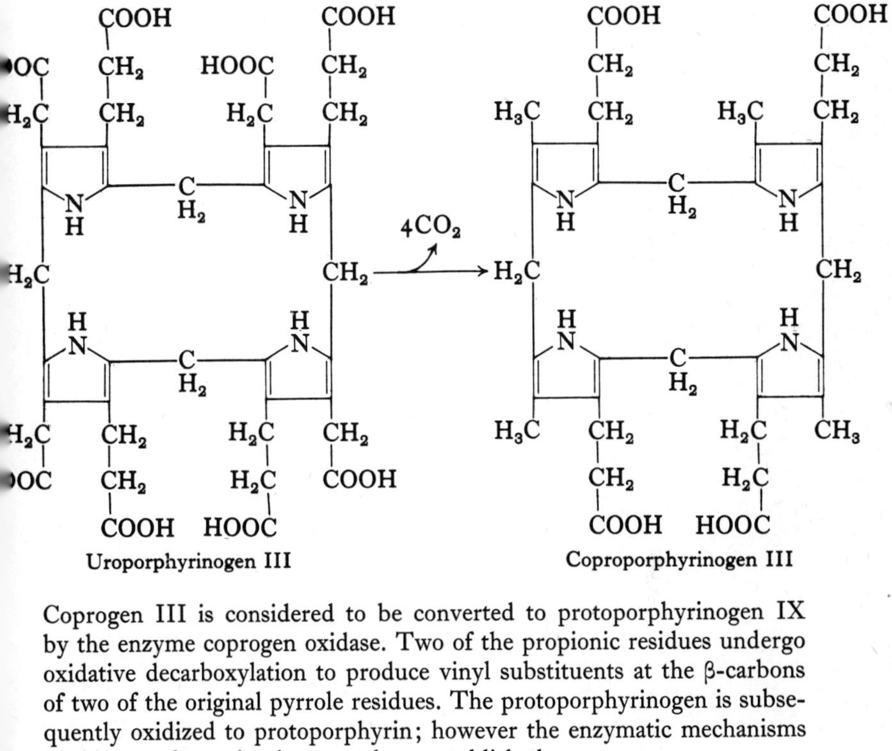

$4CO_2$

Uroporphyrinogen III → Coproporphyrinogen III

Coprogen III is considered to be converted to protoporphyrinogen IX by the enzyme coprogen oxidase. Two of the propionic residues undergo oxidative decarboxylation to produce vinyl substituents at the β-carbons of two of the original pyrrole residues. The protoporphyrinogen is subsequently oxidized to protoporphyrin; however the enzymatic mechanisms of this transformation have not been established.

Protochlorophyll ⟶ Chlorophyll

The protoporphyrin apparently subsequently chelates with iron or magnesium. The protohaem derivatives formed from the chelation of iron give rise to the cytochromes, catalase and peroxidases. Magnesium derivatives of protoporphyrin give rise to chlorophylls. The mechanisms of iron and magnesium chelation have not been established in higher plants. However, the functioning or protoporphyrin and magnesium protoporphyrin as precursors in chlorophyll biosynthesis is demonstrated by the

Protoporphyrinogen IX

Protoporphyrin IX

observed accumulation of these compounds in chlorophyll-less mutants and other conditions in which ALA is supplied to etiolated plants.[517] The proposed sequence in the transformation of magnesium protoporphyrin to chlorophyll is depicted in Fig. 3.18.

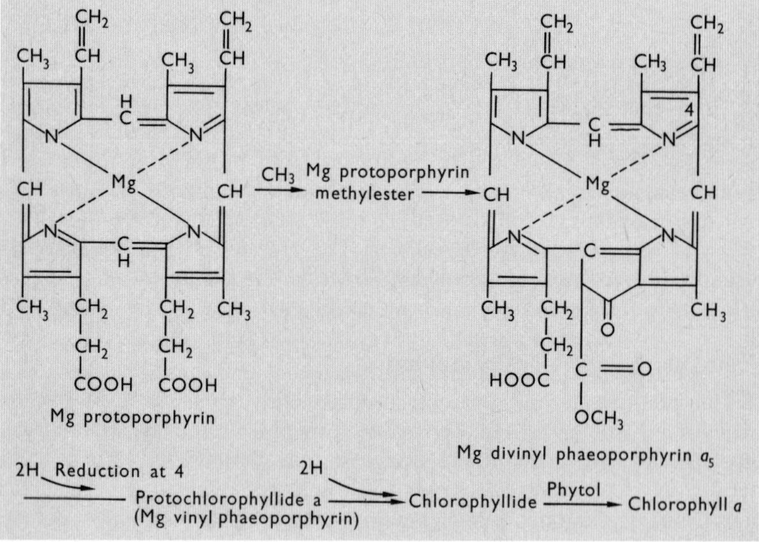

Fig. 3.18 Proposed sequence for Mg protoporphyrin conversion to chlorophyll *a*.

Control of chlorophyll synthesis

Apart from a few noteworthy exceptions such as the cotyledons of some conifer and citrus seeds, the accumulation of chlorophyll in most higher plants is a light-dependent process. When time-course studies are made of chlorophyll formation following illumination, three distinct phases are evident. An initial rapid small increase in chlorophyll content is followed by a lag phase which is followed by a period of rapid chlorophyll production. The lag period can be of varying length depending upon tissue age, species, light intensity, etc., and can in some instances extend to several hours.

It is considered that the sequence of chlorophyll accumulation can be accounted for by the following events. The initial rapid accumulation of chlorophyll apparently represents the conversion of existing protochlorophyllide to chlorophyll (Fig. 3.18). This photoconversion involves a reduction; as yet the source of reductant has not been established. It has been suggested that the condensation of –SH groups in proteins of etioplast[336] membranes with the subsequent formation of disulphide –S–S– bridges could provide the reductant. Such a mechanism for reductant generation would in part account for the observed ultrastructural changes in etioplast ultrastructure which accompany chlorophyll production following illumination.[173]

The small amount of chlorophyll accumulated immediately following illumination appears to be related to the low level of protochlorophyllide present in etiolated tissue. The lag phase of chlorophyll production has been interpreted as the time required for the production of the enzymes necessary for the sustained production of protochlorophyll. It has been shown that there is no requirement for continued illumination during the lag phase; plants given a brief illumination will produce chlorophyll without a lag when they are returned to the light after an intervening dark period.[658] Spectral studies have indicated that the triggering of enzyme production for chlorophyll synthesis is mediated by the photoreceptive pigment phytochrome.[508] Production of chlorophyll following illumination is prevented by inhibitors of protein[331] and ribonucleic acid synthesis.[58] Such observations have stimulated extensive experimentation aimed at locating the deficient enzymes which are synthesized following illumination to facilitate chlorophyll synthesis. It has been demonstrated that protochlorophyllide accumulates in excess of that present in control tissue when etiolated leaves are fed ALA in the dark.[228] This finding indicates that the enzymes necessary for ALA conversion to protochlorophyllide are not deficient in dark-grown tissue. It is suggested that under normal conditions the etiolated dark grown tissue has a reduced capacity for ALA synthesis. Illumination apparently sets in motion events leading to the production of enzymes required for ALA synthesis.[332]

BIOSYNTHESIS OF COFACTORS

Associated with the demonstration of the essentiality of vitamins in animal nutrition it was demonstrated that the requirement for many of these vitamins could be satisfied by the inclusion of plant foods in the diet. Analysis subsequently indicated that plants contained many of the vitamins in appreciable amounts and therefore were apparently capable of synthesizing them. After the demonstration that many vitamins functioned as cofactors in metabolic reactions a role for these compounds in plant metabolism was established. While the involvement of these compounds, such as biotin, tetrahydrofolic acid, thiamine pyrophosphate, pyridoxal phosphate and coenzyme A, in metabolic reactions is understood, very few studies have been performed on the mechanism of their synthesis in higher plants even though in several instances the cofactors were originally isolated from plant sources. Most investigations of cofactor biosynthesis have been performed with microbial systems. These investigations clearly implicate various amino acids in cofactor synthesis and in many instances the functioning of the cofactor in metabolic processes directly involves the nitrogen atom originally derived from the amino acids.

Biotin

This coenzyme participates in carboxylation and decarboxylation reactions. It is particularly important in lipid biosynthesis where it is required as a cofactor in the carboxylation of acetyl-CoA CH_3—CO ~ CoA to form Malonyl-CoA CH_2—CO ~ CoA by the enzyme acetyl-CoA

$$\underset{\displaystyle\text{COOH}}{\underset{|}{\text{CH}_2}}\text{—CO} \sim \text{CoA}$$

carboxylase. The reaction involves an initial carboxylation of the nitrogen atom of the biotin ring as indicated on p. 97. The resulting N-carboxybiotin subsequently functions as the carboxyl donor.

The biosynthesis of biotin, established from studies with microbial systems, involves cysteine, pimetyl CoA and carbamyl phosphate according to the scheme on p. 97.[393] Note that the nitrogen residues are derived from the amino group of cysteine and carbamyl phosphate; in plants the nitrogen in carbamyl phosphate probably originates from glutamine.[481]

Tetrahydrofolic acid

Tetrahydrofolic acid or tetrahydropteroyl (the terms are synonymous) derivatives are extremely important in reactions involving the transfer or metabolism of one carbon compounds such as formyl, hydroxymethyl or methyl groups. It will be recalled that the participation of tetrahydrofolic acid was encountered in the discussion of glycine–serine interconversions.

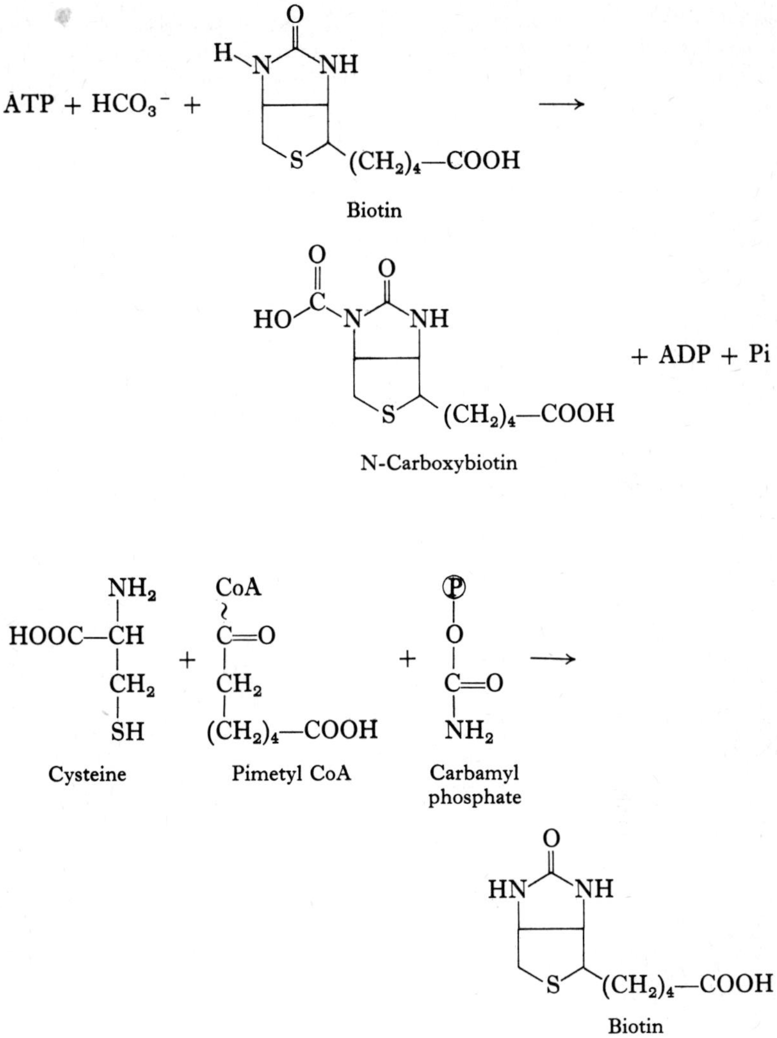

$$ATP + HCO_3^- + \text{Biotin} \longrightarrow \text{N-Carboxybiotin} + ADP + Pi$$

Biotin: H–N, NH, O, S, $(CH_2)_4$—COOH

N-Carboxybiotin: HO–C(=O)–N, NH, O, S, $(CH_2)_4$—COOH

$$\underset{\text{Cysteine}}{HOOC—CH(NH_2)—CH_2—SH} + \underset{\text{Pimetyl CoA}}{CoA\sim C{=}O—CH_2—(CH_2)_4—COOH} + \underset{\text{Carbamyl phosphate}}{Ⓟ—O—C{=}O—NH_2} \longrightarrow \text{Biotin}$$

Biotin: HN, NH, O, S, $(CH_2)_4$—COOH

Biotin: activity and biosynthesis

Folate derivatives are also involved in methionine biosynthesis and apparently participate in reactions involved in purine synthesis. Cossins and coworkers [118, 126, 585] have succeeded in extensive characterization of folate derivatives in plants and have demonstrated their localization in chloroplasts and mitochondria.

In addition to the demonstration of the occurrence of the folate derivatives it has been shown that many of the enzymes involved in C_1 interconversions are present in plant extracts.[118, 298, 615] Serine, formate and formaldehyde are the most important sources of C_1 units and the metabolism of C_1 units involving folic acid derivatives is depicted below (Fig. 3.19).

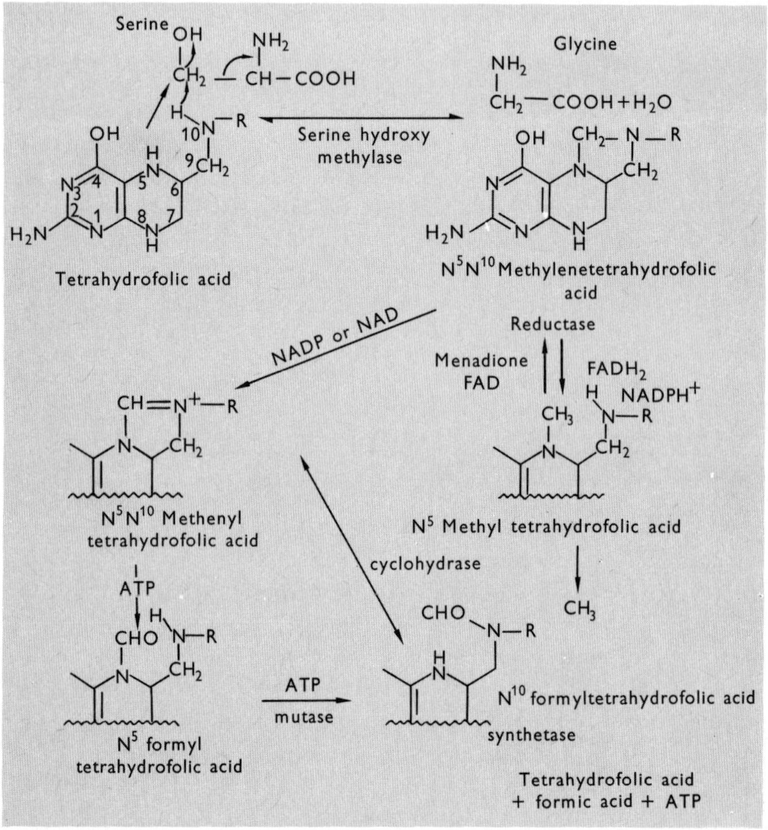

Fig. 3.19 Pathways of C_1 interconversions; note importance of nitrogen atoms 5 and 10 in this metabolism.

Although the presence of folic acid derivatives and their involvement in intermediary metabolism has been demonstrated, the mechanism of their biosynthesis has not been established in plants. Studies with microbial systems indicate that the important nitrogen residues N_5 and N_{10} originate from guanine and *p*-amino benzoic acid according to the reactions outlined in Fig. 3.20. The synthesis of guanine is discussed on p. 122;

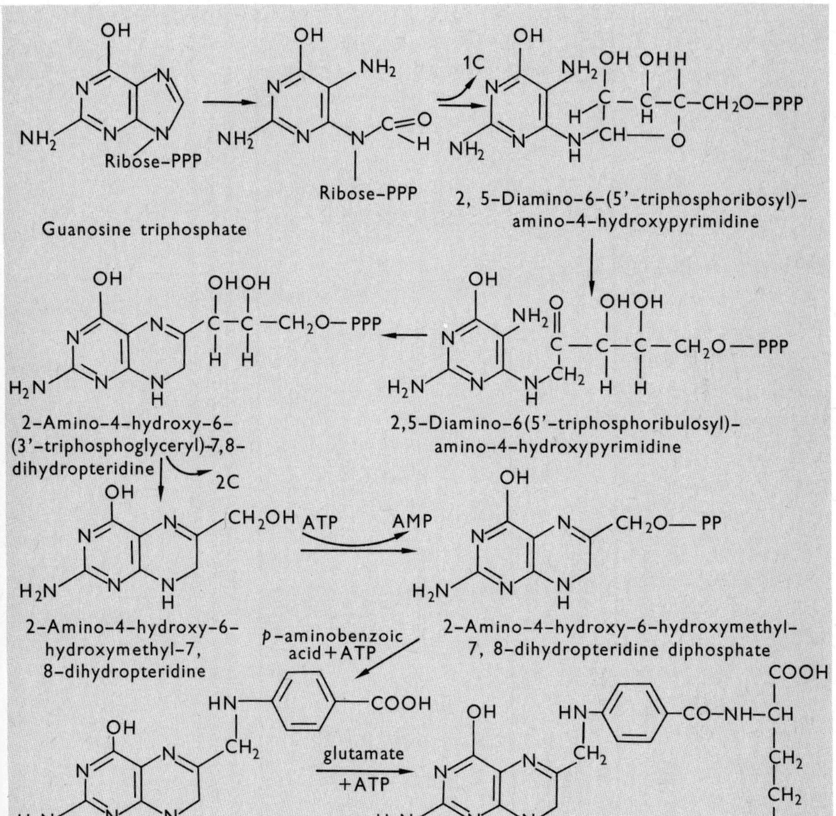

Fig. 3.20 Proposed scheme for the biosynthesis of tetrahydrofolic acid.

Table 3.5 Levels of folate derivatives in photosynthetic tissues.

Derivative	Radish (*Raphanus sativus*) cotyledons		Pea (*Pisum sativum*) leaf		Wheat (*Triticum vulgare*) leaf		Spinach (*Spinacia oleracea*) leaf	
	Before γ-GCP	After γ-GCP	Before γ-GCP	After γ-GCP	Before γ-GCP	After γ-GCP	Before γ-GCP	After γ-GCP
10-HCO-H_4PteGlu	294	553	237	160	115	238	191	218
10-HCO-H_4PteGlu$_2$	136	10 090	71	3 366	95	6 780	48	3 566
5-HCO-H_4PteGlu	101	n.d.	134	n.d.	57	n.d.	n.d.	n.d.
5-CH_3-H_4PteGlu	462	698	443	570	153	275	4 923	4 671
10-HCO-H_4PteGlu$_3$	58	n.d.	n.d.	n.d.	n.d.	861	n.d.	n.d.
H_4PteGlu	57	262	17	91	14	n.d. }		n.d.
5-HCO-H_4PteGlu$_2$	62	3 255	57	2 296	n.d.	3 807 }	191	3 207
5-CH_3-H_4PteGlu$_2$	149	15 220	493	5 394	22	1 397 }		11 140
H_4PteGlu$_2$	19	n.d.	117	1 278	n.d.	n.d. }		653
Total recovered from column	1 338	30 080	1 569	13 160	456	13 360	5 353	23 460
Range of total folates in extracts before chromatography	1 400 to 2 800	32 200	2 200 to 7 100	14 400	1 500 to 2 100	15 900	5 500 to 7 900	28 100

n.d., Not detected. Data are expressed as ng PteGlu equivalents/g dry wt for *Lactobacillus casei*. Extracts were prepared from fresh tissue and chromatographed on DEAE-cellulose before and after γ-glutamylcarboxypeptidase (γ-GCP) treatment. Fractions were assayed with *L. casei* and *Pediococcus cerevisiae*. The large bracket denotes that the pteroylglutamate activity eluted in the position of these derivatives could not be assigned to any one derivative. (From Spronk and Cossins,[585] Table 1, p. 3159).

p-amino benzoic acid could arise by amination of hydroxy benzoic acid originally synthesized from tyrosine or phenylalanine.

In the biosynthesis guanosine triphosphate loses a one-carbon fragment in the form of folic acid and produces 2,5-diamino-6-(5′-triphosphoryl)-amino-4-hydroxypyrimidine. Ring closure follows the conversion of ribosyl to the ribulosyl derivative with the resulting formation of 2-amino-4-hydroxy-6-(3′-triphosphoglyceryl)-7,8-dihydropteridine. Elimination of a 2-carbon residue in the form of glycol aldehyde produces 2-amino-4-hydroxy-6-hydroxymethyl-7,8-dihydropteridine. This latter compound, following a phosphorylation mediated by ATP, can react with *p*-amino benzoic acid to produce dihydropteroic acid. This compound can react, in the presence of ATP, with glutamate to produce dihydrofolic acid which in plants seems to be reduced by an $NADPH^+$-dependent reductase to tetrahydrofolic acid.[614] The naturally occurring folic acid derivatives contain varying numbers of glutamic acid residues. The significance of this is not clear but as indicated in Table 3.5 this diversity in glutamic acid residues also occurs in folate derivatives from plants.

Thiamine pyrophosphate

The coenzyme thiamine pyrophosphate plays an integral role in carbon metabolism. It is frequently involved in situations in which carbon residues are non-oxidatively removed such as in the conversion of pyruvate to acetaldehyde. It is also required during the oxidative decarboxylation catalysed by α-keto acid dehydrogenases. The cofactor participates in transketolase reactions in which a 2-carbon fragment from a ketose is transferred onto the aldehyde carbon of an aldose. Such transketolase activity is important in the interconversion of sugar phosphates involved in the pentose phosphate respiratory pathway and the Calvin–Benson photosynthetic cycle.

The coenzyme consists of a substituted pyrimidine linked by a methylene group to a substituted thiazole. It has been established in microbial systems that biosynthesis probably involves the amino acids methionine, alanine and aspartate. The aspartate is utilized in the formation of the pyrimidine ring derivative according to the sequence[393] on p. 102.

The 2-methyl-4-amino-5-hydroxymethyl pyrimidine is then converted to the pyrophosphate derivative which can then condense with the thiazole moiety. The synthesis of the thiazole component is believed to occur by the condensation of alanine and methionine to 4-methyl-5-β-hydroxyethyl thiazole which is then phosphorylated to the monophosphate. Condensation of the substituted pyrimidine pyrophosphate and the thiazole monophosphate produces thiamine monophosphate according to the scheme outlined on p. 103. The thiamine monophosphate is converted to thiamine

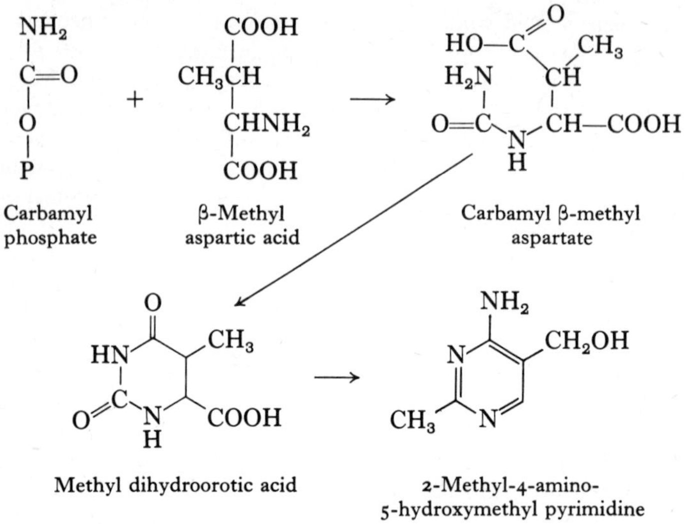

Carbamyl phosphate + β-Methyl aspartic acid → Carbamyl β-methyl aspartate → Methyl dihydroorotic acid → 2-Methyl-4-amino-5-hydroxymethyl pyrimidine

and phosphate and it is the free thiamine which receives a pyrophosphate residue from ATP to form the active coenzyme thiamine pyrophosphate.[393]

Pyridoxal phosphate

In animal systems there is an essential requirement for pyridoxal phosphate in transaminase and amino acid decarboxylations catalysed by purified enzymes. In contrast it is frequently difficult to establish an absolute requirement for pyridoxal phosphate in these reactions when partially purified enzymes extracted from plants are utilized. However plant transaminases usually show sensitivity to inhibitors known to interfere with pyridoxal phosphate function.[177] It thus appears that in plant transaminases that the pyridoxal phosphate is tightly bound to the enzyme and is not lost during purification.

The functioning of pyridoxal phosphate in transamination or amino acid decarboxylation in part involves the nitrogen atom in the pyridoxal ring. Pyridoxal phosphate forms a Schiff base with the amino acid (Fig. 3.21). The Schiff base provides a conjugated system of double bonds extending to the nitrogen atom in the pyridine ring and this provides the potential for decarboxylation or deamination of the α-carbon of the amino acid. In transamination the amino group is transferred to form pyridoxamine-5-phosphate and the free keto acid released. The amino group from pyridoxamine-5-phosphate can be transferred to an acceptor keto acid by a reversal of the sequence.

Although the metabolic role of pyridoxal phosphate in transamination has been established little is known concerning its mode of biosynthesis.

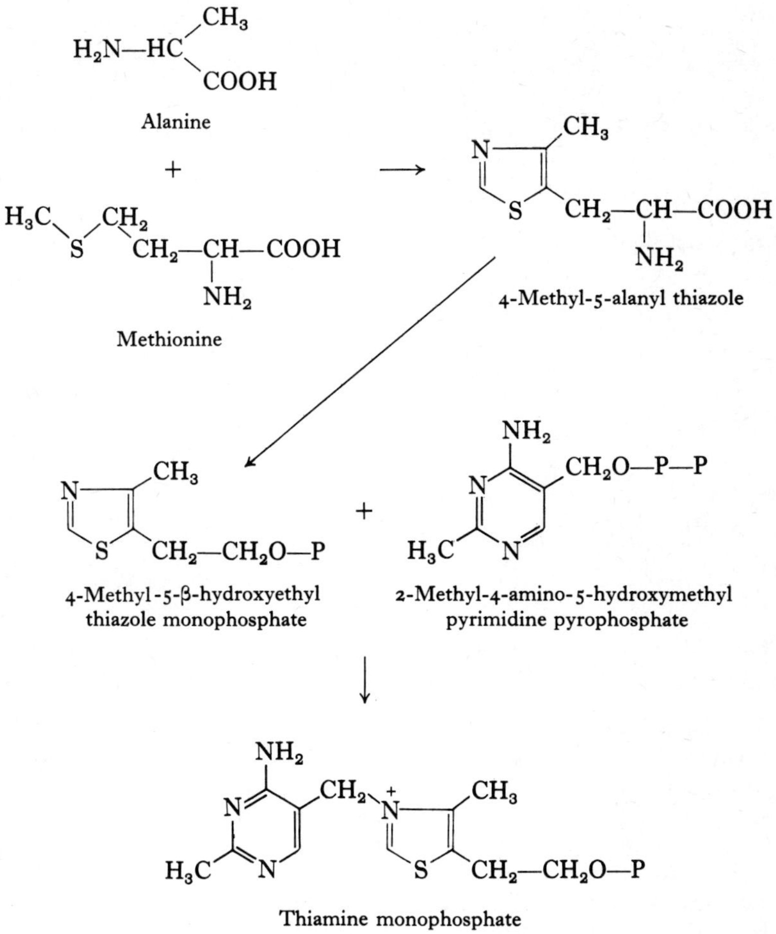

$H_2N{-}HC(CH_3)COOH$

Alanine

+

$H_3C{-}S{-}CH_2{-}CH_2{-}CH(NH_2){-}COOH$

Methionine

→

4-Methyl-5-alanyl thiazole

↓

4-Methyl-5-β-hydroxyethyl thiazole monophosphate

+

2-Methyl-4-amino-5-hydroxymethyl pyrimidine pyrophosphate

↓

Thiamine monophosphate

Biosynthesis of thiamine monophosphate

Coenzyme A

Coenzyme A functions in acyl group transfer and in this regard plays important roles in respiratory and lipid metabolism. Reactivity of coenzyme A in these reactions is dependent upon the presence of a sulphydryl group originally derived from the amino acid cysteine. Biosynthesis of the complex molecule involves participation of derivatives of valine and aspartate condensed with cysteine which in turn combines with the nucleotide ATP[393] according to the sequence outlined on pp. 105–6 (Fig. 3.22).

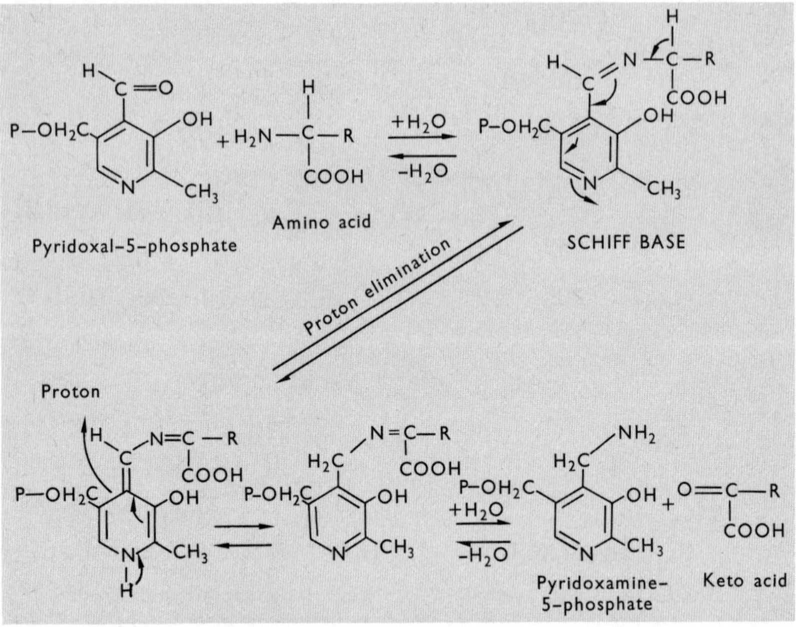

Fig. 3.21 Proposed scheme for the functioning of pyridoxal phosphate in transamination. Note the essential role of the nitrogen in the pyridoxal ring.

Few figures are available for the content of these cofactors in plants. Analyses which have been conducted have usually been directed to establishing the potential vitamin content of the plant.[532] These assays, although more recently based on chemical techniques, historically depended on feeding studies. The methods used frequently do not estimate the actual cofactor but more usually include closely related compounds. Thus data available for thiamine content do not differentiate between the free base, thiamine phosphate or the functional thiamine pyrophosphate. Thiamine (aneurine), vitamin B, occurs in highest concentrations (1 mg/100 g) in nuts and cereal grains but is present in lower levels in fruits and leaf tissue (0.1–0.2 mg/100 g).

Biotin has been detected in a whole range of plant materials; nuts and seeds are generally the best sources of the cofactor containing about 10 μg/100 g.

Estimates of vitamin B_6 or pyridoxine may not differentiate between pyridoxal, pyridoxamine or their derivatives and thus again there are few reliable figures on cofactor content. Pyridoxine content is higher in seeds than in fruit tissue but appears to be present in leaf tissue at levels comparable to that in seeds at about 5 μg/g. Tissue contents of coenzyme A do not seem to have been determined; figures are available on pantothenic acid content of plant foods but these do not differentiate between free or bound forms. Distribution of pantothenic acid appears to contrast from that of the other vitamins in that nuts have a relatively low content of this component. As is generally the case, cereal grains are good sources of pantothenic acid with contents of 15 μg/g.

Fig. 3.22

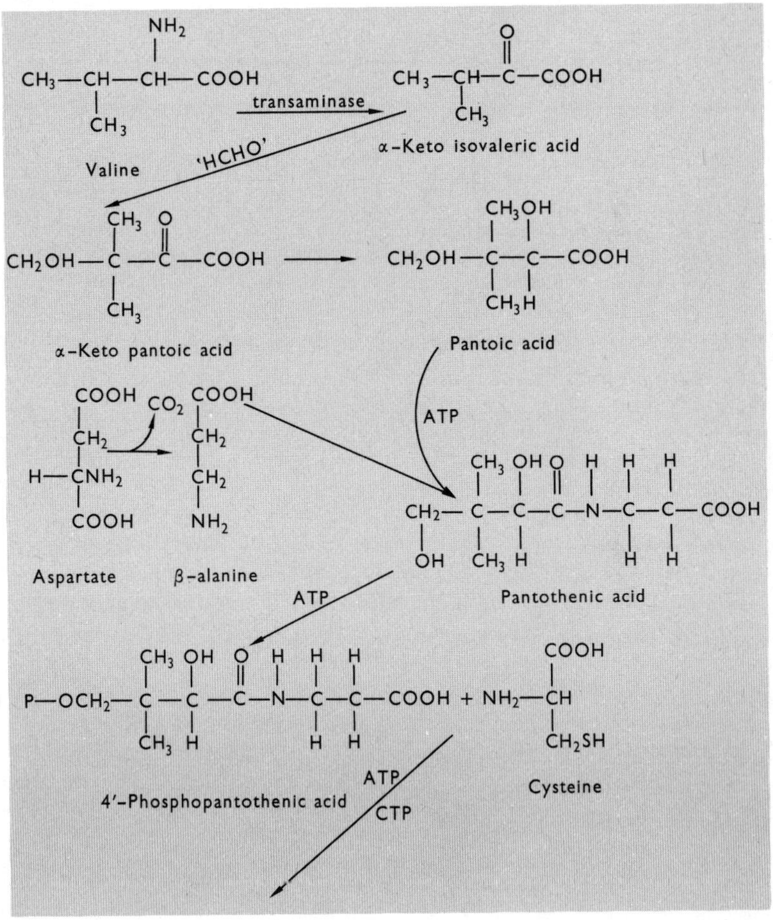

Valine
transaminase
α-Keto isovaleric acid
'HCHO'
α-Keto pantoic acid
Pantoic acid
CO_2
Aspartate
β-alanine
ATP
Pantothenic acid
ATP
4′-Phosphopantothenic acid
Cysteine
ATP
CTP

(*Continued over page*)

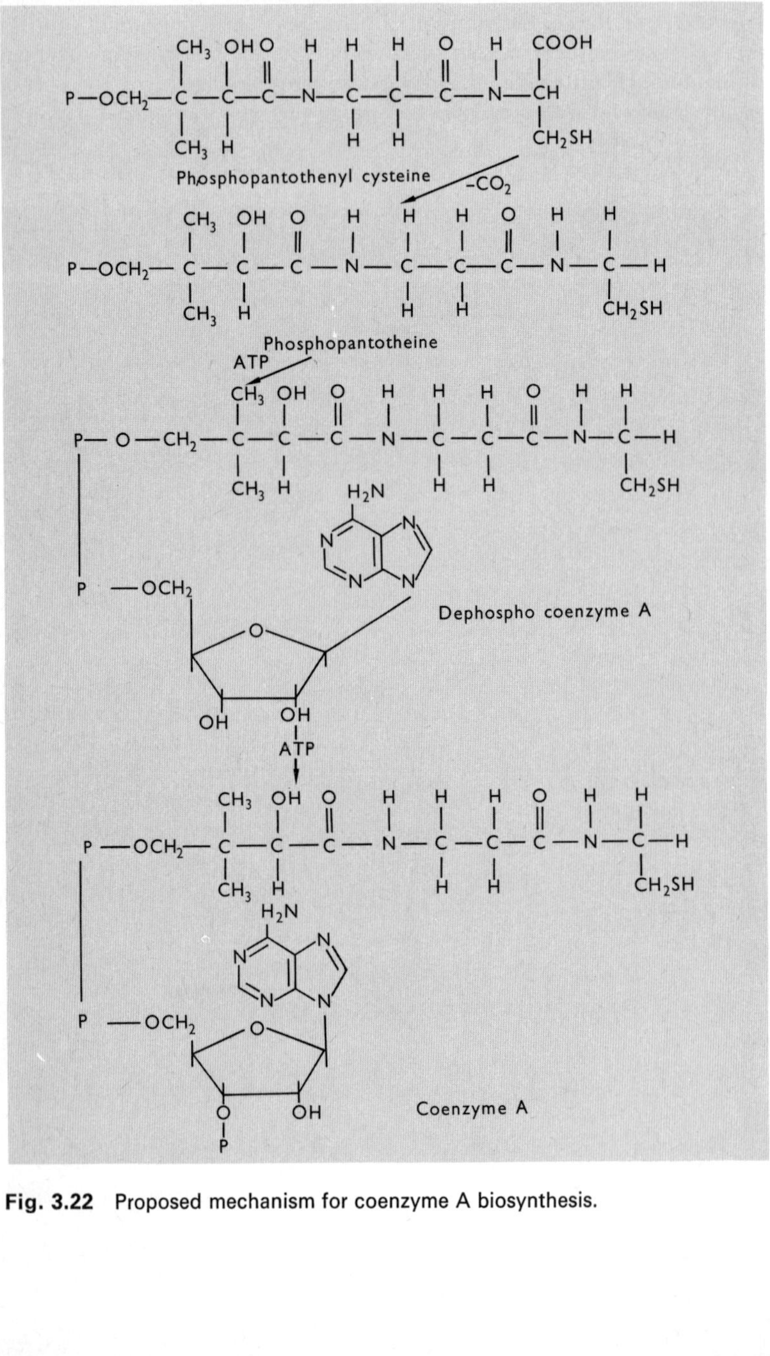

Phosphopantothenyl cysteine

$-CO_2$

Phosphopantotheine

ATP

Dephospho coenzyme A

ATP

Coenzyme A

Fig. 3.22 Proposed mechanism for coenzyme A biosynthesis.

4

Purines, Pyrimidines, Nucleosides and Nucleotides

An important series of nitrogenous constituents in plants is derived from the nitrogenous bases purine and pyrimidine.

PYRIMIDINES

The basic structure of a pyrimidine is a six-membered ring with nitrogen atoms at positions 1 and 3. The principal pyrimidines occurring in plants contain hydroxyl, amino and methyl substituents at hydrogen atoms 2, 4 and 5. The major pyrimidines are uracil, thymine and cytosine.

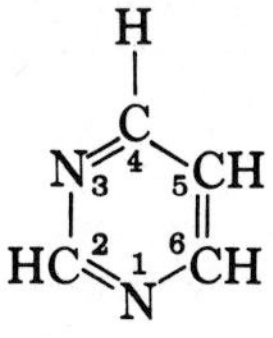

Pyrimidine	Uracil 2,4-Dioxypyrimidine	Thymine 5-Methyl- 2,4-dioxypyrimidine	Cytosine 2-Oxy-4-amino- pyrimidine

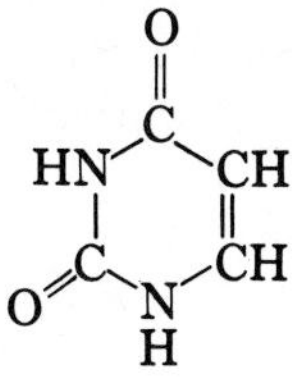

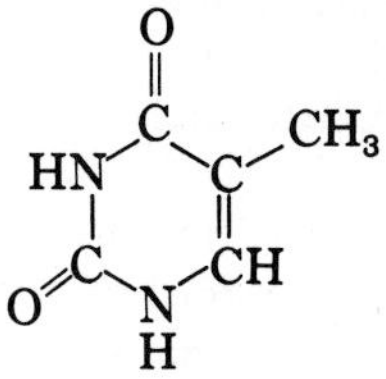

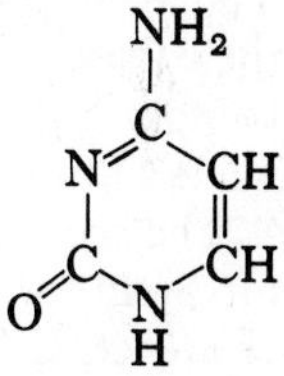

These bases seldom occur in the free state but are more usually complexed with a pentose sugar and a pentose sugar phosphate to form nucleosides and nucleotides respectively. The bases additionally are components of nucleic acids; uracil is found in ribonucleic acid, thymine is found in deoxyribonucleic acid and cytosine occurs in both types. Other pyrimidine derivatives, particularly methyl and sulphur-containing products of cytosine and uracil, occur as minor bases in transfer RNA.

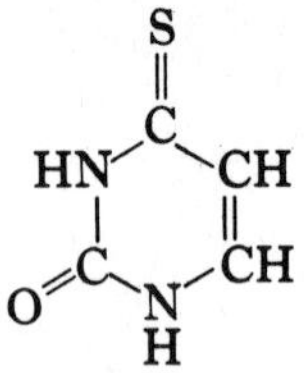

4-Thio uracil

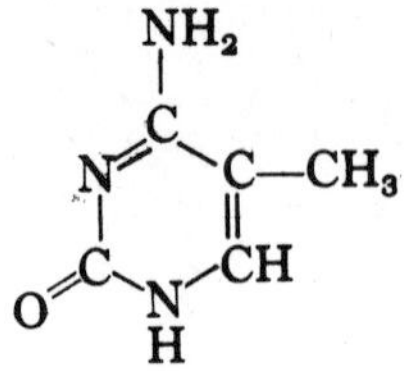

5-Methyl cytosine

PURINES

Purine bases may be considered derivatives of pyrimidines in which a five-membered imidazole component is attached to the pyrimidine to give a nine-membered ring. The labelling of the various atoms in this ring has been the subject of some debate. The currently preferred system is indicated on the left, although in older literature the nomenclature on the right is used.

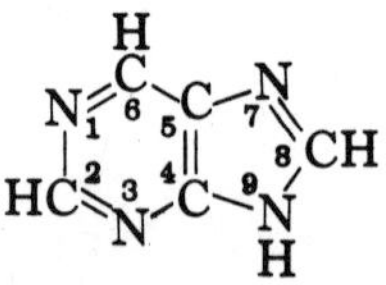

Purine ring
New system

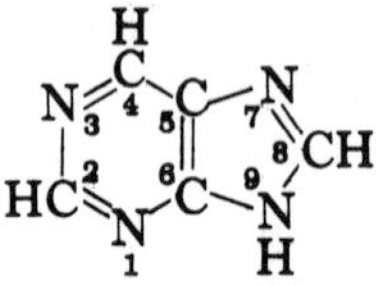

Purine ring
Old system

The major purines are adenine and guanine. As in the case of pyrimidines, these bases seldom occur free in plants but are usually complexed with pentose sugar phosphates and their principal occurrence is as bases in the nucleic acids.

The reports of the occurrence of the free bases may in part be accounted for by breakdown of the nucleotides and nucleic acids during extraction. However, there are also reports of other purines in plants which are not present in nucleic acids and these in all likelihood occur as the free base. The principal purine in this regard is xanthine and its derivatives. Xanthine is of widespread occurrence and is probably an important intermediary in purine metabolism. Other xanthine derivatives such as theophylline (1,3-dimethyl xanthine) and caffeine (1,3,7-trimethyl xanthine) occur in tea leaves (*Thea sinensis*); coffee (*Coffea arabica*) contains caffeine and theobromine (3,7-dimethyl xanthine) is found in cocoa (*Theobroma cacao*). The physiological role of these xanthine derivatives has not been established but they may be secondary metabolites analogous to alkaloids in other species. Their synthesis apparently involves progressive methylation of xanthine with methionine serving as methyl donor.

Adenine
6-Amino purine

Guanine
2-Amino-6-oxy purine

Xanthine

Theophylline

Theobromine

Caffeine

Methylated derivatives of guanine and adenine occur as constituent minor bases in nucleic acids. Additionally transfer RNA contains adenine derivatives containing isopentenyl groups. These isopentenyl substituted adenine compounds may also occur as the free base or as the ribose or ribose phosphate derivatives. These derivatives are of great interest since they show cytokinin activity. The first natural cytokinin was isolated from corn (*Zea mays*) kernels and called zeatin;[371] its chemical structure is 6-(4-hydroxy-3-methylbut-2-enyl) amino purine. A closely related adenine derivative 6-(γγ-dimethylallyl) amino purine or isopentenyl adenine has also been identified from plants.[569]

Significantly triacanthine, 3-γγ-isopentenyl adenine, which occurs in the leaves of various species[245] as the free pyrimidine base, does not have cytokinin activity. Since N^6-IPA is enzymatically degraded it has been suggested[245] that triacanthine may represent an inert storage pool of precursor for the natural cytokinins.

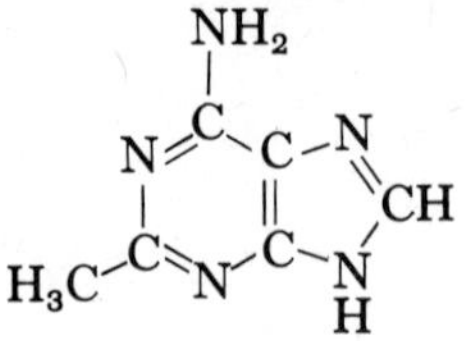

2-Methyl adenine

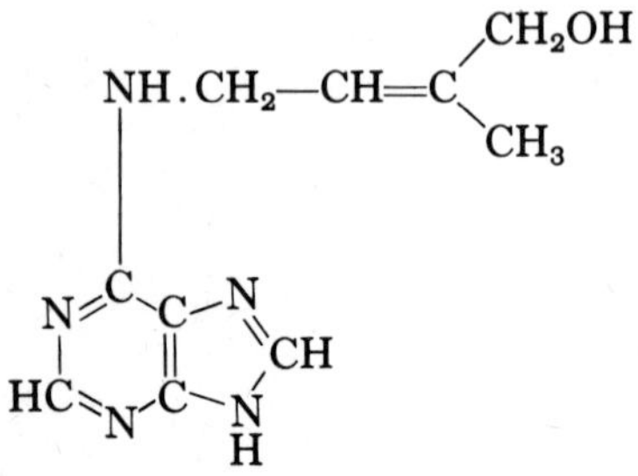

Zeatin
6-(4-Hydroxy-3-methylbut-2-enyl) amino purine

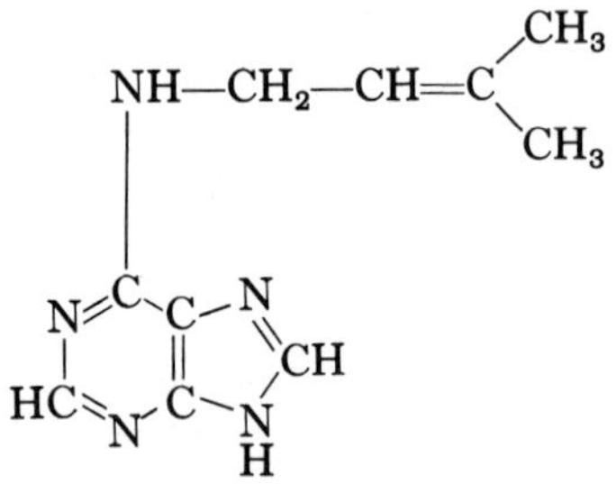

Isopentenyl adenine (IPA)
6-(γγ-Dimethylallyl) amino purine

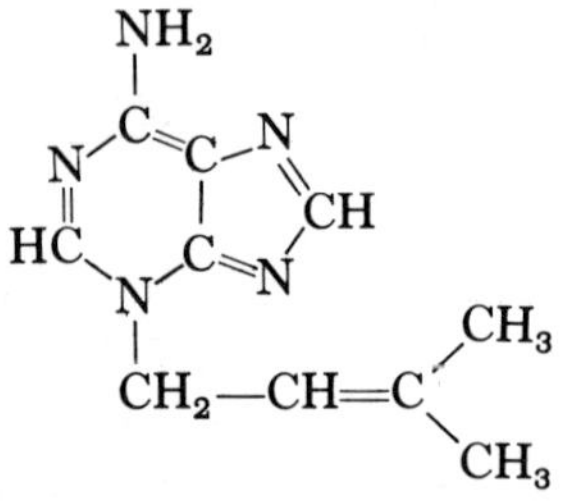

3-γγ-Isopentenyl adenine
Triacanthine

NUCLEOSIDES

The purine and pyrimidines may be complexed with a sugar, usually a ribose or deoxyribose, to produce a nucleoside. The sugar is attached in a β-glycosidic link to the N-9 of the purines adenine and guanine to form adenosine and guanosine respectively. Similar β-glycosidic linkage of ribose to the N-1 of the pyrimidines uracil and cytosine produces uridine and cytidine.

Unless designated otherwise, the sugar moiety in the nucleoside is always ribose. When deoxyribose combines with the base the resulting

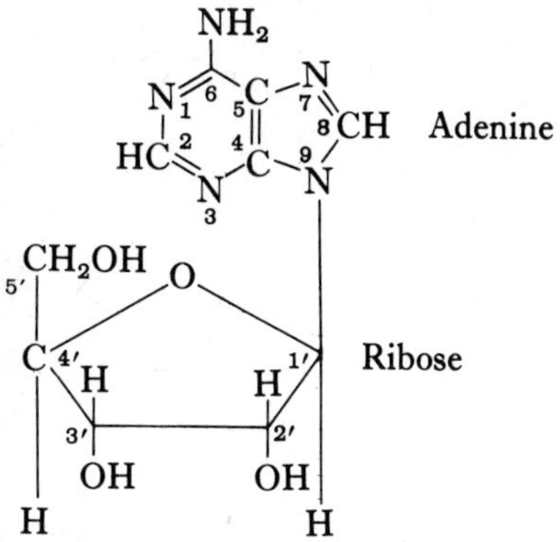

Adenine
Ribose

Adenosine (AOH)

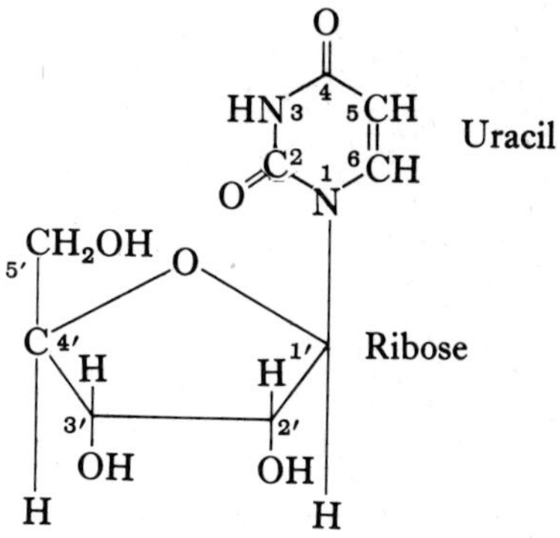

Uracil
Ribose

Uridine (UOH)

complexes are designated deoxy-adenosine, deoxy-guanosine, deoxy-cytidine. Uracil does not occur as the deoxyribose derivative. Thymine does not occur as the ribose derivative but does combine with deoxyribose to form thymidine; the deoxy prefix is redundant in this instance.

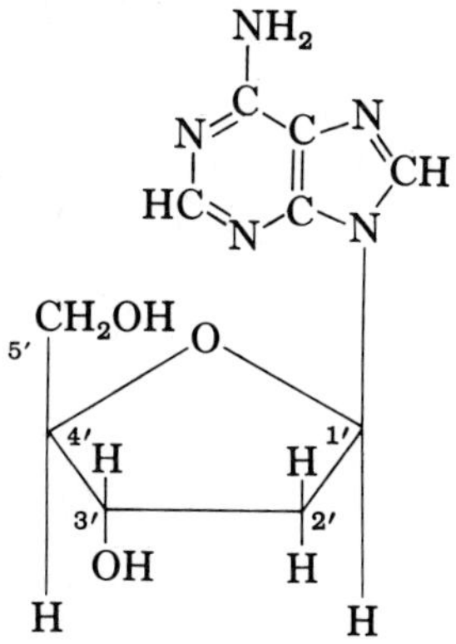

Deoxy adenosine (d AOH)

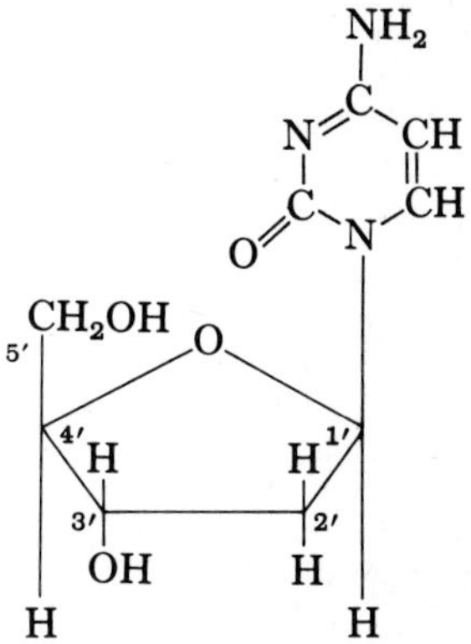

Deoxy cytidine (d COH)

The ribonucleosides occur in the free state. In dry peas (*Pisum sativum*), for example, the nucleosides occur with the relative distribution uridine > adenosine > xanthosine and > guanosine.[76] During the germination phase xanthosine increases and adenosine content declines.[77] Some other base sugar complexes are characteristic of certain species but their physiological role has not been established. Crotonoside is a free nucleoside occurring in *Croton tiglium* formed from the purine isoguanine and ribose.

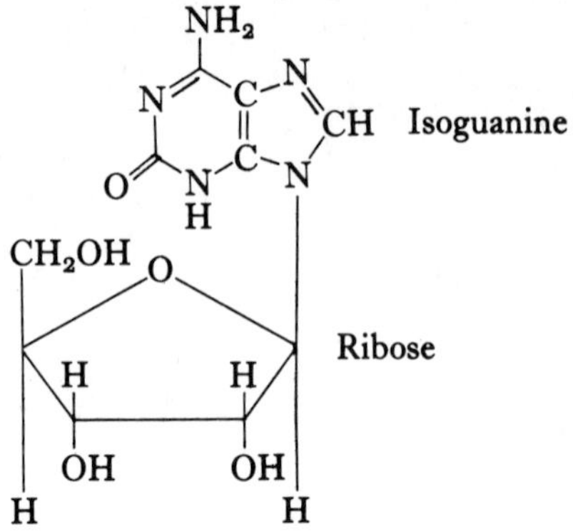

Isoguanine

Ribose

Crotonoside

Vicin which occurs in the seed of *Vicia* spp. is the glucoside of the pyrimidine divicin. However, this is not a true nucleoside since the sugar is linked by an oxygen substituent to the pyrimidine ring and not to the constituent N-1 atom.

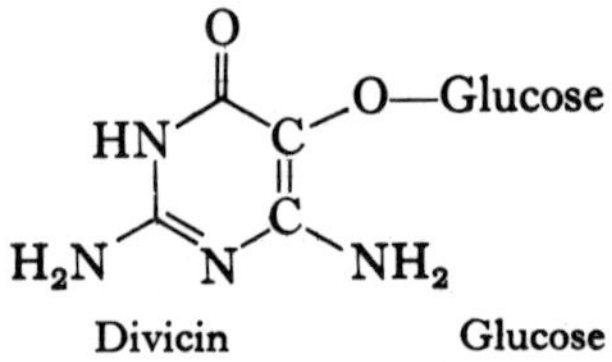

Divicin Glucose

Vicin

The isopentenyl derivative of adenine, zeatin, has been reported to occur as the riboside also.[569]

NUCLEOTIDES

The addition of a phosphate group to the sugar residue of a nucleoside produces a nucleotide. Thus, by definition a nucleotide is a Base–Sugar–Phosphate. Phosphate esterification can occur at the three hydroxyls 2′, 3′ and 5′ of the ribose moiety of the nucleosides. In deoxyribonucleosides phosphate esterification can occur only at the 3′ and 5′ position of the deoxyribose. In nature it appears that the prominent free nucleotides are the 5′ phosphate esters; however, 3′ phosphate derivatives may be produced by nucleic acid degradation. Additionally it is possible to form cyclic nucleotides such as 2′,3′-AMP and 3′,5′-AMP. The latter compound is of great interest since it has been demonstrated that in animal tissue it may function as a secondary messenger between a hormone and its induced effect; cyclic AMP also has a regulatory function in protein synthesis in bacterial systems.[534]

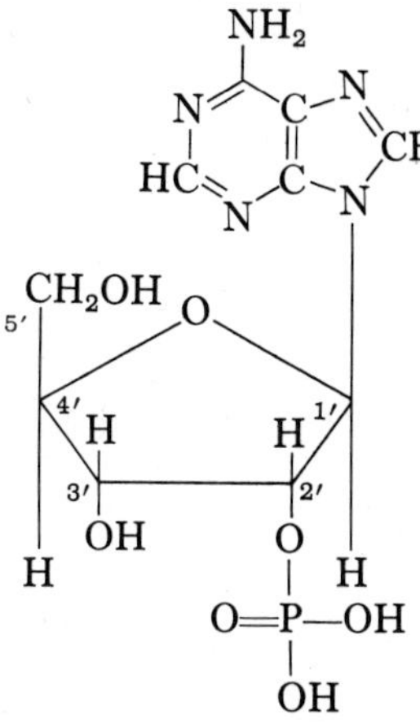

Adenosine 2′-monophosphate
2′-adenylic acid

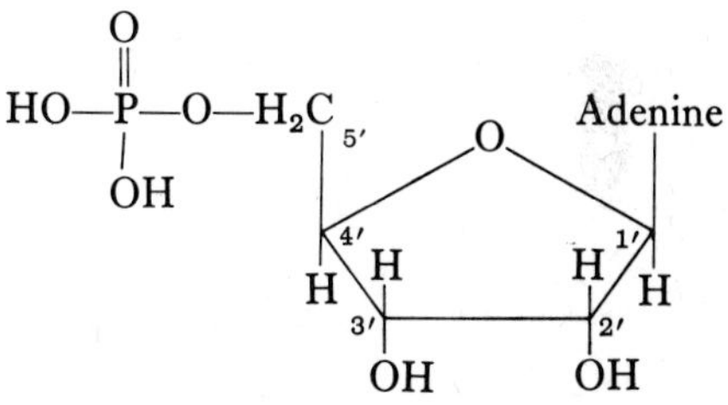

Adenosine 5′-monophosphate
5′-adenylic acid

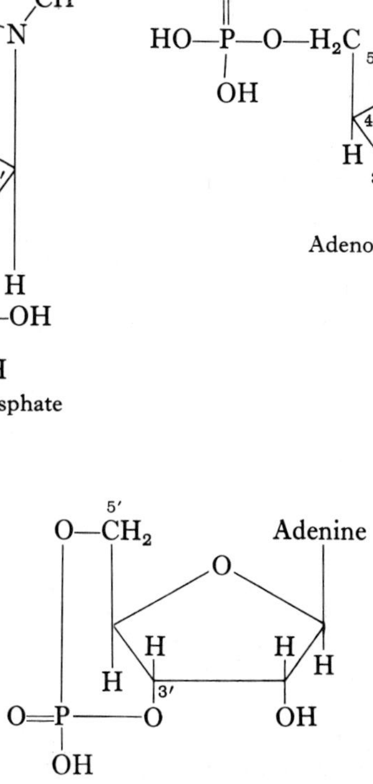

Adenosine 3′,5′-cyclic monophosphate
3′,5′-cyclic adenylic acid

Some of the effects of gibberellic acid on plant development can be mimicked by cyclic AMP,[215] giving rise to the speculation that gibberellic acid responses may involve a secondary messenger. Unfortunately, to date it has not been possible to demonstrate the presence of adenyl cyclase in plant tissues which is responsible for the production of cyclic AMP from ATP; nevertheless recent analyses using a specific immunological technique[211] have demonstrated the presence of cyclic 3′,5′-AMP in plant tissues and we can expect a series of investigations aimed at elucidating its physiological role.

The predominant naturally occurring phosphorylated nucleosides can contain one, two, or three phosphates at the 5′ position. The additional β and γ phosphates are increasingly unstable so that in assessment of the nucleotide content of plant tissues caution must be taken to prevent the cleavage of terminal phosphates from the di- and triphosphates.

General structures of mono-, di-, and triphosphates

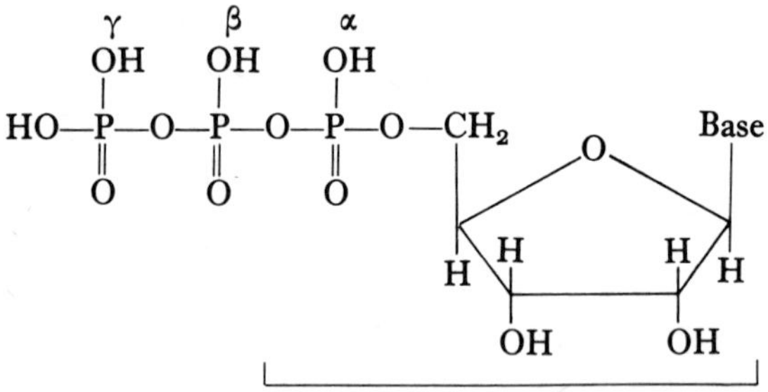

γ β α
OH OH OH
HO—P—O—P—O—P—O—CH_2 O Base
O O O
H H H H
OH OH

Nucleoside 5′-monophosphate (NMP)

Nucleoside 5′-diphosphate (NDP)

Nucleoside 5′-triphosphate (NTP)

Characteristically, plant material is extracted in cold perchloric acid, trichloracetic acid or methanolic formic acid. The resulting extract is subsequently purified to remove contaminants and the nucleotides are separated from the partially purified extract by column chromatography. When such analyses are conducted, in addition to the monophosphate, diphosphate, and the triphosphate derivatives of cytidine, uridine, adenosine and guanosine, other nucleotides and nucleotide derivatives are observed. Principal among these are the pyridine nucleotides and various carbohydrate derivatives of the nucleoside diphosphates. A typical separation of nucleotides from plant material is shown in Fig. 4.1 and major free nucleotide composition is given in Table 4.1.

The complexity of the series of products arising from the purine and pyrimidine bases raises the problem of nomenclature. Thus, a given nucleotide can be called a monophosphate of the corresponding nucleoside or it could be an acid with the name derived from the nucleoside. Since the phosphate can occupy a variety of positions it is customary to designate its location on the ribose moiety. Some terminologies and alternatives in current usage are outlined in Table 4.2.

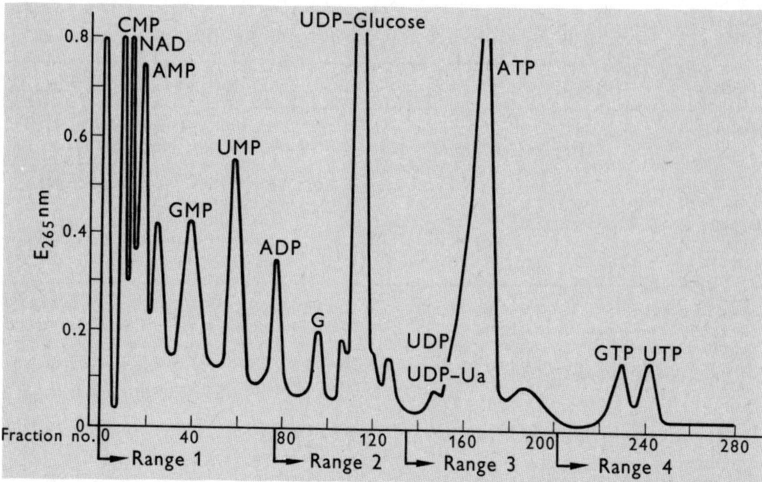

Fig. 4.1 Anion-exchange chromatogram of acid-soluble nucleotides of *Acer pseudoplatanus* cells after growth for 4 days.
(Dowex 1-Formate ; x8 ; linear-formate gradients.)
UDP-Ua = a UDP-uronic acid ; G = unidentified guanosine nucleotide.
(From Brown and Short,[79] Fig. 2, p. 1366.)

Table 4.1 Predominant free nucleotide content of Bean (*Phaseolus vulgaris*) petioles and leaf lamina. (From Brown and Mangat,[78] Tables 1 and 3, pp. 1860–1861.)

Nucleotide	Concentration (mμ moles/10 g fr. wt.) Petioles	Lamina
CMP	10	—
NAD	—	80
AMP	10	120
GMP	—	70
NADP	—	18
ADP ribose	5	30
UMP	5	150
ADP	30	180
AdR*	10	20
UDP glucose	240	240
NADPH +	20	70
UDP	—	—
CTP	10	—
ATP	330	690
GTP	60	80
UTP	90	140
GR†	10	—
Total	830	1888

*Unidentified adenine nucleotide arising during alkaline degradation of NAD (nicotinamide adenine dinucleotide).
†Unidentified guanine nucleotide.

Table 4.2 Nomenclature of products arising from purine and pyrimidine bases.

Base	Nucleoside	Nucleotide as acid	Nucleotide as phosphate
Adenine	Adenosine	2′ Adenylic acid	Adenosine 2′ monophosphate
		3′ Adenylic acid	Adenosine 3′ monophosphate
		5′ Adenylic acid	Adenosine 5′ monophosphate
		5′, 3′ Cyclic adenylic acid	Adenosine 5′, 3′ monophosphate (cyclic)
Guanine	Guanosine	5′ Guanylic acid	Guanosine 5′ monophosphate
Uracil	Uridine	5′ Uridylic acid	Uridine 5′ monophosphate
Cytosine	Cytidine	5′ Cytidylic acid	Cytidine 5′ monophosphate

Each 5′ nucleotide also occurs as the diphosphate and triphosphate. The free bases are usually designated as ACGTU, the nucleoside as AOH, etc., the 3′ nucleotide as Ap and the 5′ nucleotide as pA. 5′ Adenosine triphosphate is abbreviated to pppA or 5′ATP. Ribose is always understood unless deoxy sugars are indicated by d′, e.g. d′AOH—deoxyadenosine, d′AMP—deoxyadenylic acid.

PURINE AND PYRIMIDINE BIOSYNTHESIS

From the foregoing description of the chemical characteristics of the free purine and pyrimidine bases and their nucleoside and nucleotide derivatives it might be expected that synthetic events would involve an initial production of the various free bases which could be converted subsequently to nucleosides and nucleotides. In fact the converse is true and it appears that purine and pyrimidine 5′ nucleotides are the initial products of biosynthesis.

Pyrimidine biosynthesis

In plants most studies have been devoted to an investigation of pyrimidine biosynthesis. Studies with appropriately labelled precursors, and enzymatic investigations, indicate that the biosynthesis of uridylic acid and cytidylic acid proceeds along the following sequence.

The components of the initial reaction in pyrimidine biosynthesis are aspartic acid and carbamyl phosphate. As discussed earlier (p. 39), carbamyl phosphate is formed from glutamine, bicarbonate and ATP in plants.[481, 483]

The carbamyl phosphate reacts with aspartic acid to produce *N*-carbamyl aspartic acid in a reaction catalysed by the enzyme aspartate transcarbamylase.[482] Removal of water from *N*-carbamyl aspartic acid

$$NH_2{-}\overset{\displaystyle O}{\overset{\|}{C}}{-}O{-}\underset{\displaystyle O}{\underset{\|}{\overset{\displaystyle OH}{\overset{|}{P}}}}{-}OH \;+\; \begin{array}{c} COOH \\ | \\ H{-}C{-}H \\ | \\ H{-}C{-}NH_2 \\ | \\ COOH \end{array} \xrightarrow[\text{Aspartate trans-carbamylase}]{Pi} \begin{array}{c} \text{HO–C(=O)–}CH_2 \\ | \\ H_2N{-}C(=O){-}NH{-}CH{-}COOH \end{array}$$

Carbamyl phosphate — Aspartic acid — *N*-Carbamyl aspartic aci

catalysed by dehydroorotase produces dihydroorotic acid. Oxidation of dihydroorotic acid by an NAD-dependent dihydroorotate dehydrogenase produces orotic acid.[320]

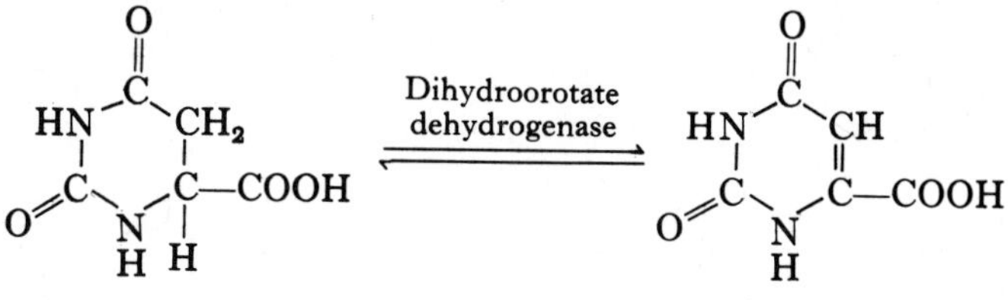

Dihydroorotic acid ⇌ (Dihydroorotate dehydrogenase) Orotic acid

Dihydroorotic acid

Orotic acid

Orotic acid is converted to orotidine 5′-phosphate by the attachment of the 5′-phosphoribosyl moiety from phosphoribosyl pyrophosphate in a reaction catalysed by the enzyme orotidine 5′-phosphate-pyrophosphorylase.[541, 702]

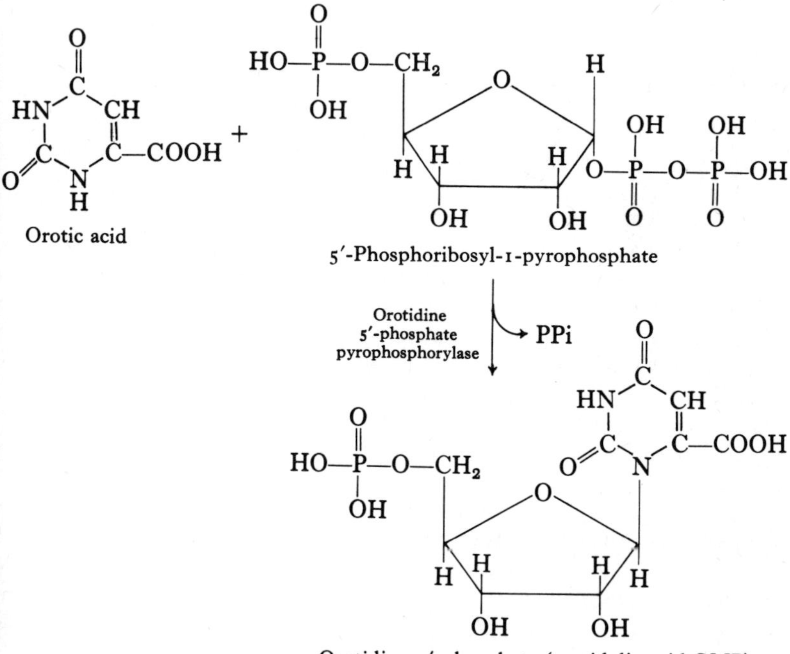

Orotic acid + 5′-Phosphoribosyl-1-pyrophosphate

Orotic acid

5′-Phosphoribosyl-1-pyrophosphate

↓ Orotidine 5′-phosphate pyrophosphorylase → PPi

Orotidine 5′-phosphate (orotidylic acid OMP)

Decarboxylation of orotidine 5′-phosphate in a reaction catalysed by orotidine 5′-phosphate decarboxylase produces uridine 5′-phosphate.[541]

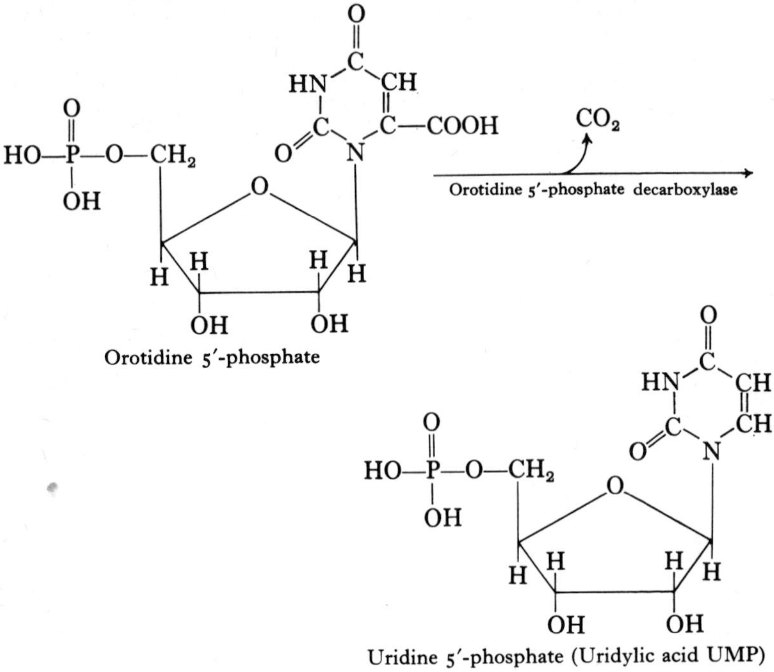

Although there are reports[84] of the decarboxylation of orotic acid and orotidine to uracil and uridine thus bypassing orotidine 5′-phosphate in some plants these findings have not been confirmed in other species and the outlined sequence appears to be the most usual pathway.

In animal and bacterial systems uridine 5′-phosphate can be converted to uridine 5′-triphosphate by the activity of the enzyme uridylate kinase. Uridine 5′-triphosphate then undergoes an amination of the 6′ position of the pyrimidine ring to yield cytidine 5′-triphosphate. Glutamine serves as the amine source in an ATP-dependent reaction.

Although the enzymes involved in these reactions have not been detected in plants, the ready conversion of uracil derivatives into cytidylic acid suggests the operation of such a pathway.[329, 469]

Purine biosynthesis

The sequence of reactions involved in purine biosynthesis established in studies with bacterial and animal systems is described below (Fig. 4.2).

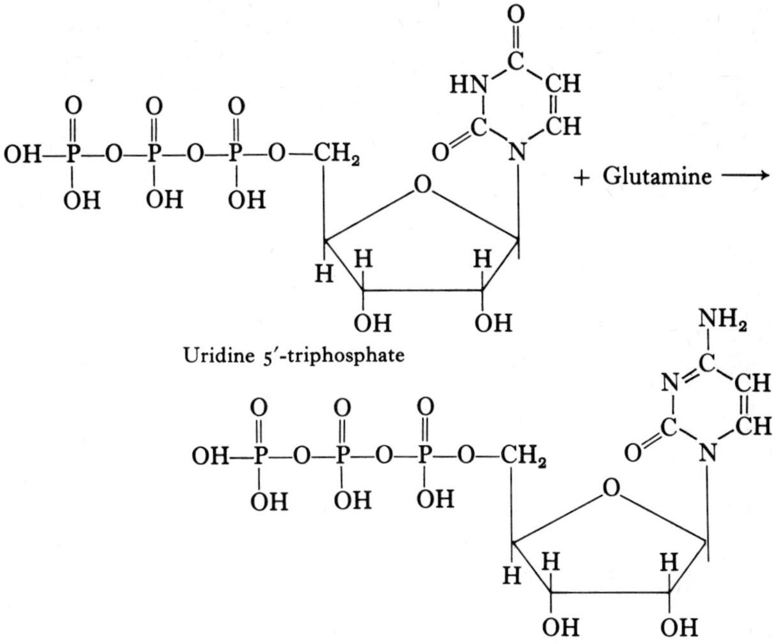

+ Glutamine ⟶

Uridine 5′-triphosphate

Cytidine triphosphate

Ribose 5′-phosphate undergoes enzymatic pyrophosphorylation at the expense of ATP to form 5-phosphoribosyl-1-pyrophosphate. In the next reaction catalysed by an amido tranferase, the 5-phosphoribosyl-pyrophosphate reacts with an amino donor and the amino group replaces the pyrophosphate with the resulting formation of 5-phosphoribosyl-1-amine. In plant systems asparagine, glutamine and carbamyl phosphate have all been implicated as amino donors, with asparagine being the most efficient.[320, 525] The 5-phosphoribosyl-1-amine reacts with glycine in the presence of ATP and the enzyme glycinamide ribonucleotide synthetase to form glycinamide ribonucleotide. In this reaction the carboxyl group of glycine reacts with the 1 amino group of the 5-phosphoribosyl-1-amine to form an amide linkage between glycine and the amino sugar.

The glycinamide ribonucleotide may be converted to *N*-formyl-glycinamide ribonucleotide by the addition of a formyl group from N^5N^{10}-methenyltetrahydrofolate. In the next reaction the amide group of glutamine is transferred in an ATP-dependent process to the carbamyl group of *N*-formyl-glycinamide ribonucleotide to yield *N*-formyl-glycinamidine ribonucleotide. This latter product following elimination

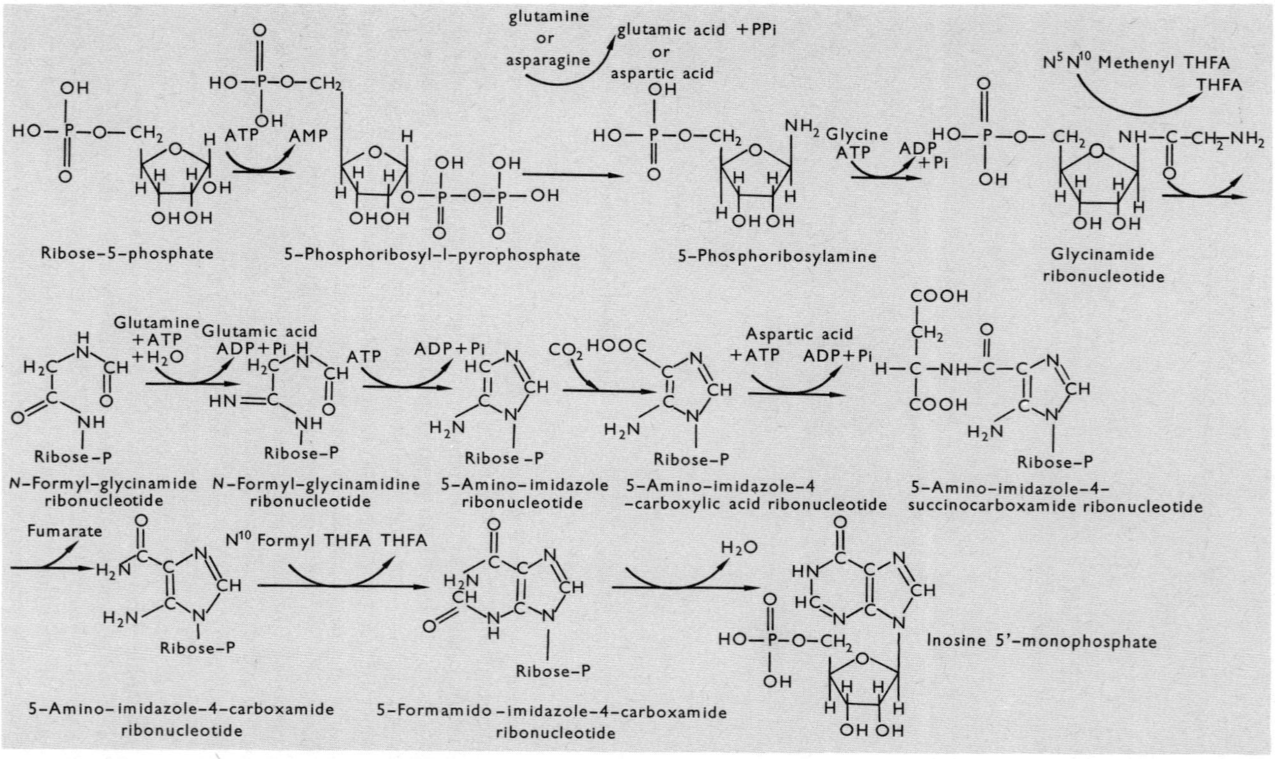

Fig. 4.2 Proposed scheme for the biosynthesis of purine nucleotides. Note the key roles of the amino acids glycine

of water produces 5-amino-imidazole ribonucleotide which contains the five-membered imidazole ring portion of purines.

5-Amino-imidazole ribonucleotide is then carboxylated to yield 5-amino-imidazole-4-carboxylic acid ribonucleotide. The next step involves the addition of the nitrogen atom which will correspond to position 1 of the purine ring. An entire aspartic acid molecule is combined in an ATP-dependent process to yield 5-amino-imidazole-4-*N*-succinocarboxamide ribonucleotide which is then converted to 5-amino-imidazole-4-carboxamide ribonucleotide and fumaric acid. Addition of a formyl group, donated by N^{10} formyl tetrahydrofolate, to the 5 amino group of the ribonucleotide yields 5-formamino-imidazole-4-carboxamide ribonucleotide. Elimination of water from this compound allows for ring closure of the purine ring and the formation of inosine 5′-monophosphate which is the precursor of adenosine 5′-monophosphate and guanosine monophosphate.

The conversion of inosine 5′-monophosphate to adenosine 5′-monophosphate involves the addition of an amino group at position 6 of the purine ring (Fig. 4.3). This is brought about by a GTP-dependent reaction of inosine 5′-monophosphate with aspartic acid to yield aspartyl inosine monophosphate (adenylosuccinic acid). Elimination of fumaric acid from this complex in a manner similar to that described previously yields adenosine 5′-monophosphate.

In the formation of guanosine 5′-monophosphate, inosine 5′-monophosphate is oxidized in an NAD-dependent reaction to xanthosine 5′-monophosphate (Fig. 4.3). This nucleotide can then be aminated in an ATP-dependent reaction with glutamine serving as an amino donor to form guanosine 5′-monophosphate.

Evidence for the operation of this biosynthetic sequence in higher plants is based upon the demonstration of some of the enzymes involved in early reactions of the scheme by Waygood and coworkers.[320, 525, 667] Additionally, Iwai *et al.*[297] have demonstrated the occurrence of glycinamide ribonucleotide in plant extracts and the accumulation of this compound in the presence of inhibitors of folate metabolism suggests its utilization in purine biosynthesis according to the depicted scheme. Labelled glycine and 5-amino-4-imidazole-carboxamide are readily incorporated into purine nucleotides[469, 525] as would be expected if synthesis proceeded as described in the above sequence.

THE SALVAGE PATHWAY OF NUCLEOTIDE BIOSYNTHESIS

The breakdown of nucleic acids and nucleotides such as might occur during mobilization of seed reserves or during normal turnover of cellular constituents has the potential of producing nucleosides and free bases. These products can be degraded further as discussed later or they can be

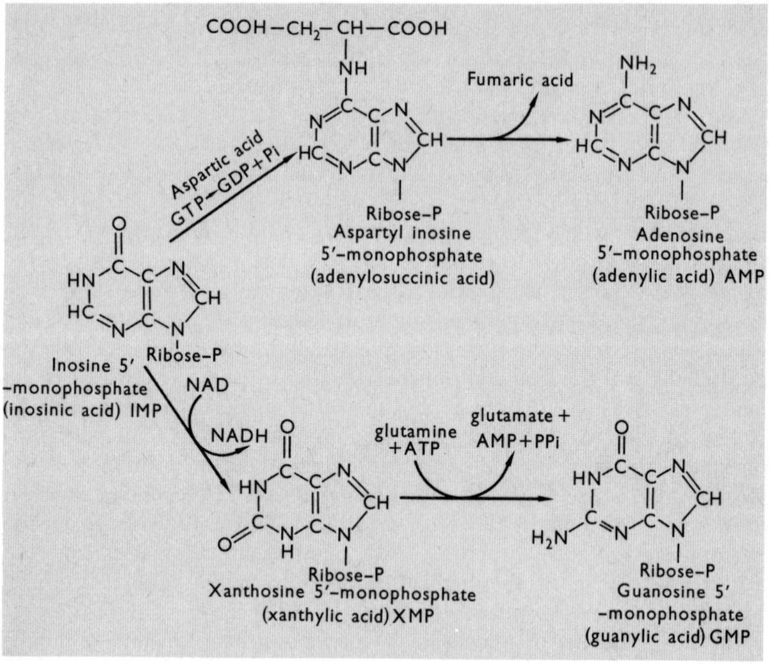

Fig. 4.3 Proposed scheme for the biosynthesis of guanosine 5′-monophosphate and adenosine monophosphate ; note the key role of the amino acid, aspartic acid and the amide glutamine as amino donors in the synthesis of these nucleotides.

interconverted into other nucleotides. This process of nucleotide synthesis from presynthesized purine and pyrimidine bases or nucleosides is referred to as the salvage pathway.

Labelling studies with pyrimidine bases and nucleosides and associated enzymatic studies indicate the operation of a salvage pathway in plants (Fig. 4.4).

Labelled uracil does not appear to be converted into uridine in plant tissues and investigations with plant extracts[413,539] have failed to detect the enzyme uridine phosphorylase which normally catalyses uracil–uridine conversion in animal and bacterial cells. In plants, exogenously supplied uracil is readily incorporated into uridine 5′-monophosphate. This conversion involves the participation of phosphoribosylpyrophosphate and the enzyme uridine monophosphate pyrophosphorylase which has been detected in plant extracts.

Uracil + Phosphoribosylpyrophosphate ⟶ Uridine 5′-monophosphate + PPi

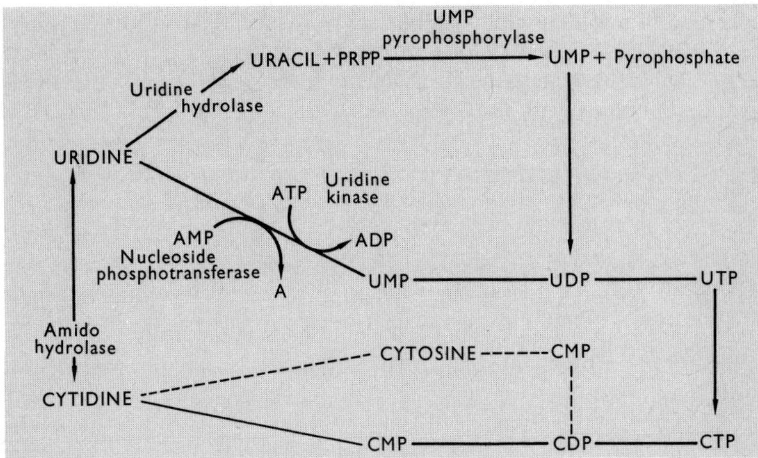

Fig. 4.4 Proposed interconversions of pyrimidine bases, nucleosides and nucleotides in higher plants. Reactions established on the basis of labelling and/or enzymatic data are indicated by continuous lines.

Uridine may be converted to uracil and ribose by a specific hydrolase; additionally it may be converted to uridine monophosphate by a uridine kinase. There are reports of a specific uridine kinase;[669] however the phosphorylation can also be carried out by a non-specific nucleoside phosphotransferase in which AMP serves as the phosphate donor.[147]

Uridine + Adenosine 5′-monophosphate ⟶ Uridine 5′-monophosphate + Adenosine

Subsequent conversion of the 5′-monophosphate to UTP is apparently mediated by a nucleotide kinase.

Cytidine may be converted to CTP by direct phosphorylation;[540] however, whether this involves nucleoside and nucleotide kinases, or the nucleoside phosphotransferase in conjunction with a nucleotide kinase, has not been established. Cytidine additionally may be converted to uridine by a deaminase (amido hydrolase).[273] This enzyme shows activity against deoxycytidine but is inactive against the corresponding nucleotides. Uridine produced by deaminase activity may be metabolized to UTP which can then also generate CTP[540] as outlined on page 119. Mechanisms for the conversion of cytidine to cytosine have not been established and no estimates have been made of cytidine monophosphate pyrophosphorylase activity.

Exogenously applied labelled adenine and adenosine are readily incorporated into ATP in a range of plant species,[469] thus indicating the operation of a salvage pathway (Fig. 4.5). Adenosine may be phosphorylated by a kinase or may be converted to adenine. This hydrolysis appears to be catalysed by a specific adenosine hydrolase (*N*-ribosyl-adenineribohydrolase), which will also hydrolyse deoxyadenosine but is ineffective

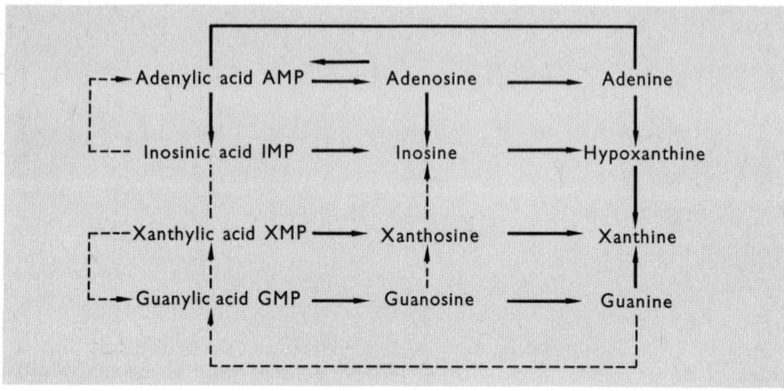

Fig. 4.5 Possible interconversions of purine derivatives. Continuous lines indicate reactions which have been tentatively established on the basis of enzymatic and/or labelling studies.

against guanidine.[120] The conversion of adenine to adenosine has not been established. It appears on the basis of enzymatic data[460] that adenine is converted directly to the nucleoside-monophosphate in a reaction catalysed by adenine monophosphate pyrophosphorylase (Fig. 4.5).

$$\text{Adenine} + \text{Phosphoribosylpyrophosphate} \longrightarrow \text{Adenosine 5}'\text{-monophosphate} + \text{PPi}$$

Labelled adenosine has been reported to be converted to inosine in some tissues.[491] In other tissues this conversion is reported to occur at the nucleotide level with adenylic acid undergoing deamination to produce inosinic acid.[566] In certain species adenine appears to give rise to hypoxanthine and in turn hypoxanthine may be oxidized to xanthine.[29] This oxidation is presumably catalysed by the enzyme xanthine oxidase. However, to date there has been no conclusive reports of the presence of this enzyme in higher plants.

Guanine derivatives appear to be metabolized less extensively than those derived from adenine. In pea (*Pisum sativum*) seedlings guanylic acid in addition to being converted into guanosine and guanine also was metabolized to β-2,6-dehydroxy-pyrimidinyl-alanine.[80] In certain tissues guanine is metabolized to uric acid[493] in the sequence described (Fig. 4.7); in these instances it appears that guanine can be deaminated to xanthine.

However, enzymes involved in this interconversion have not been established in higher plants.

PYRIMIDINE AND PURINE DEGRADATION

In addition to functioning as precursors for nucleotide biosynthesis in the salvage pathway free bases may be degraded. By the utilization of labelled thymine and uracil it has been established[179] that breakdown of pyrimidines proceeds through dihydropyrimidines and β-ureido amino acids to β-aminoisobutyric acid in the case of thymine and to β-alanine in the case of uracil (Fig. 4.6). Enzymes involved in the catabolic sequence

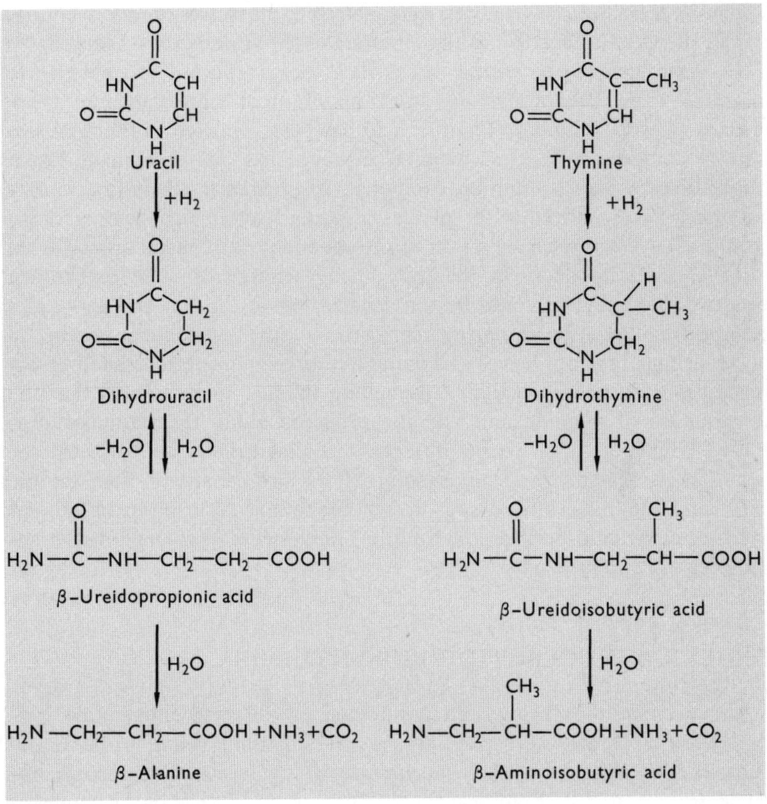

Uracil → $+H_2$ → Dihydrouracil ⇌ ($-H_2O$ / H_2O) → $H_2N—C(=O)—NH—CH_2—CH_2—COOH$ β-Ureidopropionic acid → H_2O → $H_2N—CH_2—CH_2—COOH + NH_3 + CO_2$ β-Alanine

Thymine → $+H_2$ → Dihydrothymine ⇌ ($-H_2O$ / H_2O) → $H_2N—C(=O)—NH—CH_2—CH(CH_3)—COOH$ β-Ureidoisobutyric acid → H_2O → $H_2N—CH_2—CH(CH_3)—COOH + NH_3 + CO_2$ β-Aminoisobutyric acid

Fig. 4.6 Proposed degradative sequence for uracil and thymine.

and the nature of the reductant required in the first site of the degradative process have not been determined. ^{14}C-labelled cytosine was not degraded by rape (*Brassica napus*) seedlings in which most of the catabolic studies have been conducted.[645] It was suggested that degradation of this base may be dependent upon its prior conversion to uracil. Enzymatic studies, however, indicate that this conversion does not occur at the level of the free base in plants but the nucleoside cytidine could be deaminated to uridine which could then give rise to uracil by the action of a specific hydrolase.[645]

The β-alanine produced during uracil degradation may be utilized in Coenzyme-A biosynthesis or it may undergo a transaminase reaction to produce formyl acetate which on conversion to acetate can be metabolized in the Krebs' tricarboxylic acid cycle or utilized in lipid biosynthesis. A β-alanine transaminase has been demonstrated in extracts of rape seedlings which shows specificity for α-ketoglutarate as the amino group acceptor.[645]

In some species the purine bases have been reported to give rise to allantoic acid. This conversion may be of great significance in those species where allantoin or allantoic acid has been reported to function as a storage component.[60, 519] A detailed appraisal has not been made but it might be expected that aerobic catabolism of purines to allantoic acid may be primarily associated with ageing senescing leaves. Such a breakdown would allow for the conservation of nitrogenous components arising from nucleic acid catabolism in the leaf during senescence. The breakdown product allantoic acid could be transported out of the leaf and serve as a reserve for utilization in spring regrowth.[519] Purine catabolism seems to occur primarily in old leaves and storage reserve endosperm tissue but not in embryos or young seedlings. Labelling studies indicate[29, 493] that the degradation of guanine and adenine proceeds along the sequence described (Fig. 4.7). The recent observations that uricase and allantoinase occur in peroxisomes and glyoxysomes[622, 630] are consistent with aerobic catabolism of purines occurring in leaf tissue and endosperm tissue and provide enzymatic verification for the operation of this degradative sequence.

DEOXYRIBONUCLEOTIDE BIOSYNTHESIS

Very few studies have been conducted on deoxyribonucleotide biosynthesis in plants. However, recent investigations using plant tissue cultures have demonstrated the occurrence of low levels of free deoxyribonucleotides in plants. Significantly the content of deoxyribonucleotides in the tissue cultures is highest just prior to the period of most active growth suggesting that the synthesis of deoxyribonucleotides coincides with or immediately precedes the period of deoxyribonucleic acid synthesis which occurs prior to cell division (Fig. 4.8).

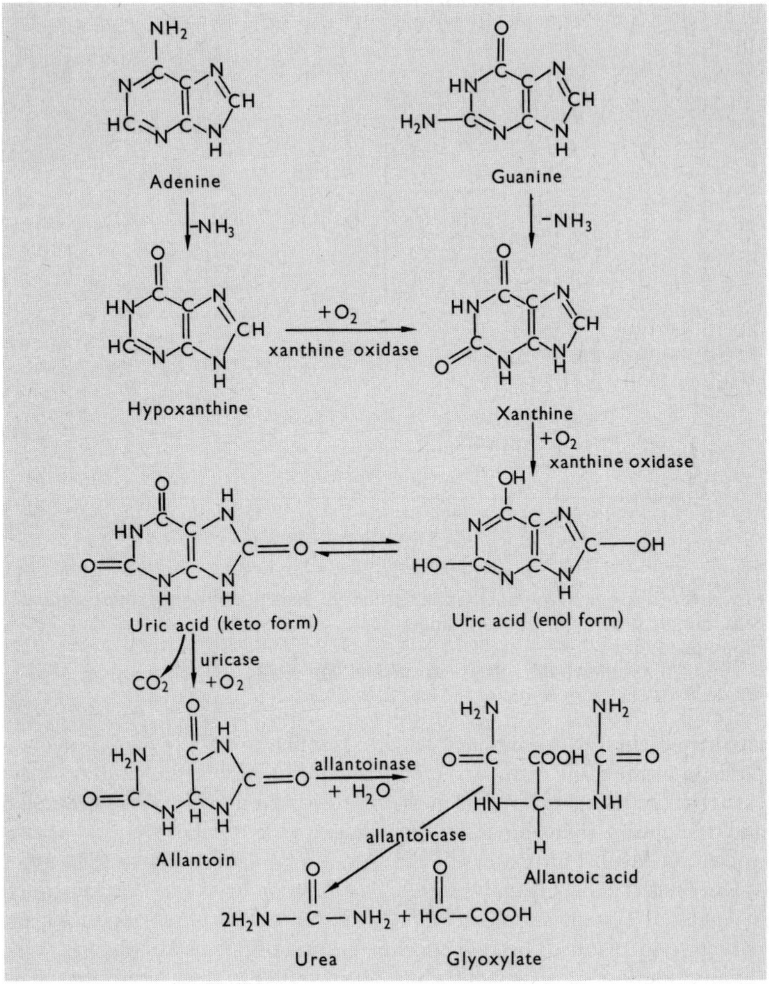

Adenine
$-NH_3$
Guanine
$-NH_3$
Hypoxanthine
$+O_2$
xanthine oxidase
Xanthine
$+O_2$
xanthine oxidase
Uric acid (keto form)
Uric acid (enol form)
uricase
CO_2
$+O_2$
Allantoin
allantoinase
$+ H_2O$
Allantoic acid
allantoicase
$2H_2N-C(=O)-NH_2 + HC(=O)-COOH$
Urea
Glyoxylate

Fig. 4.7 Aerobic catabolism of purines.

Since deoxyribonucleotides differ from ribonucleotides only in the presence of 2-deoxyribose instead of ribose it might be expected that they may be formed by analogous reactions involving 2-deoxyribose derivatives rather than phosphorylated ribose precursors. However, when ^{14}C-cytidine labelled in both the pyrimidine residue and ribose moiety was supplied to bean (*Vicia faba*) roots, the resulting dCMP and dTMP iso-

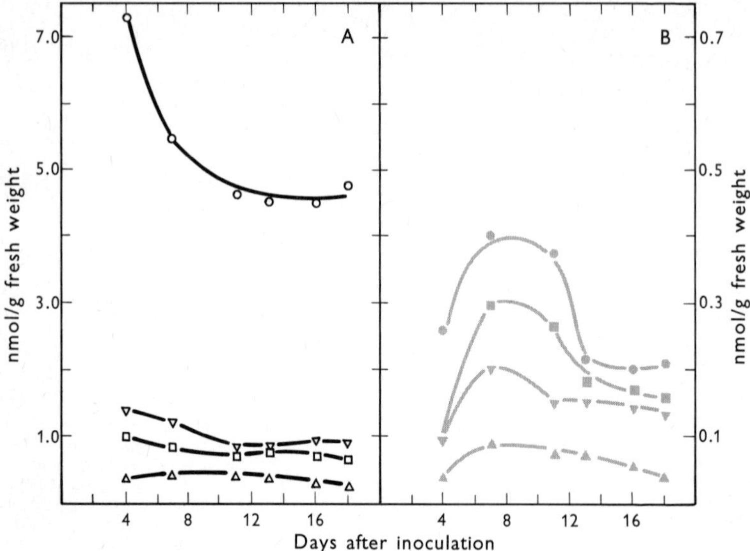

Fig. 4.8 Changes in the nucleotide content of sycamore (*Acer pseudoplatanus*) cells during growth in agar medium. Rapid growth commences at day 10. A. Ribonucleotides: ○——○ ATP; □——□ GTP; △——△ CTP; ▽——▽ UTP. B. Deoxyribonucleotides: ●——● dATP; ■——■ dGTP; ▲——▲ dCTP; ▼——▼ dTTP. From Nygaard,[468] Fig. 2, p. 31.)

lated from the DNA contained label in the pentose and pyrimidine components in the same ratio as the original cytidine.[476] These findings indicate that the ribose is not detached from the pyrimidine prior to reduction and it appears that conversion of ribose to deoxyribose occurs at the nucleotide level (Fig. 4.9).

Enzymes capable of catalysing such a reduction have been demonstrated in bacterial systems but there are to date no reports of ribonucleotide reductase in plants. The reduction in bacterial systems commences with ribonucleotide 5′-diphosphate which is converted to the 2′-pyrophosphate derivative with ATP serving as the pyrophosphate donor. Removal of the pyrophosphate with a dehydrogenation is followed by a hydrogenation to produce the deoxyribose derivative. Although $NADPH^+$ serves as the initial hydrogen donor it appears that other flavoproteins are also involved in the reductive step.

Since plants contain free thymidine triphosphate and also contain thymine as a constituent base of DNA they possess the capability of synthesis of this pyrimidine base. No studies have been reported on thymine biosynthesis in plants. It might be expected, as is the case with the

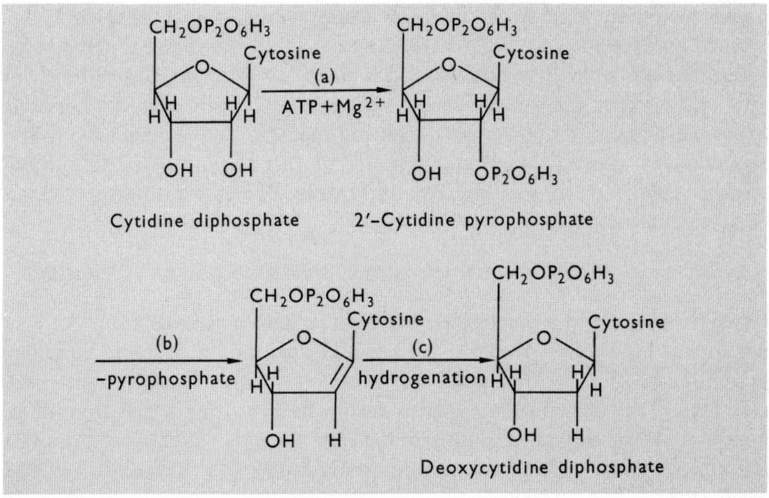

Fig. 4.9 Proposed sequence for the conversion of ribonucleoside diphosphate to deoxyribonucleoside diphosphate.

other pyrimidine bases, that its initial synthesis occurs at the nucleotide or more specifically at the deoxyribonucleotide level. In bacterial and animal systems it has been established that deoxyuridine 5′-monophosphate undergoes methylation and reduction in a reaction catalysed by thymidylate-synthetase to produce thymidylic acid dTMP. This is an interesting reaction since the methyl group is formed from hydrogen and methylene, arising from methylene-tetrahydrofolate and the end product is dihydrofolate rather than tetrahydrofolate.

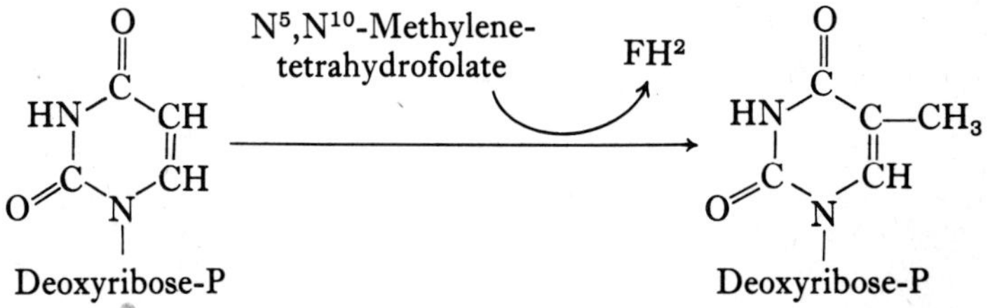

Deoxyuridine 5′-phosphate   Thymidine 5′-phosphate

A salvage pathway for thymidylic acid biosynthesis from thymidine has been demonstrated in various plant tissues.[250] Enzymes involved in this pathway were originally thought to be thymidine kinase and thymidine monophosphate kinase. Activity of these enzymes increased and decreased coincident with the period of active DNA synthesis in tissue cultures,[250] algal cells[549] and in germinating maize (*Zea mays*) seedlings.[669] The thymidine kinase activity from maize[669] and peanuts (*Arachis hypogaea*)[551] also shows activity towards uridine, deoxyuridine, cytidine and deoxycytidine. It has been suggested[147] that the measured thymidine kinase is in fact a non-specific nucleoside phosphotransferase which catalyses the reaction.

$$\text{Nucleoside} + \text{AMP} \longrightarrow \text{Nucleoside 5'-monophosphate} + \text{Adenosine}$$

Capacity to carry out nucleoside monophosphate synthesis from ATP is dependent upon the presence of an ATP degrading system. The degradation product AMP can then function in the phosphotransferase reactions. This observation explains earlier findings[668, 670] that thymidine kinase could be separated into two subunits both of which were required for kinase activity. One subunit apparently is an ATP hydrolase and the other subunit the nucleoside phosphotransferase.

NUCLEOSIDE DIPHOSPHATE DERIVATIVES

In addition to functioning as precursors in nucleic acid synthesis the nucleoside triphosphates play a key role in carbohydrate and lipid metabolism.

The glycosyl donor in the biosynthesis of disaccharides, oligosaccharides and polysaccharides is a nucleoside diphosphate sugar derivative. These sugar derivatives are formed in reactions catalysed by pyrophosphorylases in which the nucleoside 5′-triphosphate reacts with a sugar-1-phosphate to produce the nucleoside diphosphate sugar. The sugar residue replaces the terminal phosphate from the 5′ position of the nucleoside triphosphate and pyrophosphate is released.

$$\text{NTP} + \text{sugar-1-phosphate} \longrightarrow \text{NDP sugar} + \text{PPi.}$$

Although the deoxynucleoside triphosphates can also participate in this reaction, their low concentration suggests that their role in sugar interconversion is minimal. The most abundant nucleoside diphosphate sugar in plants is uridine diphosphate glucose (UDP-G). This product is of great importance since in addition to functioning directly in polymerization reactions producing di-, oligo- and polysaccharides it can undergo extensive interconversions through the activity of epimerases, dehydrogenases and decarboxylases (Fig. 4.10). The products arising from these interconversions can then function as precursors for the synthesis of other cellular polymers.

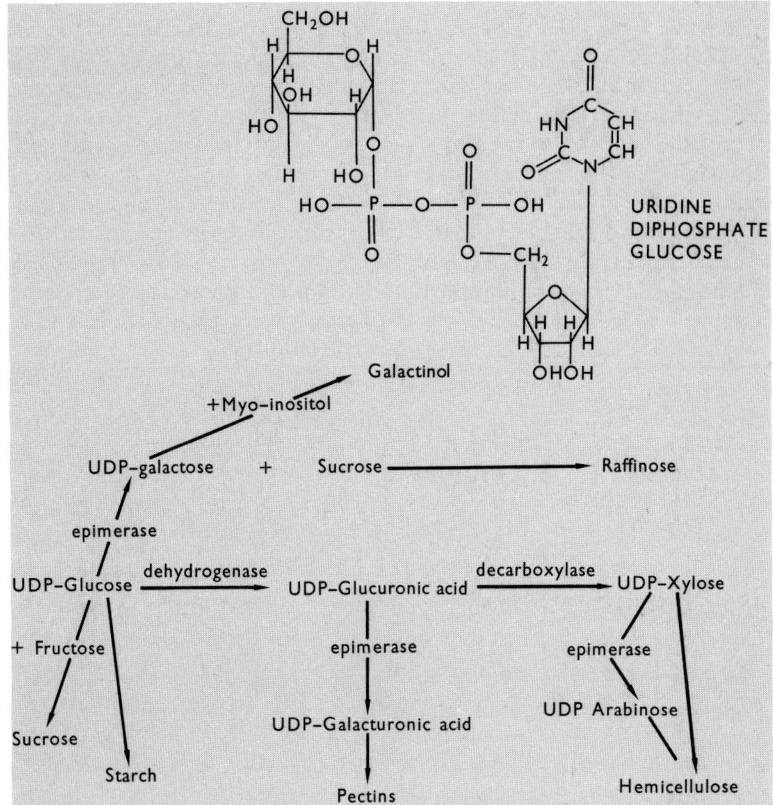

Fig. 4.10 Scheme showing role of UDP-glucose in sugar transformations.

Although UDP-glucose can serve as a precursor in starch biosynthesis, partially purified preparations of starch synthetase show a more rapid starch synthesis when ADP-glucose is the glycosyl donor. This observation appears to indicate that ADP-glucose is the prime glucose donor for starch biosyntheses *in vivo*; however, the cellular concentration of UDP-G is so much greater than ADP-G that the role of the former nucleoside diphosphate sugar in starch biosynthesis cannot be excluded.[255]

Similarly, both UDP-glucose and GDP-glucose have been reported to function as glucose donors in the synthesis of cellulosic type polysaccharides with β1-4 glucosidic linkages. Synthesis of this β1-4 polyglucan occurs *in vitro* in the isolated Golgi apparatus leading to the suggestion that cell wall biosynthesis is mediated by this organelle.[515] However, there is still considerable debate concerning the *in vivo* glucosyl donor (UDP-G or GDP-G) for cell wall synthesis.[255]

In some algae GDP-glucose is subjected to the extensive interconversions undergone by UDP-glucose in higher plants.[255]

In contrast to the functioning of the derivatives of uridine, adenosine and guanosine triphosphates, in carbohydrate metabolism, cytidine triphosphates appear to be involved principally in lipid biosynthesis. Although research in this area is fragmentary, preliminary evidence indicates that in plants CTP functions in the synthesis of phospholipids in an analogous manner to that established in bacterial and mammalian systems. Essentially two mechanisms appear to operate. In the biosynthesis of phosphatidyl-serine, phosphatidyl-inositol and phosphatidyl-glycerol-phosphate (Fig. 4.11), CTP reacts originally with phosphatidic acid to produce CDP-diacylglyceride (see formula on p. 133).

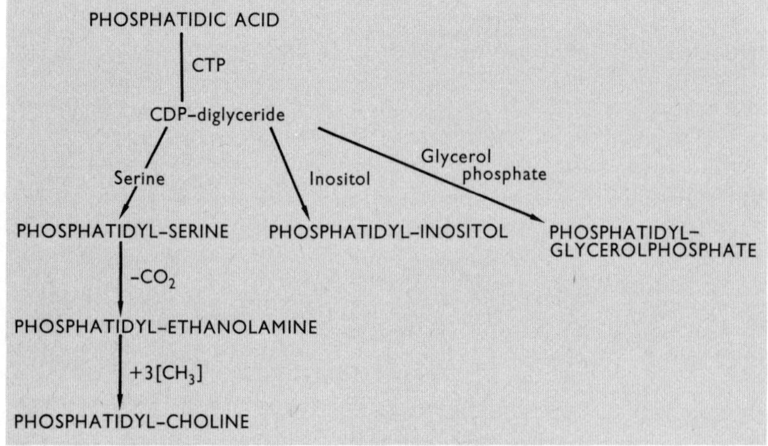

Fig. 4.11 Proposed sequence for synthesis of phospholipids.

Enzymes catalysing this reaction have been demonstrated in plants.[612] In subsequent reactions the cytidine monophosphate is replaced by serine, inositol or glycerol phosphate respectively. Phosphatidyl-serine may be decarboxylated to phosphatidyl-ethanolamine which upon substitution of three methyl groups produces phosphatidyl-choline.

In an alternative pathway, phosphatidyl-choline appears to be synthesized by the condensing of a diglyceride with a choline residue derived from CDP-choline.[412] The enzymes involved in these reactions and the mechanism of CDP-choline formation have been studied in higher plants.[389, 437] Phosphorylcholine occurs as a constituent in plant extracts and the formation of CDP-choline and phosphatidyl choline follow the following sequence.

```
H2C—O—C—R
 |     ‖
 |     O
HC—O—C—R'
 |     ‖
CH2    O
 |
 O
 |
HO—P=O
 |
 O
 |
HO—P=O
 |
 O
 |
CH2
```

Cytidine diphosphate diacylglycerol

$$\text{Choline}+\text{ATP} \longrightarrow \text{Phosphorylcholine}+\text{ADP}$$
$$\text{CTP}+\text{Phosphorylcholine} \longrightarrow \text{Cytidine diphosphate choline}+\text{PPi}$$
$$\text{CDP-Choline}+1,2\text{-Diacylglycerol} \longrightarrow \text{Phosphatidyl-choline}+\text{CMP}$$

PYRIDINE AND FLAVIN NUCLEOTIDES

Reference to Table 4.1 indicates that principal components of the acid soluble nucleotides extracted from higher plants are the pyridine nucleotides nicotinamide adenine dinucleotide (NAD) and nicotinamide adenine dinucleotide phosphate (NADP) and their reduction products $NADH^+$ and $NADPH^+$. (In older literature and in some current journals these nucleotides are referred to diphosphopyridine nucleotide, DPN, and triphosphopyridine nucleotide, TPN.) These pyridine nucleotides along with the flavin nucleotides function primarily in oxidation-reduction reactions. The pyridine nucleotides occur in the free state whereas the flavin nucleotides appear most frequently as enzyme complexes in flavoproteins.

Pyridine nucleotides

The pyridine nucleotides are both in fact dinucleotides derived from the pyridine nicotinamide and the purine adenine. Although little has been done on the enzymology of the biosynthesis of these complex molecules, labelling studies[222, 666] indicate that the pyridine moiety is synthesized from nicotinic acid which condenses with phosphoribosyl pyrophosphate to produce nicotinic acid mononucleotide and pyrophosphate (see Chapter 3). Nicotinic acid mononucleotide then condenses with ATP to produce des-amido-nicotinic acid adenosine dinucleotide. Addition of an amino group donated from glutamine into the des-amido derivative produces nicotinamide adenosine dinucleotide (NAD). Phosphorylation of NAD in a reaction in which ATP functions as the phosphate donor produces NADP. This NAD kinase has been detected in plants;[478] additionally specific phosphatases converting NADP → NAD and $NADPH^+$ → $NADH^+$ have been detected in plant extracts.[197]

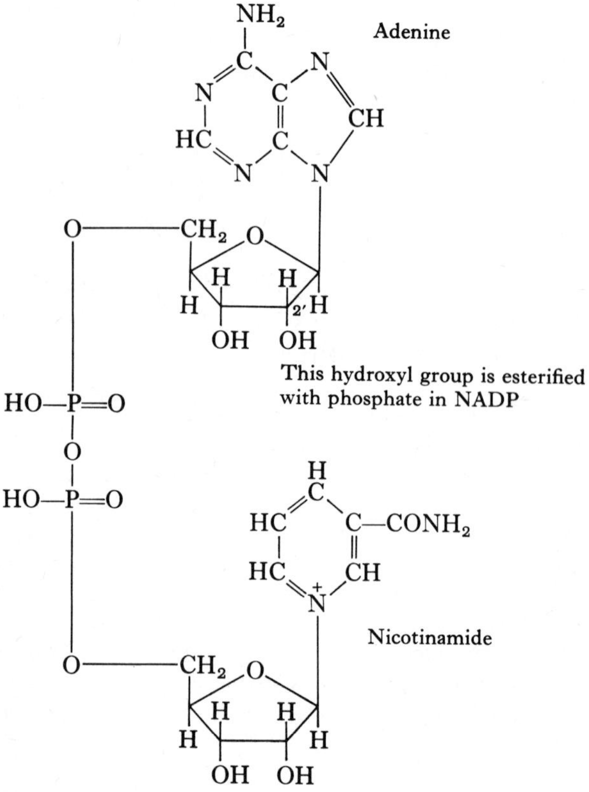

Adenine

This hydroxyl group is esterified with phosphate in NADP

Nicotinamide

Structure of Nicotinamide adenine dinucleotide (NAD). In Nicotinamide adenine dinucleotide phosphate (NADP) a phosphate is attached to the 2′ carbon of the adenosine ribose.

The pyridine nucleotides occur in both the oxidized and reduced state *in vivo* and the ratios of NAD, NADP, and $NADH^+$ and $NADPH^+$ fluctuate with environmental conditions. Illumination leads to an increase in $NADPH^+$ level presumably as a result of photosynthetic activity. However, there also appears to be a light stimulated interconversion of NAD → NADP.[478] Photoreduction of NADP occurs in the chloroplast. Since the chloroplast membrane seems to be relatively impermeable to pyridine nucleotides[533] it appears that the $NADPH^+$ produced is utilized solely for reductions within the chloroplast.

The reduction of both NAD and NADP involves the pyridine ring of nicotinamide. During reduction a hydride ion from the substrate is transferred to the 4 position of the pyridine ring and the other hydrogen from the substrate is released as a proton. Addition of the hydride ion at position 4 of the ring results in a change of bonding configuration around the nitrogen atom of the pyridine ring. These changes are accompanied by alterations in spectral characteristics resulting in strong absorption at 340 nm (Fig. 4.12). Since the reduction of NAD and NADP both involve reduction of the nicotinamide component, oxido-reduction reactions involving these constituents can be easily monitored by following absorbance change at 340 nm in a spectrophotometer. NAD and NADP both in

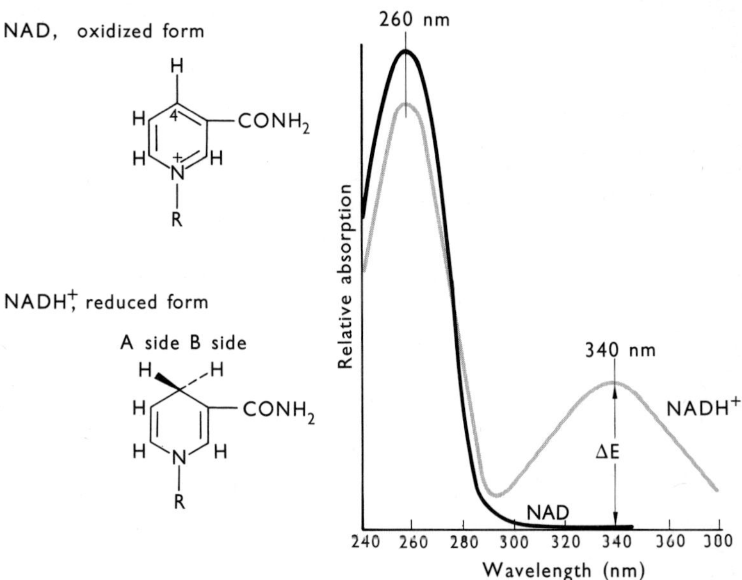

Fig. 4.12 Reduction of pyridine nucleotides. Note change in valence configuration in the nicotinamide component which leads to characteristic changes in spectral properties.

the oxidized and reduced form show an additional absorbance peak at 260 nm associated with the adenine residue.

Flavin nucleotides

The flavin nucleotides occur as prosthetic groups of flavoproteins. The strength of attachment of the nucleotides to the proteins varies and in some instances the involvement of flavin nucleotides in enzyme reactions is readily demonstrated, whereas in others it is only inferred on the basis of inhibitor studies. An illustration of this point is provided by the enzyme nitrate reductase. Crude and partially purified enzyme preparations from soybean (*Glycine max*) show increased nitrate reduction on the addition of exogenous flavin nucleotides whereas the enzyme prepared from maize (*Zea mays*) shows only marginal enhancement.[41]

Two types of flavin nucleotides have been characterized which function in oxido-reduction reactions: flavin mononucleotide (FMN) and flavin adenine dinucleotide (FAD). The former is not a true nucleotide since it consists of a nitrogenous base isoalloxazine complexed to a pentahydric alcohol phosphate (ribitol phosphate) rather than the customary ribose phosphate of the previously encountered nucleotides. Flavin adenine dinucleotide consists of flavin mononucleotide complexed through a phospho diester linkage to adenosine 5′-monophosphate (similar to the linkage encountered in NAD).

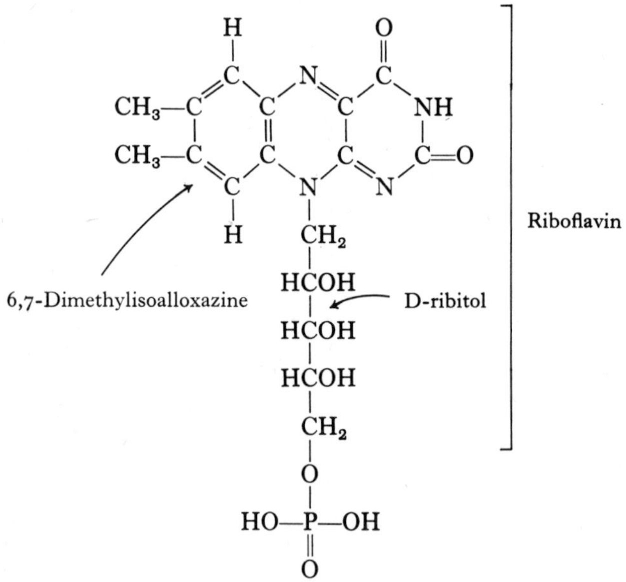

6,7-Dimethylisoalloxazine

D-ribitol

Riboflavin

Structure of Flavin mononucleotide

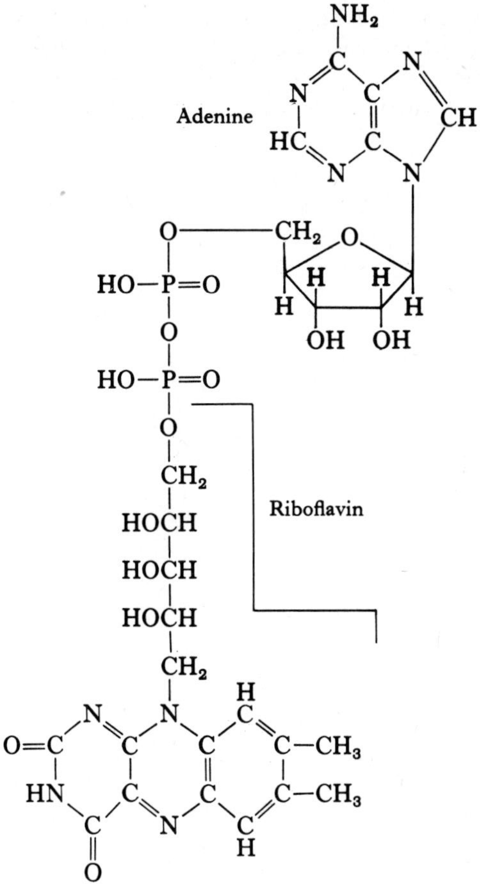

Adenine

Riboflavin

Structure of Flavin adenine dinucleotide

There have been only limited reports on studies of flavin biosynthesis in plants. Evidence from microbial studies[393] suggests that the synthesis proceeds by way of riboflavin which can readily be detected in plants. Guanine as the ribose or ribotide derivative apparently functions as the precursor in the biosynthesis. Cleavage of the imidazole ring of the purine by the action of the enzyme cyclohydrolase accompanied by a deamination and oxidation produces the compound 4-ribitylamino-5-amino-uracil. The origin of the ribitol group is not clear but it probably originates from ribose by reduction. 4-Ribitylamino-5-amino-uracil reacts with a four carbon diacetyl unit to form 6,7-dimethyl-8-ribityl-lumazine. This compound is converted directly to riboflavin. Two molecules of 6,7-dimethyl-8-ribityl lumazine are necessary and during the formation of riboflavin 4-ribitylamino-5-amino-uracil is released (Fig. 4.13).

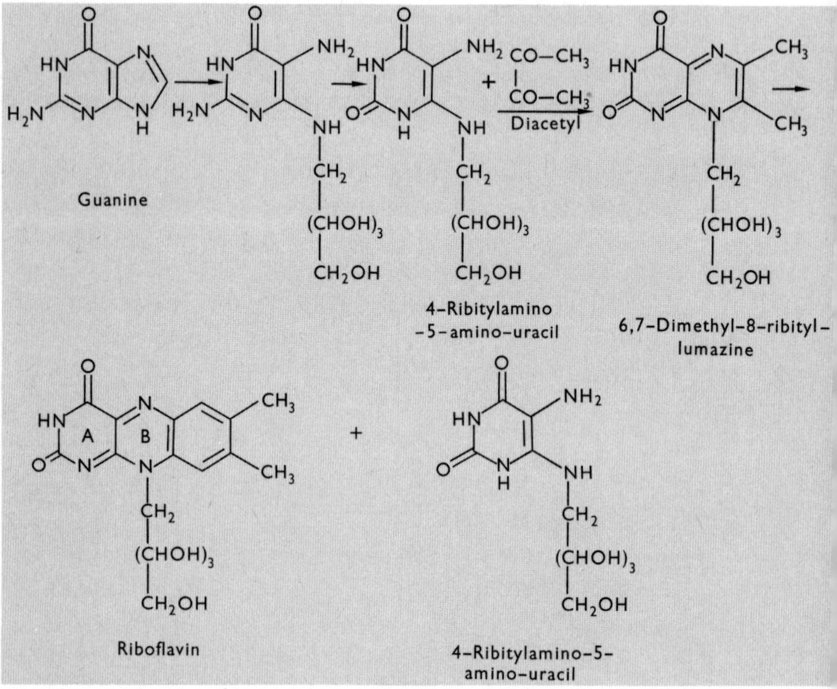

Fig. 4.13 Proposed sequence for riboflavin biosynthesis.

Phosphorylation of riboflavin in an ATP-dependent reaction catalysed by flavokinase produces flavin mononucleotide. Combination of FMN with ATP is catalysed by flavin adenine nucleotide synthetase leading to the production of FAD. Presence of these two enzymes has been demonstrated in a range of plant species.[218, 219]

The functioning of flavin nucleotides in oxido-reduction reactions depends on the capacity of the isoalloxazine ring to undergo reduction. Although the sequence is not completely established, it appears that a pair of hydrogen atoms transferred from the substrate reacts with the nitrogen atoms in the isoalloxazine ring to yield the reduced flavin nucleotides. In both FMN and FAD, reduction occurs in the isoalloxazine residue to produce $FMNH_2$ and $FADH_2$ respectively. Reduction of the flavin nucleotides results in a bleaching of the yellow coloration characteristic of the oxidized flavins. This bleaching and formation of leucoflavins can be followed spectrophotometrically as the reduction of the oxidized flavo nucleotides is accompanied by a loss of absorption in the 375 and 450 nm range (Fig. 4.14).

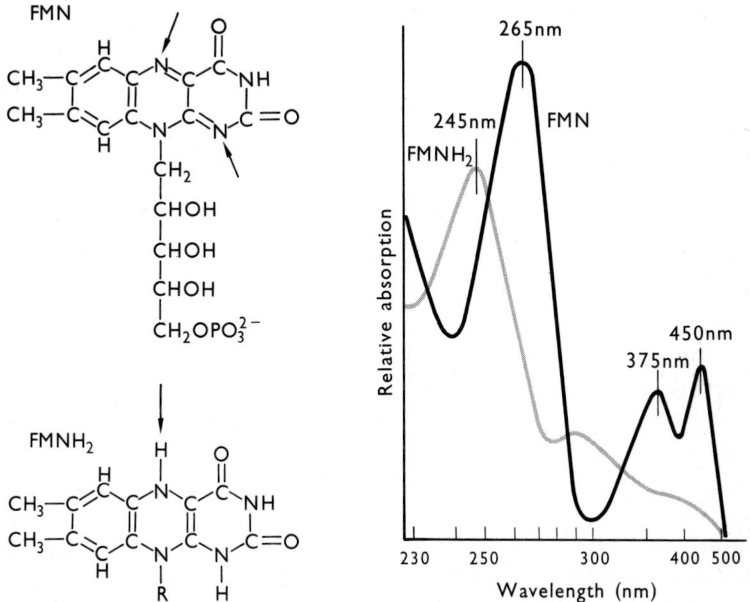

Fig. 4.14 Reduction of Flavin mononucleotide. Note attachment of hydride ion to the nitrogen atoms in the isoalloxazine ring and the accompanying bonding changes which result in characteristic spectral changes.

5

Nucleic Acids

COMPOSITION OF NUCLEIC ACIDS

Historically, studies of nucleic acids were conducted on calf thymus and yeast. These investigations indicated that calf thymus nucleic acid was composed of phosphoric acid, adenine, guanine, cytosine, thymine and a sugar derivative originally identified as laevulinic acid but later identified as deoxyribose. In contrast nucleic acid from yeast was found to contain uracil instead of thymine and ribose instead of deoxyribose. These early characterizations and other studies of nucleic acids from animal and plant sources led to the concept that animal nucleic acids differed in composition from plant nucleic acids. However, with more refined analyses and the development of the histochemical **Feulgen** dye specific for deoxyribonucleic acid, it became evident that both animals and plants contain two classes of nucleic acid: deoxyribonucleic acid (DNA) and ribonucleic acid (RNA).

RNA contains phosphate residues, ribose, the purine bases adenine and guanine and the pyrimidines cytosine and uracil. DNA contains phosphate, deoxyribose, the purine bases adenine and guanine and the pyrimidine bases cytosine, methyl cytosine and thymine.

Dependent on hydrolysis conditions, the nitrogenous bases can be released as the nucleosides or nucleotides. Dilute alkali hydrolysis of RNA produced mixtures of 2′,3′-cyclic ribonucleoside monophosphate, 2′-ribonucleoside monophosphate and 3′-ribonucleoside monophosphate. DNA is not degraded by dilute alkali. Enzymatic degradation of RNA may produce 3′-ribonucleotides or 5′-ribonucleotides depending upon enzyme source. Such observations led to the conclusion that in ribonucleic acid the bases were combined to ribose phosphate and these nucleotides were condensed in linear fashion by phosphodiester linkages between 3′ and 5′ groups of adjacent ribose residues. Similar investigations indicated that the bases must be arranged in similar manner in DNA.

Since drawing the structures of the component phosphate sugars and

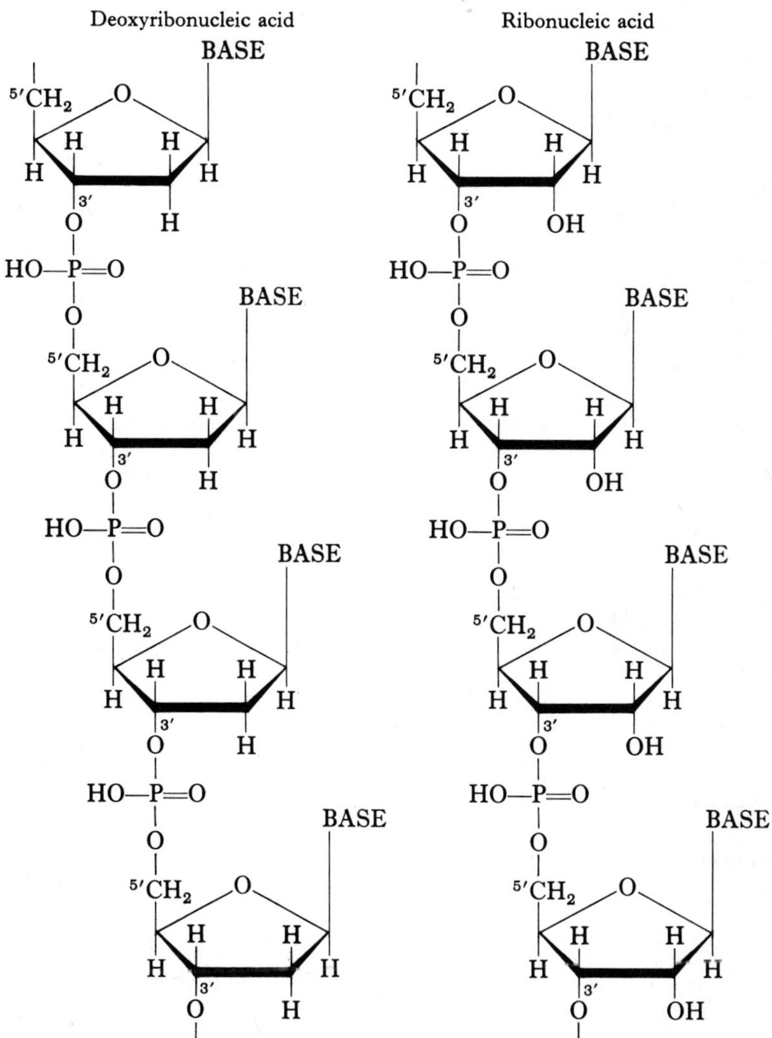

Polynucleotide structure of DNA and RNA showing phosphodiester linkage

bases in RNA and DNA is extremely cumbersome it is customary to use various alternative forms such as:

```
         A             A             U             C
         |             |             |             |
        C2'           C2'           C2'           C2'
(a)      |             |             |             |
        C3' α         C3' α         C3' α         C3'
  p      |     p  β    |     p  β    |     p  β    |    OH
        C5'           C5'           C5'           C5'
```

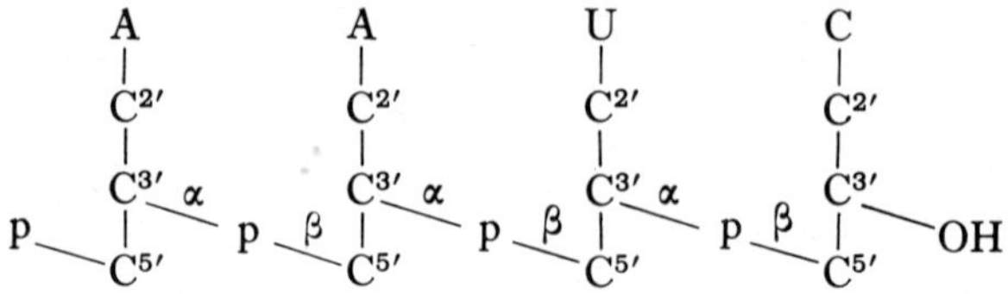

(b) pApApUpC

Dilute sodium hydroxide and some enzymes degrade the β bond to produce hydrolysis products 3′ phosphate Ap, etc. Other enzymes attack the α bond to produce 5′ phosphate pA, etc.

In analyses of the hydrolysis products of DNA from several sources there is a constancy in the ratio of adenine : thymine and similarly the ratio of guanine to cytosine and methyl cytosine is 1. In contrast there is no constancy in the base ratios of RNA (Table 5.1).

Table 5.1 Base Composition of DNA and RNA from different plant sources. Note A :T and G :C+MC ratio of 1 in DNA and lack of constancy in RNA.

Origin	Base composition of DNA as molar percentages				
	Guanine (G)	Adenine (A)	Cytosine (C)	5-Methyl-cytosine (MC)	Thymine (T)
Daucus carota (leaf)	23.2	26.7	17.3	6.0	26.8
Pinus sibirica (seed)	20.8	29.2	14.6	4.9	30.5
Cucurbita pepo (seed)	21.0	30.2	16.1	3.7	29.0
Phaseolus vulgaris (seed)	20.6	29.7	14.9	5.2	29.6

	Base composition of RNA as molar percentages			
	Guanine (G)	Adenine (A)	Cytosine (C)	Uracil (U)
Pinus sibirica (seed)	31.3	25.1	24.3	19.3
Cucurbita pepo (seed)	30.6	25.2	24.8	19.4
Phaseolus vulgaris (seed)	31.4	24.9	24.1	19.6

The constancy of the A:T=1 and G:C=1 ratios is of great importance as it is encountered in all types of DNA and gave rise to the now famous Watson & Crick model for DNA (Fig. 5.1). In this model two parallel chains of polynucleotides, in which the nucleotides are held together by diester linkage with the bases projecting from the side chain,

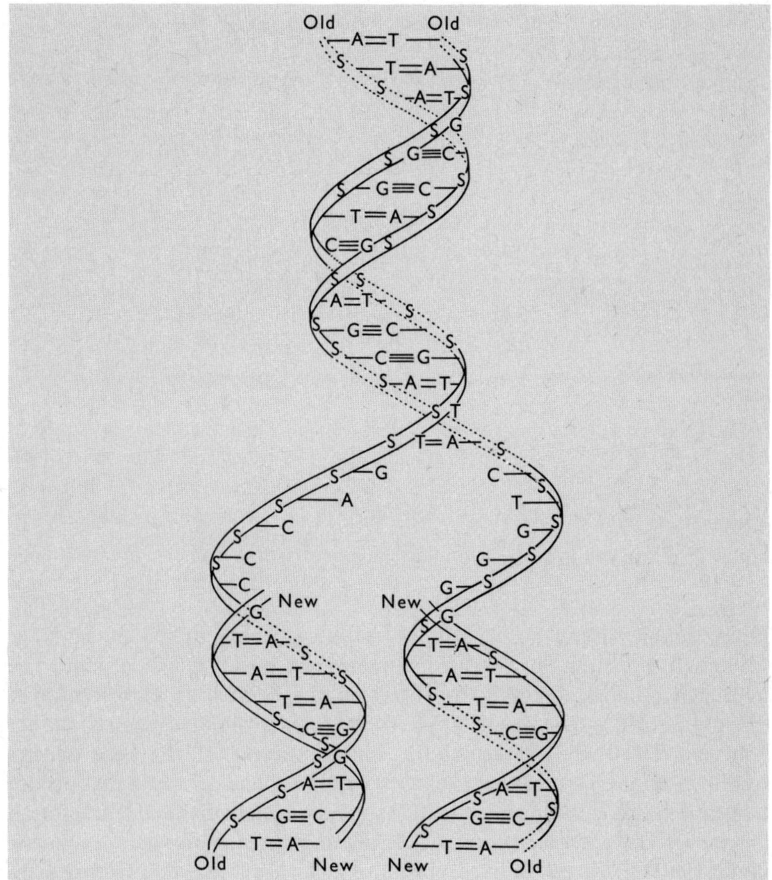

Fig. 5.1 Model of DNA as suggested by Watson and Crick. During replication the strands separate and each forms a template for the synthesis of a new complementary strand.

are wound in a helix. For each adenine of one chain a thymine projects from the other chain and the bases are held together by hydrogen bonding (Fig. 5.2). In a similar way, cytosine or methyl cytosine forms hydrogen bonds with guanine in the parallel chain.

Hyperchromicity

The double stranded nature of DNA is an important characteristic and confers on the molecule certain diagnostic properties. Solutions of DNA absorb ultraviolet light as do the component nucleotides. However, owing to the double helical structure and associated base pairing, the

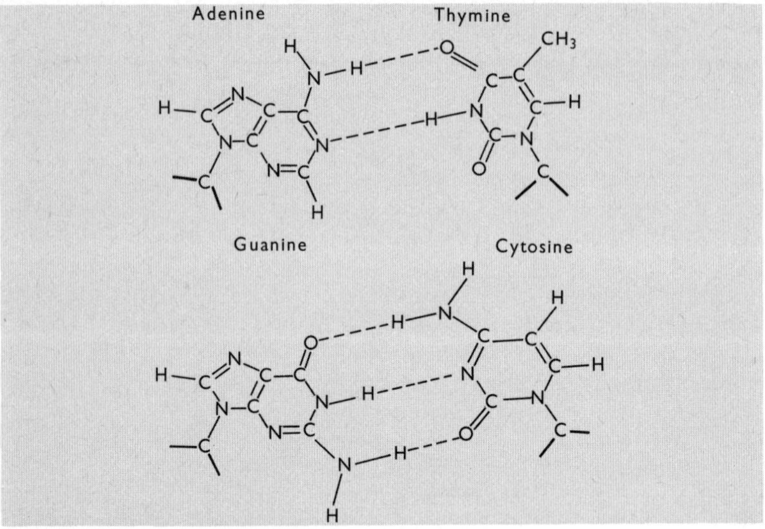

Fig. 5.2 Proposed hydrogen bonding between bases in complementary strands of DNA.

DNA double strand absorbs light less efficiently than the nucleotides. When a DNA solution is heated, a temperature is reached at which the hydrogen bonding between the bases in the complementary strands is disrupted. More energy is required to break the hydrogen bonding between G:C than between A:T. The disruption of the base pairing results in an increase in absorptive capacity of the constituent nucleotides and thus there is an increase in the optical density of the DNA solution (Fig. 5.3). This increased optical density caused by a disassociation or melting of the DNA is called hyperchromicity. The melting temperature (Tm temperature) at which half of all the hyperchromicity has been achieved depends upon ionic strength of the solution and on the base composition of the DNA. The greater the GC content the higher the Tm. If a heated solution of DNA is cooled then there is a reversal of the hyperchromic shift and the optical density decreases as the separated strands reassociate, **anneal**, as the temperature is lowered.

Similar hyperchromicity occurs in transfer RNA, described later, indicating the presence of some base pairing in this RNA species.

Hybridization

The capacity of base pairing means that chains of polynucleotides can 'recognize' other chains whose molecular composition is similar to their own: if strands of DNA are separated then under appropriate conditions

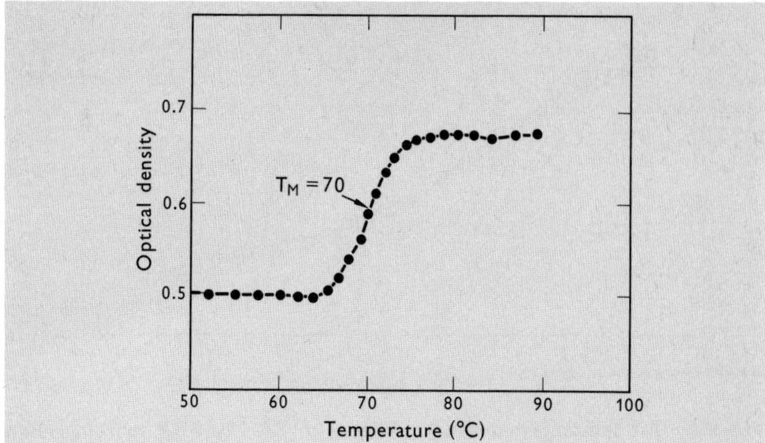

Fig. 5.3 Melting profile of pea (*Pisum sativum*) DNA. The optical density of the DNA solution is plotted as a function of temperature to which the solution has been heated. The temperature at which half of all the hyperchromicity has been observed is 70°C.

they will reassociate or hybridize with polynucleotide strands having complementary base sequences.

If DNA is dissociated by heat and then allowed to reassociate during a period of lowered temperatures it is found that the reassociation follows a specific time course. If the partially reassociated mixtures are passed through columns of calcium phosphate (hydroxyapatite) then the single stranded dissociated DNA passes through while the annealed double stranded DNA formed by base pairing between complementary strands is retained. The reassociation of a pair of complementary strands results from their collision and therefore the rate of reassociation depends on their concentration. Thus the factors controlling the completeness of reassociation are concentration of DNA and time. For convenience these factors can be expressed as Cot which is the product of DNA concentration and time (values are usually expressed as moles of nucleotide × seconds/litre).[75]

When experiments are conducted on DNA reassociations from higher plants it is found that there is a rapid reassociation of a certain portion of the DNA followed by a prolonged period of less rapid reassociation (Fig. 5.4). This rapid reassociation involves the association of the most abundant nucleotide sequences whereas less common nucleotide sequences can pair less rapidly.[427] The rapid reassociation of certain DNA sequences indicates that certain nucleotide sequences must be repeated many times within the DNA. Although the significance of these repeated sequences

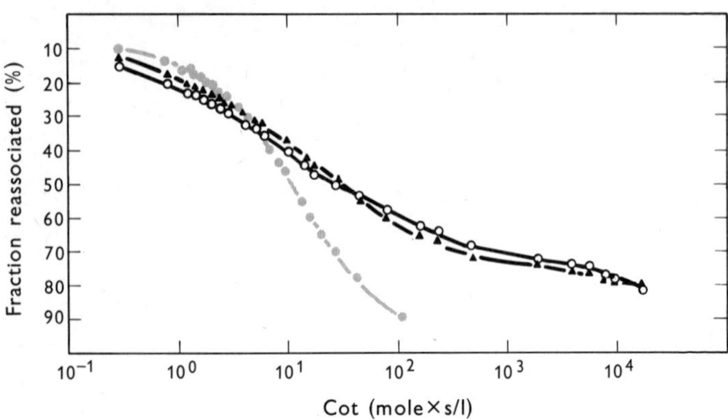

Fig. 5.4 Comparison of reassociation kinetics of *Vicia faba* DNA from cotyledons in the cell division phase of growth (▲) with that from cotyledons in the cell expansion phase of growth (○). *Escherichia coli* DNA (●) was used to check for fluctuations in reassociation conditions. The rates of reassociation were measured at 37°C in 5 × SSC, 50 % formamide by the optical method. (SSC = 0.15 M NaCl, 15 mM Sodium citrate). (From Millerd and Whitfeld,[427] Fig. 4, p. 1008.)

is not completely understood it is believed that the repetitive sequences of polynucleotides may provide a mechanism for forming many copies of a particular gene.

In addition to hybridizing with complementary strands of DNA, dissociated strands of DNA can also hybridize with polynucleotides of RNA containing complementary base sequences (Fig. 5.5). In the case of RNA/DNA hybrids the base uracil takes the place of thymine as the partner of adenine. The demonstration of RNA/DNA hybridization can be performed by various techniques; perhaps the most convenient is that using nitrocellulose filters.[552, 627] Undegraded DNA is dissociated by heating or alkaline treatment before being filtered through nitrocellulose filters. The single strands of DNA are retained on the filters which can be air-dried. These DNA filters can then be incubated with labelled RNA under appropriate conditions; after a suitable time the filters are washed to remove the unassociated labelled RNA. Subsequently the amount of radioactivity retained on the filter can be determined and used as an index for estimating the degree of complementarity between the DNA and various applied RNA species and *vice versa* (Fig. 5.6).

NUCLEIC ACID CONTENT OF PLANT MATERIALS

The assay of nucleic acids in plant materials essentially involves the removal of other cellular components by a series of solvents and then determining the nucleic acid content of the residue. Theoretically, the

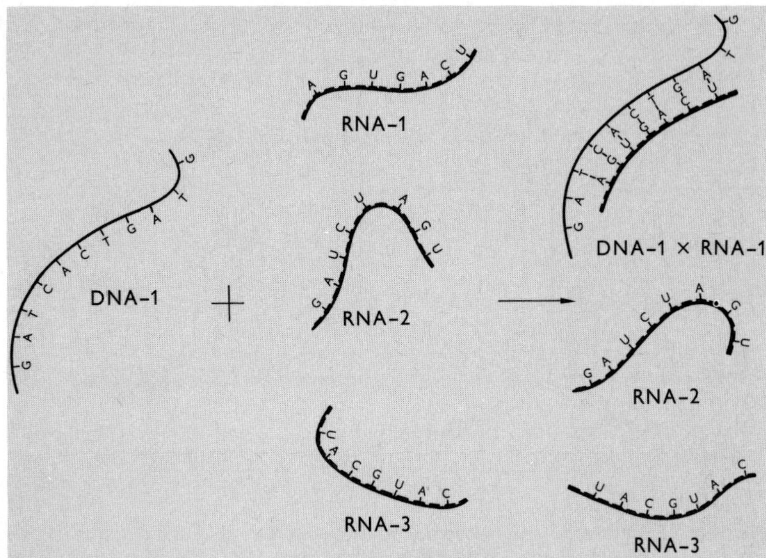

Fig. 5.5 Diagrammatic representation of DNA–RNA hybridization. Only base sequences complementary in both DNA and RNA are able to hybridize.

nucleic acid determination could involve an estimation of sugar residues since the colorimetric assays for ribose and deoxyribose are available; alternatively phosphate content could be determined or the base content could be determined by measuring the absorbance at 260 nm on a spectrophotometer. Unfortunately, such assays are not always reliable since other plant constituents frequently contaminate the nucleic acids and produce erroneous results. The most reproducible method[270, 284] involves treatment of the plant material with ethanol, trichloracetic acid and ethanol-ether to remove soluble constituents and lipids. The nucleic acids in the residue can be separated into DNA and RNA on the basis of their reactivity with alkali. The residue is usually treated with potassium hydroxide which hydrolyses the RNA and solubilizes DNA. The DNA is precipitated when the alkaline solution is acidified and the supernatant contains the nucleotides from the hydrolysis of RNA. Frequently such extracts are pigmented and this supernatant extract is purified by ion exchange chromatography. The RNA content of the eluant following chromatography is estimated by measuring absorbance at 260 nm. Other methods give variable results owing to contamination with non-nucleic acid components.

DNA content can be determined on the precipitate produced by acidification of the alkaline hydrolysate. The DNA in the precipitate is hydrolysed by hot acid and estimated by colorimetric assay of deoxyribose.

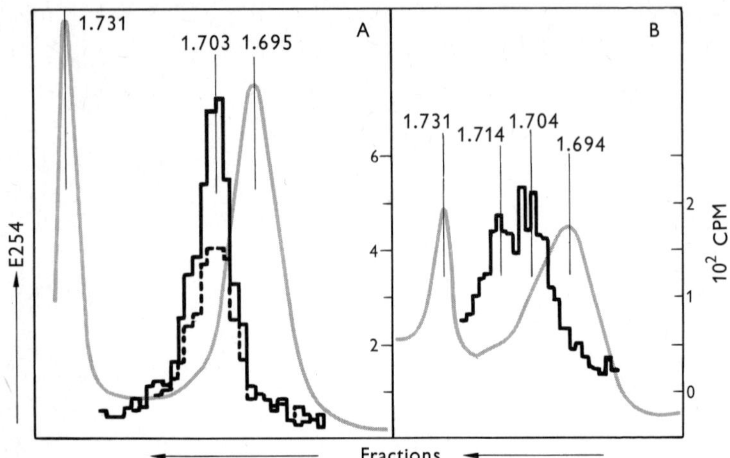

Fig. 5.6 Fractionation of ribosomal RNA genes by CsCl equilibrium centrifugation. A :50 μg of pea (*Pisum sativum*) nuclear DNA (1.695 g cm^{-3}) plus *Micrococcus lysodeikticus* DNA (1.731 g cm^{-3}) were fractionated (continuous scan), and each fraction was fixed to a millipore filter. The filters were halved, and one set was hybridized with 1.3×10^6 (continuous histogram), and the other with 0.7×10^6 mol.wt. pea rRNA (dotted histogram). B :100 μg of Swiss chard (*Beta vulgaris*) total DNA (1.694 g cm^{-3}) were sheared by sonication to 0.35×10^6 daltons and fractionated with *Micrococcus lysodeikticus* DNA (continuous scan). Individual fractions were hybridized with 1.3×10^6 mol. wt. swisschard rRNA (histogram). Note that RNA hybridizes only with a limited fraction of DNA, i.e. that with complementary base sequences. (From Scott and Ingle,[552] Fig. 2, p. 679.)

Alternatively DNA content can be determined on the residue produced by ethanol, trichloracetic acid, ethanol-ether extraction.

When these rigorous extraction and purification techniques are applied to plant materials it is found that the RNA and DNA content varies considerably between tissues and actively growing tissues have an RNA to DNA ratio of about 10:1 (Table 5.2).

The extraction of plant materials with alkaline or acidic solutions produces degraded nucleic acids. Other methods are available which, although not quantitative, provide undegraded or native nucleic acids. These methods involve homogenizing the plant material in buffers containing detergents, and inhibitors of nuclease activity, in the presence of aqueous phenol. Each research laboratory has its own favourite ingredients to be used in the original homogenizing step. Subsequent centrifugation separates two phases; the nucleic acids separate into the upper aqueous phase while proteins and other contaminants are retained in the phenol phase.

Table 5.2 DNA and RNA content of some plant tissues.

		DNA	RNA	Ref.
Peanut (*Arachis hypogaea*) cotyledons (4 days after germination)	μg/g fresh wt.	240	1360	106
Corn (*Zea mays*) embryos (48 h after germination)	μg/part	17.2	142.5	290
Rye (*Secale cereale*) embryos (48 h after germination)	μg/part	3.25	30.8	270
Pea (*Pisum sativum*) stem	μg/10 mm segment	7.3	44.0	53
Corn (*Zea mays*) endosperm (48 h after germination)	μg/part	14.0	28.5	290
Xanthium leaf	μg/g fresh wt.	340	1052	488

Providing suitable precautions are taken during the extraction, undegraded RNA and DNA can be recovered from the aqueous phase by ethanol precipitation. DNA can be recovered in the absence of RNA by digesting the RNA with ribonuclease.

The DNA and RNA recovered can then be characterized by various techniques.

DNA

The customary method of characterizing DNA is by equilibrium centrifugation in caesium chloride solution. During the centrifugation the caesium chloride forms a gradient of density and the DNA floats to the region of the gradient corresponding to its own buoyant density (Fig. 5.7). The distribution of DNA in the gradient can be determined by obtaining the optical density at 260 nm of fractions collected from the gradient. The buoyant density is a function of the base composition of the DNA. Since G–C pairs are paired by three hydrogen bonds they are more compact and hence less buoyant than A–T pairs which are paired by two hydrogen bonds. Thus buoyant density can be utilized to differentiate between DNA's with different base composition.

When buoyant density centrifugation is performed on total DNA from leaf tissue from plants it is found that the DNA profile is asymmetrical with shoulders or peaks, more dense than the main peak. These shoulders or minor peaks are referred to as representing 'satellite' DNA's, and they are more or less distinct depending on the species from which the DNA was prepared. Although originally it was thought that DNA was confined to the nucleus associated with the chromosomes, evidence accumulated in the past decade from ultrastructural and analytical studies indicates

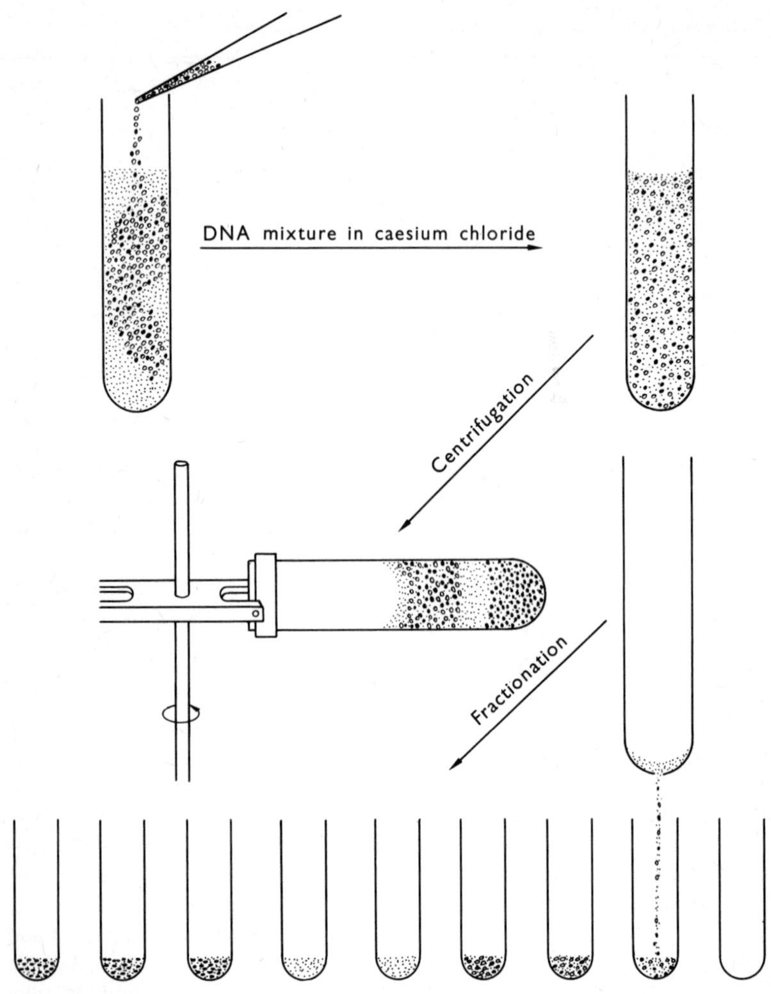

Fig. 5.7 Diagrammatic representation of caesium chloride density gradient centrifugation.

that DNA also occurs in such cellular organelles as chloroplasts and mitochondria. It is the presence of the DNA from the organelles, in addition to the nuclear DNA, which produces the asymmetric profiles (see Fig. 5.6B). If nuclei, mitochondria and chloroplasts are prepared from leaf tissue and the DNA prepared from the organelles subjected to caesium chloride centrifugation it is found[685] that each DNA fraction possesses

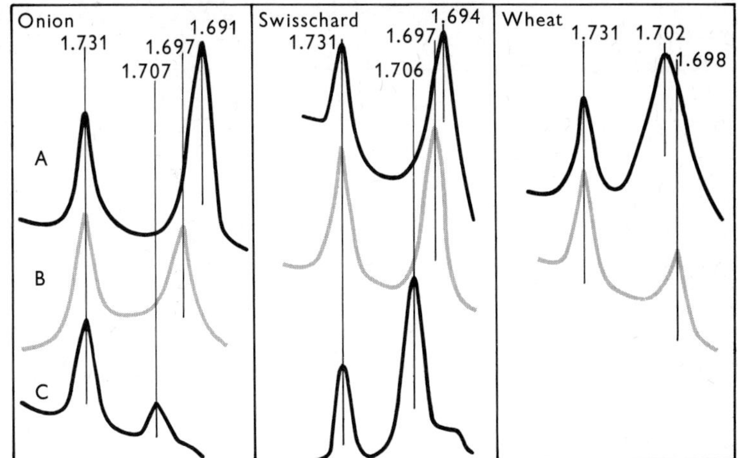

Fig. 5.8 Microdensitometer tracings of ultraviolet photographs of cell organelle DNA's. The marker DNA from *Micrococcus lysodeikticus* has a buoyant density of 1.731 g cm^{-3}. A :Nuclear DNA's from onion (*Allium cepa*), 1.691 ; Swiss chard (*Beta vulgaris*), 1.694 ; and wheat (*Triticum vulgare*), 1.702 ; B. chloroplast DNA's from onion, 1.697 ; Swiss chard, 1.697 ; and wheat, 1.698 ; C : mitochondrial DNA's from onion, 1.707 ; and Swiss chard, 1.706. (From Wells and Ingle,[685] Fig. 1, p. 178.)

distinct buoyant densities (Fig. 5.8). Nuclear DNA from different species is found to have a variable buoyant density; and even though prepared from nuclei apparently free from chloroplasts and mitochondria it may still contain satellites. A recent compilation indicates the occurrence of these nuclear satellite DNA's in many plants and it is considered that they represent DNA of the nucleolar organizer region.[128] Although there was some original dispute,[708] most recent evidence[685] indicates that chloroplast DNA has a uniform buoyant density of 1.697 and mitochondrial DNA has a buoyant density of 1.706 g cm^{-3}.

The base composition of nuclear and chloroplast DNA is found to differ in one important respect—nuclear DNA contains methyl cytosine whereas this base is absent in chloroplast DNA (Table 5.3). The buoyant density data indicate that the mitochondrial DNA is enriched in GC residues. Hyperchromicity and hybridization studies (Fig. 5.9) indicate that chloroplast DNA contains many more repetitive sequences than mitochondrial DNA while nuclear DNA contains a preponderance of non-repetitive sequences.[684]

The satellite DNA's represent only a few percent of the total DNA. In addition to differing in base composition they presumably differ in

Table 5.3 Base composition (mole/100 moles) of nuclear and chloroplastic DNA. (From Wells and Birnsteil.[684])

	Chloroplast DNA	Nuclear DNA
Adenine	31.1	28.5
Guanine	22.5	21.0
Cytosine	21.4	17.2
Thymine	25.8	28.3
5-Methyl cytosine	Not detectable	5.0

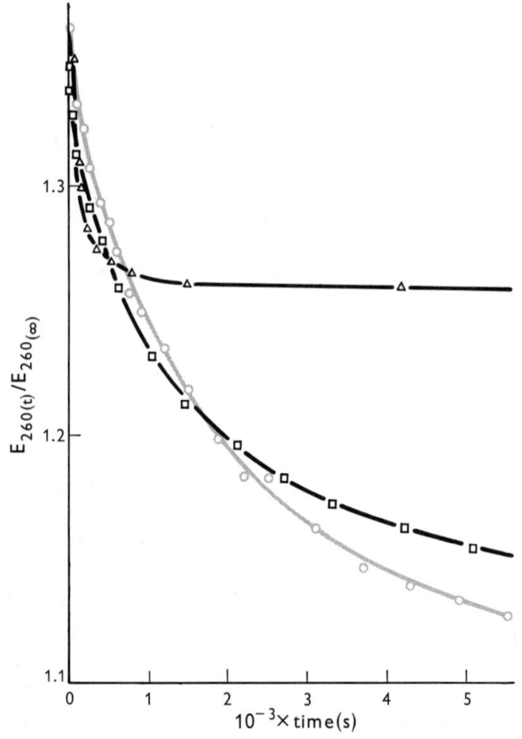

Fig. 5.9 Renaturation curves for lettuce (*Lactuca sativa*) DNA fractions. Denatured DNA (30–40 μg ml^{-1}) was incubated at (Tm$-$25)° in SSC(=0.15 M NaCl, 15 mM sodium citrate, pH 7.2). After initial renaturation of reiterated nucleotide sequences the annealing reaction in nuclear DNA (△) becomes very slow. Chloroplastal DNA (○) renatures rapidly with a 70% recovery of hyperchromicity. Mitochondrial DNA (□) shows a somewhat lower rate of renaturation than does chloroplastal DNA, with a similar recovery of hyperchromicity. (From Wells and Birnstiel,[684] Fig. 4, p. 782.)

molecular weights. The molecular weight of DNA from plants has been estimated at $5\text{–}7 \times 10^6$; however *in vivo* it is probably many-fold greater than this and the low figures are caused by the shearing of the large DNA molecules during the extraction process.

The nuclear DNA is associated with the proteins, predominantly histones, and a small amount of RNA, in the chromosomes. Chloroplast and mitochondrial DNA appear to be attached to the membranes of the organelles but devoid of associated proteins.

RNA

Undegraded RNA can be characterized by various means. Originally separations were based on sedimentation in sucrose density gradients. When such separations were performed on RNA from plants three components were identified and on the basis of their sedimentation were characterized as 25s, 18s and 4s (Fig. 5.10). Since similar 25s and 18s RNA components were obtained from ribosomal preparations these were designated heavy ribosomal RNA and light ribosomal RNA respectively and the 4s component was considered to represent soluble RNA.

Total nucleic acids can be fractionated by chromatography on methylated albumin Kieselguhr (MAK) columns. The nucleic acids which bind to the methylated albumin can be eluted with progressively increasing salt concentrations. When such fractionations are conducted several nucleic acid components can be identified (Fig. 5.10) which can be characterized as soluble RNA_1 and soluble RNA_2, DNA, light ribosomal RNA and heavy ribosomal RNA. When labelled nucleic acids are fractionated a further peak of radioactivity, not associated with a U.V. absorbing peak, is eluted at the higher salt concentrations. The species so eluted has been variously called mRNA (messenger RNA) or D-RNA since its base composition is similar to that of the DNA. Not all of the applied radioactivity is recovered by elution of methylated albumin Kieselguhr columns with high salt concentrations; if the column is subsequently treated with sodium hydroxide or detergent, methylated albumin is stripped from the column and the associated remaining radioactive material is simultaneously eluted. This fraction has been termed tenaciously bound RNA (TB-RNA). When MAK column chromatography is conducted on RNA prepared from leaf extracts resolution of peaks in the ribosomal RNA region is much poorer than that observed in extracts from non-green tissue. Although the resolution of MAK column chromatography is not as great as the currently preferred technique of gel electrophoresis it does provide a preparative capability and allows for the collection of RNA components which can be subjected to further analyses. Base analyses conducted on the RNA fractions from various tissues indicate that the components have

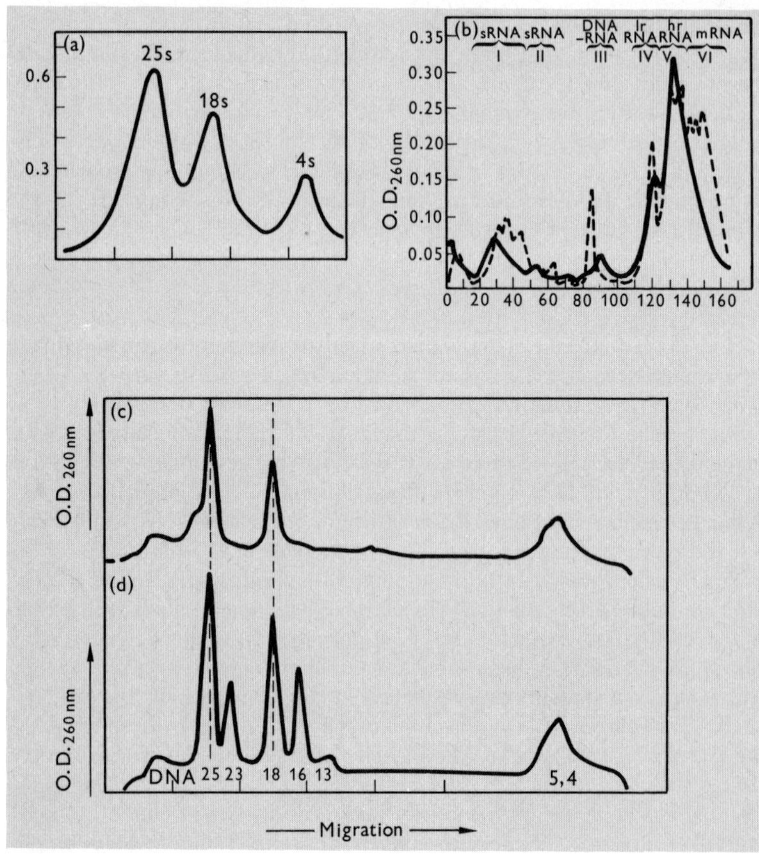

Fig. 5.10 Fractionation of plant RNA by (a) sucrose density gradient 5–20 % sucrose; (b) Methylated albumin Kieselguhr column chromatography; (c) acrylamide gel electrophoresis of RNA from root tissue; (d) acrylamide gel electrophoresis of RNA from leaf tissue.

different base compositions which appear to differ between species (Table 5.4).

These analyses do not demonstrate the presence of methylated nucleotides. However, methylation of ribosomal RNA has been shown to occur in plant tissue.[102] Significantly it is found that in chloroplast ribosomal RNA methylation occurs in the base component of the nucleotide (see p. 108) as is typical in bacterial systems. In the cytoplasmic ribosomal RNA in contrast, methylation occurs predominantly in the ribose moiety of the

Table 5.4 Base composition of RNA fractions separated by MAK column chromatography. a, Peanut (*Arachis hypogaea*) cotyledons[117] b, Soybean (*Glycine max*) hypocotyl.[293]

Fraction		Percent composition			
		CMP	AMP	GMP	UMP
Sol. RNA1	a	27.5	18.0	34.1	20.4
	b	40.9	15.9	26.4	17.1
Sol. RNA2	a	30.0	19.8	31.4	18.8
	b	31.2	19.4	28.2	21.2
DNA	a	26.3	21.9	31.0	20.9*
Light rRNA	a	25.1	22.3	32.1	20.5
	b	20.7	24.5	29.8	25.0
Heavy rRNA	a	20.0	25.4	32.7	21.7
	b	22.4	24.4	33.0	20.3
mRNA	a	18.5	28.2	29.6	23.7
or D-RNA	b	22.6	25.5	30.8	21.1

* TMP in DNA.

nucleotide.[523] Also the above analyses fail to show the occurrence of the many substituted bases characteristic of tRNA.

The D-RNA and TB-RNA are characterized by having a proportionately high adenine content and Key and coworkers[327] have drawn attention to this fact. It has been demonstrated that all mammalian messenger RNA's except those coding for histones are enriched in adenylic acid due to the attachment of polyadenylic acid sequences to the messenger RNA's. This feature has facilitated preparation of mammalian mRNA since the A-rich component can hybridize with polyuridylic acid supported on glass filters or hydroxyapatite[231] columns or can selectively bind to Millipore filters.[71] Thus A-rich mRNA can be fractionated from the other cellular RNA. Preliminary evidence[397, 679] indicates that RNA extracted from polysomes prepared from plant tissue can be fractionated by the Millipore retention technique suggesting, along with the data of Key's group,[327] that some plant messenger RNA's may be enriched in poly (A). By using the affinity of poly (A) containing mRNA to associate with oligo thymidine (dT) supported in cellulose it has been possible[655a] to isolate the mRNA for soybean leghaemoglobin. This protein it will be recalled is produced in nodules and is an important component in symbiotic nitrogen fixation (Chapter I). In contrast to the situation in mammals the A-rich segments have not been demonstrated in bacterial cells. If this situation is found to hold true it may provide a useful tool for identifying mRNA produced by the chloroplasts since their nucleic acid metabolism in many aspects closely resembles that encountered in bacteria.

The currently preferred technique of separation of RNA is by gel electrophoresis (Fig. 5.10). In this method the nucleic acids are separated on the basis of molecular weight. The original investigations of Loening and Ingle[387] indicated that nucleic acids prepared from leaf tissue contained four high molecular weight components. In contrast, the RNA from roots contained only two heavy molecular weight components and the partially resolved low molecular weight fractions. Analyses indicate that the additional RNA components in extracts from leaf tissue are due to the presence of chloroplast ribosomal RNA. On the basis of their mobility during gel electrophoresis[386] the heavy 25s and 18s light cytoplasmic ribosomal RNA have been determined to have molecular weights of 1.3×10^6 and 0.7×10^6 respectively and chloroplast heavy 23s and light 16s ribosomal RNA have molecular weights of 1.1×10^6 and 0.56×10^6. The ratio of heavy to light ribosomal RNA is 2:1; however during extraction the heavy chloroplast ribosomal RNA is frequently degraded to lower molecular weight fragments and appropriate corrections have to be made in calculation of these ratios.[285] The demonstration of chloroplast ribosomal RNA by gel electrophoresis confirmed the earlier speculation of the occurrence of RNA in chloroplasts which had been based on staining and autoradiographic techniques. Furthermore it partially accounted for the fact that ribosomal preparations from chloroplasts had a sedimentation value of 70s in comparison to 80s of cytoplasmic ribosomes.

When RNA prepared from material receiving a short exposure to labelled nucleic acid precursor is fractionated by gel electrophoresis it is found that radioactivity is dispersed in the ribosomal region of the gel. This high molecular weight heterodisperse material may correspond to mRNA.

The partially resolved low molecular weight fractions (Fig. 5.10) are designated as 5s and 4s respectively. The 4s component is composed of the cellular transfer RNA (tRNA) while the 5s fraction originates from both the chloroplast and cytoplasmic ribosomes. In addition, under appropriate conditions which involve heating and rapid cooling, a 5.8s RNA can be released from the cytoplasmic 80s ribosomes.[499] The molecular weights of the 5.8s, 5s, and 4s RNA have been estimated as 50 400, 37 900 and 25 000 respectively.

The 4s or tRNA is heterogeneous in nature containing tRNA components complementary to each of the twenty amino acids usually encountered in proteins. It is customary to identify the different tRNA's on the basis of their amino acid acceptance; thus the tRNA species accepting arginine is written $tRNA^{arg}$ while that accepting leucine would be designated $tRNA^{leu}$. If the tRNA is charged with an amino acid then the designation is arginyl-$tRNA^{arg}$ or leucyl-$tRNA^{leu}$, etc. Techniques are available for isolating tRNA and fractionating tRNA species specific for a particular amino acid. Several tRNA's specific for various amino acids have been isolated from microbial sources and the nucleotide sequence determined. Phenylalanyl-tRNA ($tRNA^{Phe}$) has been purified from wheat germ[163] and its nucleotide composition determined. It shows much similarity to yeast $tRNA^{Phe}$ and in keeping with other characterized tRNA's it can be accommodated in the so-called clover leaf configuration with Dihydrouridine, T ψ C and anticodon loops (Fig. 5.11).

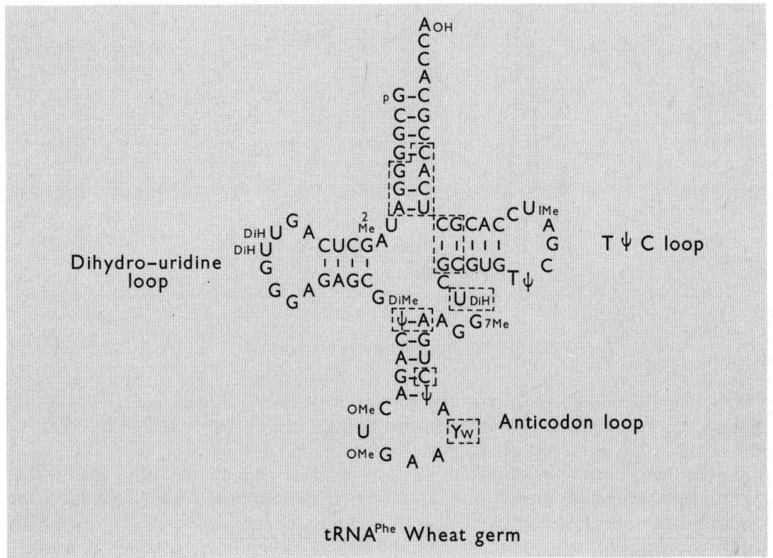

Fig. 5.11 Structure of wheat germ phenylalanyl-tRNA (from Dudock *et al.*,[163] Fig. 4, p. 944).

Other analyses[354] indicate that tRNA from plants in common with those from animals and bacteria are enriched in many substituted bases and all possess a terminal CCA. These terminal nucleotides are added to the incomplete tRNA molecule at the 3′ terminus by the enzyme CTP ATP terminal nucleotidyl transferase.

Separation of tRNA by reversed phase chromatography (Fig. 5.12a) demonstrates the occurrence of different tRNA fractions capable of accepting the same amino acid.[17] These are called iso-accepting tRNA's. Iso-accepting $tRNA^{leu}$, $tRNA^{tyr}$ and $tRNA^{met}$ have been demonstrated in plants. The occurrence of iso-accepting tRNA species might be predicted on the basis of a redundant genetic code (see Table 6.7). There are several codons for each amino acid and thus tRNA species with the appropriate anticodon might be expected. However at present there is no evidence to indicate that the separation of iso-accepting species by reversed phase chromatography or any other method is related to differences in

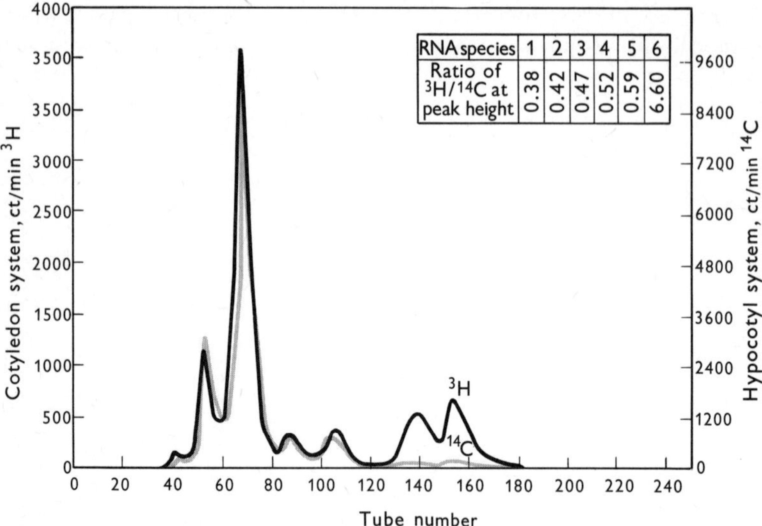

RNA species	1	2	3	4	5	6
Ratio of ³H/¹⁴C at peak height	0.38	0.42	0.47	0.52	0.59	6.60

Fig. 5.12a Separation of leucyl-tRNA^{leu} from soybean (*Glycine max*) by reversed phase chromatography. (From Anderson and Cherry,[17] Fig. 1, p. 205.)

Transfer RNA was extracted from the cotyledon (———) or hypocotyl (tint) and incubated with ^{3}H-L-leucine or ^{14}C-L-leucine respectively using amino acyl synthetases prepared from the cotyledons or hypocotyls. After incubation the reaction mixtures were combined and the acylated tRNA recovered. The acylated-tRNA was applied to a Freon column and eluted with a linear gradient of sodium chloride. The different iso-accepting leucyl-tRNA^{leu} elute at different salt concentrations, producing the different peaks of radioactivity. Note that certain leucyl-tRNA^{leu} species present in the cotyledon are absent in the hypocotyl.

bases in the anticodon region. Some of the different iso-accepting tRNA species may be associated with specific organelles. Merrick and Dure[418] have reported that iso-accepting tRNA's associated with chloroplasts differ from those occurring in the cytoplasm.

A particularly significant series of iso-accepting tRNA's^{met} are encountered in plant tissues.[369] Fractionation of tRNA from leaf tissue on benzoylated DEAE-cellulose indicates the presence of three methionyl tRNA's (Fig. 5.12b). The two major species eluted at low salt concentration (i, m) correspond to the two cytoplasmic species. The first peak (i) contains the tRNA^{met} functional in protein chain initiation (see Chapter 6). However unlike the situation in bacteria and animals this cytoplasmic initiator methionyl tRNA is not formylatable. The third tRNA^{met} peak (f), which elutes at high salt concentrations, is formylatable and is the major methionyl tRNA component of chloroplasts. The enzyme involved in the formylation, transformylase, is present in the chloroplasts. The formyl methionyl

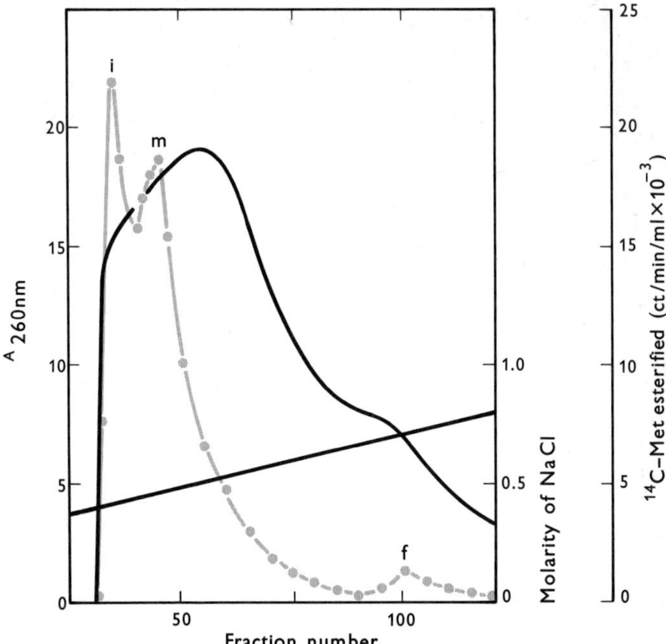

Fig. 5.12b Separation of tRNA^{met} by benzoylated DEAE-cellulose chromatography. (From Leis and Keller,[369] Fig. 1, p. 891.)

Transfer RNA was extracted from wheat germ and applied to a benzoylated DEAE-cellulose chromatography column. The column was eluted with a linear gradient of sodium chloride. The eluted RNA was recorded by measuring A260 nm. Samples of eluate were collected and assayed for methionine acceptor activity (——●——) using amino acyl synthetase prepared from wheat germ. Reprinted with permission from *Biochemistry* (1971), **10**. Copyright by the American Chemical Society.

tRNA may function in protein chain initiation by chloroplast ribosomes and thus resemble the initiating system in the cytoplasm of prokaryotes.

NUCLEIC ACID SYNTHESIS

Since the bulk of DNA is confined to the cell nucleus it is to be expected that for the most part DNA synthesis will be restricted to and occur most actively in the meristematic regions of the plant where the cells are undergoing mitosis or in the reproductive parts of the plant in cells undergoing meiosis and mitosis. The DNA in the nucleus is confined to the chromosomes; thus DNA synthesis and chromosome replication are closely related but not contemporaneous events. The synthesis of DNA during chromosome replication and mitosis has been extensively studied and Stern's group[590] has followed the course of DNA synthesis during meiosis.

In contrast to the restricted period of DNA synthesis it appears that capacity for RNA synthesis is maintained for prolonged periods in plant cells.

Timing of DNA synthesis

Cell division may be conveniently divided into phases. During periods

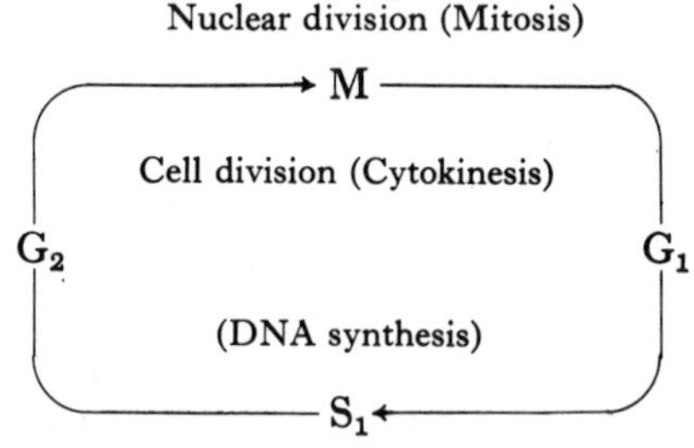

G_1 and G_2 cell growth proceeds; however during this time DNA synthesis occurs in the S phase without an interruption of cell growth. The period of cell growth preceding DNA synthesis is termed G_1 and the growth phase following DNA synthesis is G_2. At the termination of G_2 the cell nucleus undergoes the characteristic changes of mitosis which can be divided into: (a) **prophase** during which time the chromosomes become visible and at the end of which the nuclear membrane breaks down; (b) **metaphase** during which the chromosomes align at the equator of the cell on the metaphase plate; (c) **anaphase** when the chromosomes separate and begin to move towards the pole and enter (d) **telophase**, when chromosome structures disappear, the nuclear membrane reforms and the nucleoli reappear. The period between telophase and the onset of prophase is **interphase**. Feulgen staining and autoradiography indicate that prophase nuclei contain twice as much DNA as the telophase nuclei. Telophase nuclei are customarily said to have the diploid or 2C content whereas prophase nuclei are described as 4C. The DNA content of nuclei is doubled during interphase at the S phase between G_1 and G_2. This doubling of DNA content during the interphase can be confirmed by the demonstration that DNA precursors are incorporated into chromosomal DNA at this time. DNA synthesis usually stops during G_2 before the onset of mitosis. Other labelling studies indicate that the histone proteins with which the DNA is associated in the chromosome also duplicates during the S phase. However the arrangement of the DNA and histones in the chromosome and the interrelationship of their metabolism are the subject of much discussion far beyond the scope of the current text. Synthesis of DNA does not occur at a uniform rate in all the chromosomes or at the same time along the length of the same chromosome; some chromosome parts are labelled early whereas other parts are labelled later in S phase. Each chromosome can be considered as a series of linked DNA sequences each of which is duplicated at a specific time.

The factors controlling the onset of DNA synthesis have not been resolved. Various hormones, IAA and kinetin for example, have been shown to stimulate cell division in cell tissue cultures.[136] Deoxyribonucleotides have been shown to accumulate in cell tissue cultures[468] and the enzymes involved in thymidylic acid synthesis[250] increase immediately preceding DNA synthesis leading to the speculation that the level of precursor molecules may in some undefined manner determine the onset of DNA synthesis.

DNA synthesis usually terminates in G_2 before mitosis. However, the termination of DNA replication is not the trigger for entry into G_2 and the onset of mitosis. Instances are known where DNA synthesis is continued without nuclear division. Such is the case in the cells of the suspensor formed during early embryogenesis.[661] A similar DNA synthesis without nuclear division occurs during the development of cotyledons of many seeds.[427, 548, 570] This sustained synthesis of DNA could represent a mechanism for the amplification of specific genes if only certain DNA sequences were synthesized. However on the basis of hybridization studies Millerd *et al.*[427] conclude that in the bean (*Vicia faba*) cotyledon the total nuclear DNA complement is replicated and thus in this instance the endomitosis or polyteny represents a means of amplification of the total genome.

Mechanism of DNA synthesis

The linear arrangement of nucleotides in the double helical DNA molecules specifies the genetic information contained in that DNA. Since this genetic information must be passed on to each daughter cell as the nuclei and cells divide it is imperative that the DNA is replicated in such a manner as to yield identical daughter molecules so that genetic continuity is maintained. The Watson and Crick model for the DNA molecule provides a working hypothesis as to how such a replication could occur. The two strands of a DNA double helix are complementary and thus replication of each to form complementary daughter strands produces two daughter duplex DNA molecules identical to the parent, each of which contains one strand from the parental DNA. This type of **semi-conservative replication** was originally demonstrated by the elegant experiments of Meselson and Stahl[420] using the bacterium *Escherichia coli*. Filner[185] has utilized similar techniques to demonstrate that DNA is similarly replicated semi-conservatively in tobacco (*Nicotiana tabacum*) cells (Fig. 5.13). Cells were grown in ^{15}N nitrate and then transferred to ^{14}N nitrate and the DNA isolated from such cells during the exponential phase of growth where generation time was two days. The isolated DNA was fractionated by caesium chloride density centrifugation. One day after transfer of the tobacco cells from ^{15}N nitrate to ^{14}N nitrate the DNA sedimented as a uniform band. However, the fully ^{15}N labelled DNA band disappeared between days 2 and 3. Accompanying the disappearance of the ^{15}N labelled band there was an appearance of an intermediate band due to hybrid DNA. Totally light ^{14}N DNA appeared on day 5, i.e. during the second replication. During subsequent replications as more ^{14}N DNA was produced the hybrid continued to persist but became a progressively smaller fraction of the total. These changes in DNA density are in agreement with the semi-conservative mode of replication.

DNA polymerase

The enzyme catalysing DNA replication is DNA polymerase; the overall reaction may be represented as follows.

$$\begin{matrix} n_1\ \text{dATP} \\ n_2\ \text{dGTP} \\ n_3\ \text{dCTP} \\ n_4\ \text{dTTP} \end{matrix} \xrightarrow[\text{Mg}]{\text{Preformed DNA}} \text{DNA}-\begin{bmatrix} \text{dAMP}n_1 \\ | \\ \text{dGMP}n_2 \\ | \\ \text{dCMP}n_3 \\ | \\ \text{dTMP}n_4 \end{bmatrix} + [n_1 + n_2 + n_3 + n_4]\ \text{PPi}$$

A capacity for plant extracts to incorporate deoxyribonucleotides into acid insoluble material (DNA) has been demonstrated in a number of

---

Fig. 5.13 Semi-conservative replication of DNA in tobacco (*Nicotiana tabacum*) cells. DNA isolated at 24-hour intervals after transfer from ^{15}N to ^{14}N, and banded in caesium chloride. Band a: Reference (*Micrococcus lysodeikticus*) DNA; Band b: ^{15}N DNA; Band c: Hybrid DNA; Band d: ^{14}N DNA. Top and bottom frames show the resolution of a mixture of reference, ^{15}N and ^{14}N DNA. (From Filner,[185] Fig. 3, p. 37.)

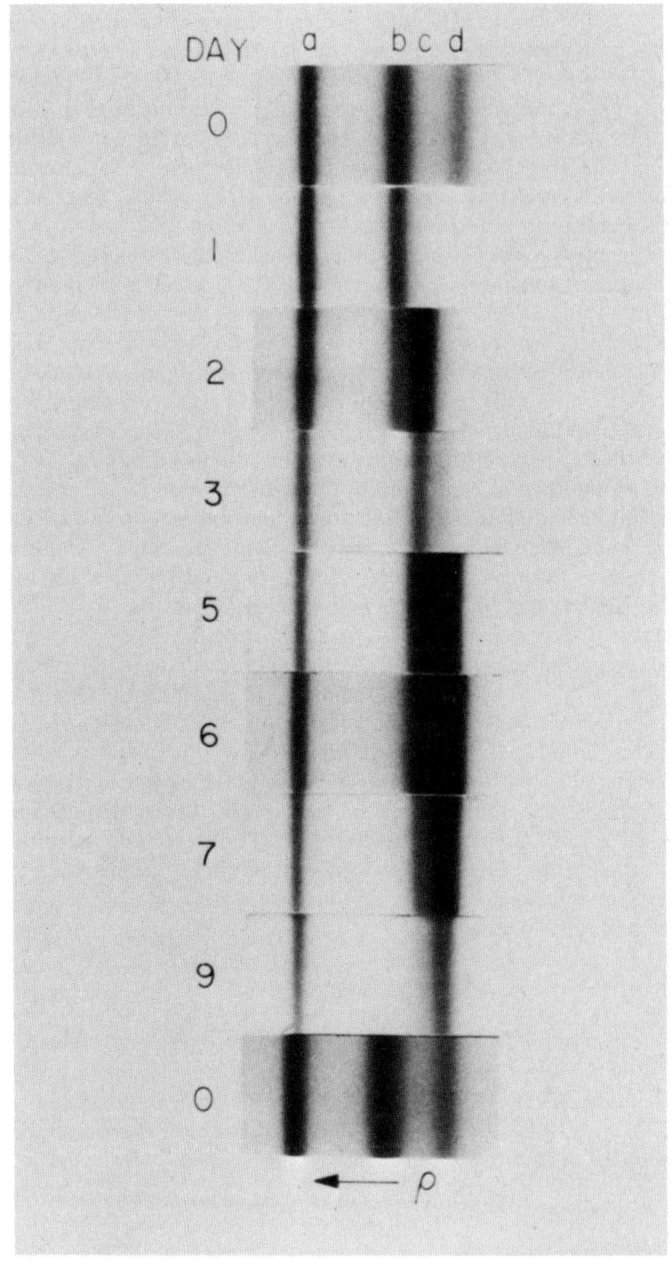
DAY
a
b c d
0
1
2
3
5
6
7
9
0
ρ

laboratories.[165,250,444,604] The active component in the extracts may be nuclei, chromatin, or a soluble enzyme. Soluble DNA polymerases from corn (*Zea mays*)[604] and wheat (*Triticum aestivum*)[444] have been studied and the DNA polymerases released from the chromatin prepared from sugar beet (*Beta vulgaris*) have recently been characterized. The demonstration of DNA polymerase activity is dependent upon the presence of all the deoxyribonucleotides and exogenous DNA which functions as a template and primer in the reaction.

Fractionation of the polymerase solubilized from sugar beet[165] chromatin by column chromatography demonstrates the presence of four protein peaks catalysing DNA synthesis. The different peaks show varying preferences for DNA as templates. In some instances single stranded DNA was the more effective template whereas in other fractions double stranded DNA is more effective. The activity of the extracted polymerases was inhibited by Actinomycin D which inhibits both DNA and RNA synthesis through its capacity to bind to the guanine residues of DNA.

The significance of the multiple DNA polymerase is not clear at this time. However a similar multiplicity of polymerases is encountered in bacterial and mammalian cells. DNA polymerase activity increased in callus tissue cultures prior to cell division[250] and in sugar beet during washing treatments which lead to a rejuvenation of the tissue.[165] Such preliminary observations suggest that DNA synthesis may be regulated by the level of polymerase; however, in the callus tissue[250] the polymerase level does not decline at the termination of DNA synthesis and similarly in *Chlorella* DNA replication could not be related to the amount of DNA polymerase in the cell.[549]

Isolated chloroplasts have been shown to incorporate deoxyribonucleotide triphosphates into a DNA product which bands with chloroplast DNA in caesium chloride gradients.[579,621] The enzymes involved in this synthesis have not been characterized and the timing of chloroplast DNA synthesis relative to nuclear DNA synthesis in the cell cycle has not been studied in higher plants.

RNA SYNTHESIS

The capacity for RNA synthesis in plants is conveniently measured by incubating the plant material with a labelled nucleic acid precursor such as ^{32}P phosphate or ^{14}C or ^{3}H labelled orotic acid, or labelled bases or nucleotides. RNA is subsequently extracted from the plant material and the amount of radioactivity incorporated determined. Most plant parts are able to incorporate precursors into RNA and thus, unlike DNA synthesis which is confined to dividing cells, RNA synthesis occurs generally throughout the plant.

The incorporation of label into RNA is prevented by Actinomycin D, demonstrating the requirement for a functional DNA molecule for RNA synthesis. Substituted bases have varying effects on RNA synthesis; some are very inhibitory while others appear to have only a marginal effect (Table 5.5).

Table 5.5 The effect of Actinomycin D and base analogues on RNA synthesis in soybean (*Glycine max*) hypocotyl.

Material	Counts per minute in RNA
* 0	10 550
8-Azaguanine (2.5×10^{-3} M)	9 380
Actinomycin D (10 μg ml^{-1})	1 270
† 0	14 500
2-Thiouracil (2.5×10^{-3} M)	12 700
2-Thiocytosine	14 900
6-Azauracil	15 100
5-Fluorouracil	8 450
6-Methyl Purine	1 900

*Tissue (0.8 g) was incubated with 0.5 μCi ATP-8-^{14}C (from Key [323]).
†1 g tissue incubated with 0.5 μCi adenosine-8-^{14}C (from Key[324]).

It is not clear how the inhibition of RNA synthesis by base analogues is achieved. It may be expected that the base analogues are converted to nucleotides and replace the normal bases during RNA synthesis. Insertion of a wrong base could foreseeably result in a termination of the incomplete RNA chain. Alternatively chain elongation might proceed but the RNA produced may be non-functional due to the presence of the base analogues. Such a mode of action of the analogues, however, fails to explain why the pyrimidine analogues except 5-fluorouracil do not influence RNA synthesis in the soybean hypocotyl, for example, whereas the purine analogues were found to be more inhibitory. Similarly incorporation of the base analogues into RNA fails to explain the selective effect of 5-fluorouracil which inhibits ribosomal RNA and transfer RNA synthesis by 90% and 75% respectively while D-RNA synthesis was reduced by only 10%.[324]

Rifamycin or its related compounds have been shown to inhibit effectively RNA synthesis in bacterial but not mammalian systems. Since chloroplasts resemble bacteria in their ribosomal RNA and organization of DNA it might be expected that rifamycin could preferentially inhibit chloroplast RNA synthesis. Experiments have been conducted to study this using intact plants; however, the effects are extremely variable with some authors[59] claiming inhibition while others[580] observe little effect in the intact plant. Similar discrepancies are observed when the inhibitor is applied to isolated chloroplasts. It is suggested that the variable results might be due to differential uptake of the inhibitor by the chloroplasts.

Regulation of RNA synthesis

In 1961 Jacob and Monod[303] formulated their classical hypothesis concerning the regulation of enzyme synthesis. Briefly, according to their proposal the genetic information coding for the synthesis of a particular enzyme is contained in the region of the DNA molecule designated as the structural gene (Fig. 5.14). Production of RNA complementary to the DNA sequence (transcription) is controlled by the operator. In turn functioning of the operator is regulated by a repressor molecule. The information for the synthesis of the repressor molecule is dictated by the nucleotide sequence in the DNA region R (regulator).

The regulator gene is transcribed to form mRNA coding for the synthesis of a repressor molecule which can bind specifically to the operator region and prevent transcription of the structural gene. The repressor molecule has two specific binding sites—one for the operator locus and one for the inducer site. The binding of the inducer apparently causes conformational change in the repressor molecule and it no longer binds to the operator. Release of the repressor from the operator allows for transcription of the structural gene and the production of the enzyme coded for by that gene.

In the 1960's it was speculated that many of the effects of plant hormones may involve a similar mechanism in which the hormone (inducer) combined with the repressor to alleviate the block at the operator gene, thus facilitating the synthesis of products from the operator gene and allowing for the expression of the hormonal effects. The result of these speculations was the publication of a mass of literature demonstrating the influence of plant hormones on RNA synthesis. Key[325] and Trewavas[635] have comprehensively reviewed this literature and while it is apparent that hormonal treatments do in many instances result in an increased RNA synthesis there appears to be an overall RNA synthesis rather than a stimulated production of certain species. As yet there is no clear cut evidence of plant hormones acting as inducers in the classical sense of Jacob and Monod.

Attempts have been made to account for the enhanced nucleic acid synthesis. From the section on RNA polymerase it is clear that the rate of RNA synthesis will be dependent on the level of polymerase and availability of DNA template. The mechanism of controlling the availability of DNA template is not known at this time. Bonner's group[62] originally indicated that histones probably functioned to control gene expression (DNA template availability). The basis for this reasoning was as follows. If chromatin (the interphase chromosomes) was isolated from pea (*Pisum sativum*) nuclei it could serve as a template for RNA synthesis catalysed

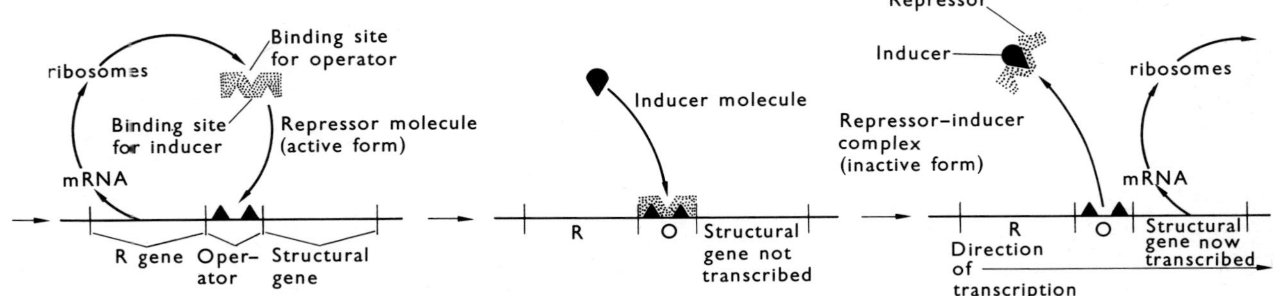

Fig. 5.14 Diagrammatic representation of the regulation of gene expression by inducer molecules.

by an RNA polymerase from *Escherichia coli*. If the histones were removed from the chromatin then the remaining DNA was a much more effective template for RNA synthesis (Table 5.6).

Table 5.6 Activity of chromatin of developing pea (*Pisum sativum*) cotyledons and of pea DNA in support of RNA synthesis by *Escherichia coli* RNA polymerase. (From Bonner, Huang and Gilden.[62])

System	RNA synthesis (pmoles nucleotide incorporated)
125 μg DNA as whole chromatin, no added polymerase	17
125 μg DNA as whole chromatin + *E. coli* polymerase	760
125 μg deproteinized DNA + *E. coli* polymerase	2220

Similar observations have been made by other workers,[306] but there is currently some hesitancy in ascribing to histones the proposed regulatory function. It is now generally conceded that the histones are insufficiently diverse to control the expression of the whole range of genes transcribed during growth and development of eukaryotic organisms. In addition to histones and DNA, chromatin contains other non-histone proteins and these may function to regulate the accessibility of DNA to the RNA polymerase.

Of course some of the non-histone proteins are the DNA polymerase and RNA polymerase.

RNA polymerase

The demonstration that isolated nuclei and chloroplasts had the capacity to incorporate ribonucleotides into RNA indicated that these organelles possessed the enzymes required for RNA synthesis. Subsequently it has been demonstrated that chromatin (the interphase chromosomes consisting of nucleic acid protein complex with a composition of 36.5% DNA, 37.5% histone, 9.6% RNA, 10.4% non-histone protein) is able to catalyse the incorporation of ribonucleic acid precursors into RNA (Table 5.7).

In contrast to the chromatin bound enzyme(s) catalysing RNA synthesis a soluble enzyme was extracted from maize (*Zea mays*) seedlings and characterized.[605] Its properties were in many respects similar to those of the enzyme originally isolated from *Escherichia coli* which had been termed RNA polymerase and catalysed the following reaction.

$$\begin{matrix} n_1ATP \\ n_2UTP \\ n_3GTP \\ n_4CTP \end{matrix} \xrightarrow{\text{DNA Template}} \underset{\left[\begin{matrix} AMPn_1 \\ UMPn_2 \\ GMPn_3 \\ CMPn_4 \end{matrix}\right]}{RNA} + (n_1 + n_2 + n_3 + n_4)\,PPi$$

Table 5.7 Requirements for RNA synthesis by soybean (*Glycine max*) hypocotyl chromatin. (From O'Brien, Jarvis, Cherry and Hanson.[475])

System	pmoles ^{3}H-UMP incorporated per 100 μg DNA
Complete	218
−Mg	95
−Mn	143
−ATP	5
−GTP	32
−CTP	15
−CTP, GTP, ATP	9

The complete system contained 0.2 μmoles GTP,CTP and ATP, 0.013 μmoles ^{3}H-UTP, 1 μmole $MgCl_2$, 0.25 μmoles $MnCl_2$, 1 μmole dithiothreitol, 20 μmoles Tris pH 8.0 and chromatin (3–6 μg DNA) in a final volume of 0.4 ml. The reaction was inhibited by ribonuclease and Actinomycin D. In the complete system addition of *Escherichia coli* RNA polymerase greatly enhanced ^{3}H-UMP incorporation into RNA indicating that only a small percentage of the potential RNA synthesis is expressed by the endogenous polymerase present in the chromatin.

Recently it has been possible to solubilize the RNA polymerases from chromatin from various sources or prepare them directly in the soluble form.[166, 235, 271, 307] When such solubilized extracts are fractionated it is found that polymerase activity is associated with several chromatographic peaks (Fig. 5.15). A similar diversity of polymerases was originally demonstrated in mammalian tissues. The polymerases in plant extracts have different properties (some of them similar to those originally used to establish the identity of mammalian polymerases). The principal diagnostic characteristic is sensitivity to the compound α-amanitin. It has been established in the mammalian studies that an insensitive enzyme Polymerase I is located in the nucleolus whereas the Polymerase II enzyme located in the nucleoplasm is inhibited by α-amanitin.[382] By analogy the two polymerases isolated from plant nuclei or chromatin may also be considered as originating in the nucleolus and nucleoplasm.

In addition to the DNA-dependent RNA polymerases associated with nuclei or chromatin it has been possible to demonstrate a DNA-dependent RNA polymerase in chloroplasts. The enzyme shows variability in its response to rifamycin depending on the source of the chloroplasts.[64, 65] The polymerase activity is generally insoluble and appears to be bound in

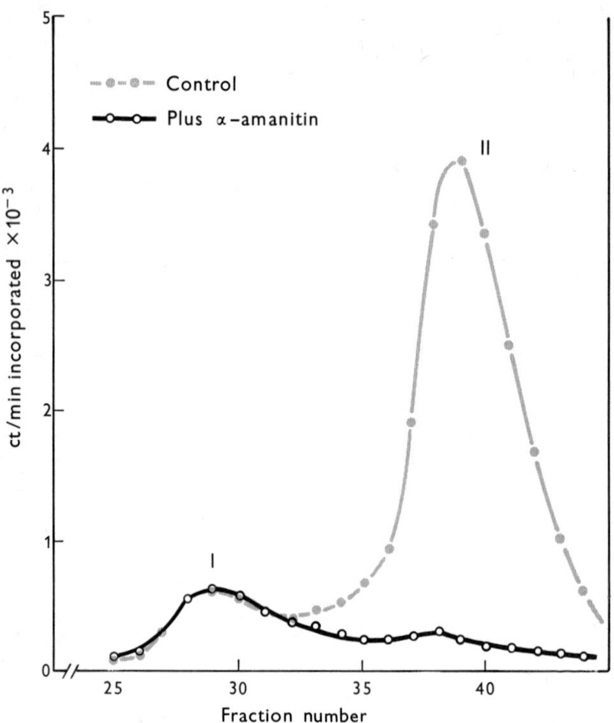

Fig. 5.15 DEAE-cellulose fractionation of RNA polymerases from wheat germ. Note α-amanitin inhibition of polymerase activity associated with peak II. (From Jendrisak and Becker,[307] Fig. 2a, p. 52.)

association with the chloroplast DNA to the chloroplast membrane.[64] Indications that there may be more than one chloroplastic polymerase is provided by the recent demonstration that some of the polymerase may be solubilized from the organelle,[64] whilst the bulk of the activity remains bound.

TRANSCRIPTION AND GENE PRODUCTS

The synthesis of RNA by DNA-dependent RNA polymerase is obviously a complex process which requires that the genetic information in the DNA is faithfully transcribed into RNA molecules. The process involves binding of the polymerase to a DNA strand, initiation of synthesis, propagation and termination. These processes are presumably accompanied by an unwinding and rewinding of the double helix.

In bacterial systems the RNA polymerase consists of a core protein with various associated polypeptides which are thought to function in determining specificity, etc. To date similar factors have not been demonstrated in higher plants; however, the finding[446] that polymerase II from maize is composed of several polypeptides may be indicative of a multienzyme complex. Additionally, the difficulty experienced in purifying polymerase I may indicate dissociation of essential accessory factors with progressive purification.[271]

Since the base sequence of the DNA is faithfully transcribed into the RNA molecule the product can hybridize with DNA. By capitalizing on this property it has been shown that chloroplast ribosomal RNA hybridizes preferentially but not exclusively with chloroplast DNA.[294] Cytoplasmic ribosomal RNA hybridizes preferentially with nuclear DNA or more specifically the nuclear satellite DNA probably corresponding to the nucleolar organizer.[552] Transfer RNA has been shown to hybridize with both chloroplast and nuclear DNA.[696] Cytoplasmic 5s RNA hybridized with a satellite DNA probably corresponding to the nucleolar organizer region while tRNA hybridized with the main band DNA.[627]

It is possible to measure the percentage of cellular DNA which is complementary to a particular RNA species by incubating DNA with excess RNA until all the complementary sites are saturated. By knowing how much of the applied DNA is hybridized to a specific RNA and the amount of DNA present in a nucleus or cell, the number of copies of the gene per nucleus can be calculated. Such calculations by Ingle and Sinclair[295] demonstrate that there are thousands of copies of genes coding for ribosomal RNA and it is estimated that there are 3900 copies of 5s RNA and 8000 copies of tRNA per nucleus in *Cucurbita pepo*.[627]

The gene product

Although certain RNA species can be isolated in an apparently undegraded state it is not clear whether this represents the form in which they were transcribed or whether they are modified subsequent to transcription. It is known that base modification in terms of methylation, thiosubstitution, etc., occurs after transcription but it also seems that the initial gene product may be modified in other ways.

Short term labelling studies followed by isolation of the RNA and separation of the products by gel electrophoresis indicate that the gene product for cytoplasmic ribosomal RNA has a molecular weight of about 2.5×10^6 daltons.[287, 356, 535] This initial product is converted to the 0.7×10^6 rRNA and a 1.4×10^6 fraction which in turn yields the characteristic 1.3×10^6 rRNA. Other species show additional rapidly labelled high molecular weight products whose kinetics of decay suggest their participation in the production of ribosomal RNA. The initial gene product for ribosomal RNA in chloroplasts has a molecular weight of 2.7×10^6 daltons which gives rise to 1.2×10^6 and 0.6×10^6 components which are then converted into the 1.05×10^6 and 0.56×10^6 chloroplast ribosomal RNA.[252]

The nature of the gene product for mRNA in plants is not known. The identification of A-rich mRNA or D-RNA indicates that the original product may undergo modification. This post-transcriptional modification of mRNA involves the addition of polyadenylic acid sequences to the 3′ end of the mRNA or its precursor. The mechanism of this polyadenylation is not completely resolved. It does not appear to involve transcription of polydeoxy thymidine residues in the DNA which could then become associated with mRNA; instead the poly (A) residues seem to be assembled without a template. An enzyme capable of poly (A) synthesis in this manner has been purified and characterized from maize (*Zea mays*)[397a] and two poly (A) polymerases have been isolated from wheat (*Triticum vulgare*).[86a] One of the poly (A) polymerases from wheat is associated with the chloroplasts which in addition contain an enzyme poly (G) polymerase capable of synthesizing polyguanylic acid residues. This association of polymerizing enzymes with the chloroplasts suggests that mRNA of organelles may be subject to post transcriptional modification, in contrast to the situation encountered in bacteria.

RNA TURNOVER AND DEGRADATION

In studies with excised plant parts it is frequently observed that the incorporation of precursors into RNA is occurring at a time when the tissue content of RNA is either declining or remaining constant. These observations where RNA synthesis is occurring without an increase in RNA content demonstrate that there must be an accompanying RNA degradation. Although it is possible that some of the lack of the accumulation of RNA in excised tissues may be attributable to increased RNA degradation associated with injury occurring during tissue preparation this is not an entirely satisfactory explanation. RNA synthesis and degradation have been shown to occur simultaneously in intact tissues and in fact RNA appears to be in a constant state of turnover. At steady state conditions when RNA content of the tissue is constant, synthesis is offset by an equal rate of degradation.

The most detailed studies of this turnover have been conducted by Trewavas[636] using the duckweed, *Lemna minor*. By appropriate short term labelling and sequential analysis it was found that DNA showed little detectable turnover, chloroplast ribosomal RNA had a half life of 15.3 days while cytoplasmic ribosomal RNA had a half life of 4.3 days. The soluble RNA which included 4s and 5s RNA of both cytoplasmic and chloroplast origin had a half life of 5.3 days. These turnover rates were effected by nutritional status of the *Lemna* and hormonal treatments and it was indicated that under different situations alteration in turnover rate could be due to changes in either synthesis or degradation.

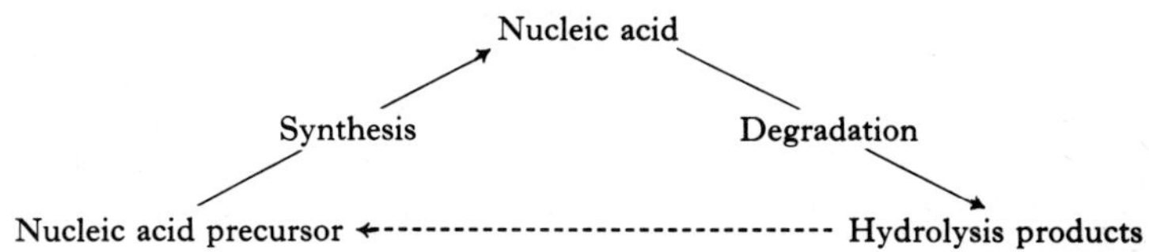

Although it is possible that the hydrolysis products could function as precursors for subsequent synthesis the data of Trewavas[636] indicate that this does not occur extensively and thus synthesis occurs from newly synthesized precursors. It is significant that Trewavas recorded a low rate of turnover of chloroplast ribosomal RNA; when RNA synthesis is followed in leaf tissue it is found that incorporation of precursors into chloroplast RNA occurs over a very limited time span whereas precursors are incorporated into cytoplasmic components even though the leaf may be declining in RNA content.[572]

The mechanism of RNA breakdown in plants has not been followed in great detail. It can be easily demonstrated that plants contain enzymes capable of hydrolysing nucleic acids. Assays for nucleic acid degrading enzymes involve the incubation of a suitably buffered plant extract with a nucleic acid substrate followed by the addition of perchloric acid or trichloracetic acid to precipitate the undegraded nucleic acids. The hydrolysis products remain in the acid solution and can be determined at 260 nm on a spectrophotometer. Assays of this type indicate that plants contain enzymes capable of hydrolysing both RNA and DNA and the activity of the enzymes in the extracts is sufficient to degrade all the cellular nucleic acids in a short period of time. It appears that the nucleic acid degrading enzymes are sequestered or located at a part of the cell distant from their substrate. Ribonuclease has been suggested to be located in the vacuole[405] and additionally has been demonstrated to occur in the chloroplast,[241] and in association with other particulate fractions.[699]

Various workers have demonstrated that the nucleic acid degradation by plant extracts is a function of several enzymes.[161] Wilson[699] has made a detailed characterization of the enzymes from corn (*Zea mays*) (Table 5.8) and indicates that his classification will accommodate many of the enzymes characterized from other sources.

The enzymes show further heterogeneity when fractionated by gel electrophoresis suggesting that there may be more nucleases.[700]

Changes in ribonuclease level have been associated with changes in growth and development; however the physiological function of the specific nucleases has not been established.

Table 5.8 Classification of corn (*Zea mays*) nucleases. (From Wilson,[699] Table 1, p. 1340.)

	RNase I	RNase II	Nuclease I
1. Substrate	RNA	RNA	RNA, DNA, 3′-Nucleotides?
2. Mode of action	Endonuclease	Endonuclease	Endonuclease
3. Products formed	2′, 3′-Cyclic nucleotides	2′, 3′-Cyclic nucleotides	5′-Nucleotides
3a. Secondary products	3′-Purine nucleotides	3′-Nucleotides (nucleosides)	
4. Base specificity	Purine, pyrimidine	Purine, pyrimidine	
5. Mol. wt.	23 000	17 000	31 000
6. pH optimum	5.2	6-7	6.2
7. Location in cell	Soluble	Microsomes (Ribosomes?)	Large particles, membranes
8. Enzyme commission classification	2.7.7.17	2.7.7.17	3.1.4.x and 3.1.3.6?

6

Proteins

PROTEIN CONTENT

Relatively few studies have been conducted to establish the true protein content of plant tissues. More usually estimates have been made of 'Crude protein' content. In these estimates plant material is digested with concentrated sulphuric acid in the presence of various catalysts with the result that nitrogenous components are converted to ammonia which in the sulphuric acid solution forms ammonium sulphate. The ammonia can be released by alkaline solutions and recovered by steam distillation and determined colorimetrically or by titration. This technique of acid digestion originally developed by Kjeldahl estimates the total nitrogen content of the plant tissue. The figure obtained is usually multiplied by 6.25 to give the crude protein content of the tissue. This conversion factor is derived from the observation that the average amino acid from which the proteins are synthesized contains approximately 16% nitrogen. Clearly protein determinations based on this method are an approximation since other non-protein nitrogenous compounds are converted to ammonia; however, the bulk of the nitrogen in plants is contained in protein and these crude protein figures, in addition to providing useful data for comparative purposes, to a great extent reflect true protein content.

On average leaf material contains approximately 2% protein on a fresh weight basis whereas seeds may be considerably enriched in protein (Table 6.1).

An alternative, more time-consuming but more accurate method of determining true protein content is to remove the non-proteinaceous nitrogenous constituents from the plant material and then determine the protein content of the residue colorimetrically or by the Kjeldahl digestion

Table 6.1 Protein content of some plant materials.

Plant		Protein (g/100 g fresh wt.)
Asparagus	*Asparagus officinalis* (shoot tip)	2.2
Beet	*Beta vulgaris* (root)	1.6
Carrot	*Daucus carota* (root)	1.2
Cauliflower	*Brassica oleracea botrytis* (curd)	2.4
Squash	*Cucurbita pepo* (fruit)	0.6
Pea	*Pisum sativum* (dry seed)	23.8
Bean	*Phaseolus vulgaris* (dry seed)	23.1
Wheat	*Triticum aestivum* (grain)	12.3

method. Usually the procedure involves sequential extraction with ethanol, trichloracetic acid and ether to remove soluble materials and lipids. The extracted lipid-free residue contains proteins and nucleic acids. The nucleic acids can be removed by hot trichloracetic acid and the residue used for Kjeldahl digestion. Alternatively the lipid-free residue can be dissolved in alkaline solution and the protein content determined colorimetrically using the method of Lowry *et al.*[391] This method takes advantage of two properties of proteins. The biuret residues $NH_2CONH.CONH_2$ produced from the protein react under alkaline conditions with copper sulphate to produce a blue colour and the tryptophan and tyrosine residues of the protein react with phosphomolybdate and phosphotungstate to produce additional blue colour.

Unfortunately, plant phenolics also react with the reagents so that it is imperative that proteins from plants are suitably purified before assays are performed by the Lowry method.

PROTEIN COMPOSITION

If plant residues are heated with hydrochloric acid the proteins are hydrolysed to amino acids. Analyses of such hydrolysates demonstrates the presence of the 20 protein amino acids. Tryptophan is destroyed during acid hydrolysis and it is customary to determine this amino acid following alkaline hydrolysis of the protein. It is generally found that plants have a fairly uniform composition with regard to the distribution of the amino acids within their proteins (Table 6.2). Notable exceptions to this, however, are the seed proteins which are characteristically enriched in glutamate, aspartate, and in some instances arginine.

PROTEIN STRUCTURE

The amino acids in the proteins are arranged in a specific sequence linked together by peptide bonds which may be considered as being

Table 6.2 Amino acid composition of plant proteins.

Amino acid	Barley (*Hordeum vulgare*) leaf [93]	Lupin (*Lupinus albus*) leaf[93]	Chinese cabbage (*Brassica chinensis*) leaf[93]	Apple peel (*Malus sylvestris*)[247]	Pea seed (*Pisum sativum*) vicilin[229]
	(g/100 g recovered amino acid)				
Aspartic acid	9.57	10.22	10.01	11.3	13.3
Threonine	5.07	5.01	5.22	6.5	3.3
Serine	4.40	4.68	4.50	9.1	7.4
Glutamic acid	11.41	11.88	11.91	6.2	17.8
Proline	4.68	4.79	4.72	6.6	5.2
Glycine	5.64	5.69	5.35	9.6	5.9
Alanine	6.71	6.21	6.10	11.2	5.2
Valine	6.37	6.27	6.06	7.2	5.2
Methionine	2.24	1.70	1.94	Not determined	0.2
Isoleucine	4.95	4.93	4.62	5.2	4.9
Leucine	9.33	9.75	9.29	8.3	9.1
Tyrosine	4.50	4.61	4.71	2.4	1.8
Phenylalanine	6.22	6.24	6.24	3.2	4.8
Lysine	6.61	6.60	7.08	4.6	8.2
Histidine	2.34	2.31	2.42	1.7	1.9
Arginine	6.89	6.35	6.46	2.0	5.9

formed by the condensation of the amino group and carboxyl residue of adjacent amino acids.

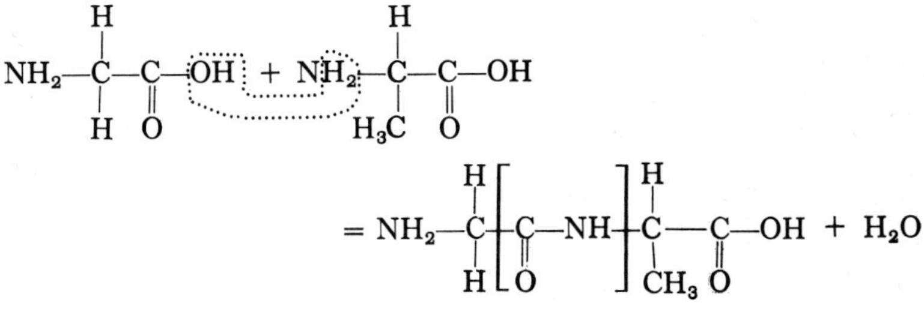

$$NH_2-CH_2-C(=O)-OH + NH_2-CH(H_3C)-C(=O)-OH$$

$$= NH_2-CH_2-\left[C(=O)-NH\right]-CH(CH_3)-C(=O)-OH + H_2O$$

Peptide bond

The linear sequence of the amino acids in the proteins confers upon the protein many of its characteristic properties. Proteins may range in molecular weight from about 6000 of ferredoxin to 2 000 000 for glutamic dehydrogenase. The average amino acid has a molecular weight of 120 so that simple proteins contain about 50 amino acid residues linked together. The sequential arrangement of the amino acids is called the **primary structure**. The primary structure has been determined for a few plant proteins and an example is given below (Table 6.3).

The properties of the peptide bond are such that the carboxyl oxygen of one peptide bond can share its electrons with the hydrogen atom

Table 6.3 Amino acid sequence of papain, a plant protease. (From Dayhoff.[142])

					5					10					15
1	Ile	Pro	Glu	Tyr	Val	Asp	Trp	Arg	Gln	Lys	Gly	Ala	Val	Thr	Pro
16	Val	Lys	Asn	Gln	Gly	Ser	Cys	Gly	Ser	Cys	Trp	Ala	Phe	Ser	Ala
31	Val	Val	Thr	Ile	Glu	Gly	Ile	Ile	Lys	Ile	Arg	Thr	Gly	Asn	Leu
46	Asn	Gln	Tyr	Ser	Glu	Gln	Glu	Leu	Leu	Asp	Cys	Asp	Arg	Arg	Ser
61	Tyr	Gly	Cys	Asn	Gly	Gly	Tyr	Pro	Trp	Ser	Ala	Leu	Gln	Leu	Val
76	Ala	Gln	Tyr	Gly	Ile	His	Tyr	Arg	Asn	Thr	Pro	Tyr	Tyr	Glu	Gly
91	Val	Gln	Arg	Tyr	Cys	Arg	Ser	Arg	Glu	Lys	Gly	Pro	Tyr	Ala	Ala
106	Lys	Thr	Asp	Gly	Val	Arg	Gln	Val	Gln	Pro	Tyr	Asn	Gln	Gly	Ala
121	Leu	Leu	Tyr	Ser	Ile	Ala	Asn	Gln	Pro	Val	Ser	Val	Val	Leu	Gln
136	Ala	Ala	Gly	Lys	Asp	Phe	Gln	Leu	Tyr	Arg	Gly	Gly	Ile	Phe	Val
151	Gly	Pro	Cys	Gly	Asn	Lys	Val	Asp	His	Ala	Val	Ala	Ala	Val	Gly
166	Tyr	Asn	Pro	Gly	Tyr	Ile	Leu	Ile	Lys	Asn	Ser	Trp	Gly	Thr	Gly
181	Trp	Gly	Glu	Asn	Gly	Tyr	Ile	Arg	Ile	Lys	Arg	Gly	Thr	Gly	Asn
196	Ser	Tyr	Gly	Val	Cys	Gly	Leu	Tyr	Thr	Ser	Ser	Phe	Tyr	Pro	Val
211	Lys	Asn													

Composition

14 Ala	A		13 Gln	Q		11 Leu	L		13 Ser	S
12 Arg	R		7 Glu	E		10 Lys	K		8 Thr	T
13 Asn	N		28 Gly	G		0 Met	M		5 Trp	W
6 Asp	D		2 His	H		4 Phe	F		19 Tyr	Y
7 Cys	C		12 Ile	I		10 Pro	P		18 Val	V

Mol. wt. = 23 426 Total no. of residues = 212

associated with the nitrogen of another peptide bond. This bond sharing can occur within the linear peptide chain and thus the chain becomes coiled into an α-helix in which 3.7 amino acid residues are present in each complete turn of the coil. This helical arrangement which occurs at intervals throughout the peptide chain is described as the **secondary structure** of the protein.

In addition to the bond sharing to produce the α-helix other interactions can occur between the component amino acids in the protein polypeptide chain. The principal interactions of this type occur between cysteine residues, in the same chain, which combine to form a disulphide –S–S– bridge thereby creating a loop in that chain. In papain, for example, the disulphide bridges are formed between cysteine residues 22 and 63, 56 and 95, 153 and 200; the seventh cysteine residue at position 25 in the peptide chain functions in the active catalytic site.[162] Additional interactions could potentially occur between the carboxyl groups of aspartate and glutamate and the amino residues of such basic amino acids as arginine and lysine. Also, the hydroxyl group of serine and threonine provide potentially active sites for the interaction with other free amino groups. The net effect of the interactions within a single peptide chain is to cause the folding and convolutions characterized as the **tertiary structure** of the protein molecule. Evidence for the three dimensional structure can be obtained by x-ray crystallography and it is obvious from Fig. 6.1 that even relatively small proteins have an extremely complex structure.

Some proteins are composed of several polypeptide chains all of which

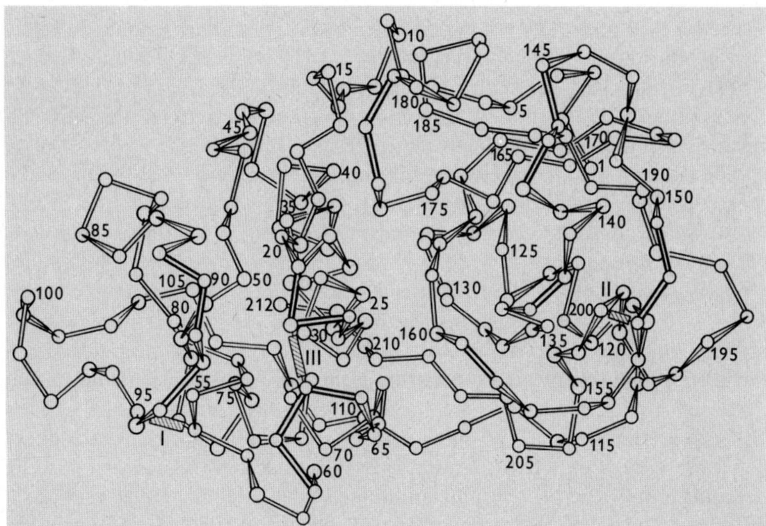

Fig. 6.1 Perspective drawing of the conformation of papain. The circles represent the α 'C' atoms of the amino acid residues. Cysteine residue at 25 and histidine residue at 158 are considered to be involved in the active catalytic site. I, II, and III in the Figure refer to locations of disulphide bridges. (From Drenth, Jansonius, Koekoek, Swen and Wolthers,[162] Figure 4, p. 932.)

are important to the expression of the properties of a particular protein. The association of the subunits represents the **quaternary structure**.

PROPERTIES OF PROTEINS

The fact that proteins are synthesized from amino acids confers upon them certain properties. Since they contain amino groups as well as carboxyl groups, proteins behave either as acids or bases and are said to be **amphoteric**. In addition to the carboxyl and amino groups, sulphydryl groups, hydroxyl groups, and imidazole groups will contribute to the ionic charge of the protein molecule. The charge on the protein at any one time will be dependent upon pH of the surrounding solution and at a certain pH the number of cations will equal the number of anions. This pH is termed the **iso-electric point.** Proteins show their minimum solubility at this point and are frequently precipitated at this pH. However, the precipitation is reversible and the protein can be solubilized by appropriate change of pH.

The amphoteric nature of proteins and their electrical charge in solution allows for their migration in an electrical field. The direction of migration of the protein molecule will depend on its relative charge which is a function of the amino acid composition and pH. Different proteins have a different charge at a particular pH and thus migrate at different rates in an electrical field. This differential rate of migration provides the basis for the separation of proteins by electrophoresis.

The large molecular weight and amphoteric nature of proteins make them quite labile and difficult to handle. During the isolation of proteins in an undegraded state certain procedures have to be followed. The plant vacuole is a repository for organic acids, hydrolytic enzymes and many secondary metabolites. The plant cell is surrounded by a tough cell wall which is disrupted only by considerable mechanical force. The plant cells contain phenolic compounds which can be converted to quinones; both the phenolics and quinones denature proteins. Part of the denaturation may involve the interaction between the quinones and the –SH groups of the protein molecule. In practice, isolation of proteins attempts to overcome these hazards. Plant material is homogenized in the cold, to prevent denaturation of the protein and minimize hydrolytic enzyme activity, in a buffer which prevents excessive fluxes of pH in the vicinity of the protein. The buffer usually contains a sulphydryl reagent such as cysteine, β-mercaptoethanol or dithiothreitol which prevents oxidation or cross linking of free sulphydryl groups in the protein. Additionally it is possible to add a compound such as polyvinyl pyrrolidine[388] to sequester phenolic compounds. The homogenate is centrifuged to remove cellular debris and used as a source of protein, usually an active enzyme.

Proteins may be associated with specific cellular organelles and if it is desired to locate a protein with an organelle a gentle homogenization is usually performed in an isotonic sucrose solution to prevent organelle disruption. The organelles are recovered by differential centrifugation.

The proteins in plant homogenates can be fractionated by various methods; usually fractionation of the protein is preceded by a concentration process. This concentration is frequently achieved by precipitation with ammonium sulphate. Proteins do not occur in true solution but exist as emulsoid or colloidal sols. Such dispersions are hydrophilic; however when salts are added there is a competition for water to form a solution of the added salt and as water availability is decreased the proteins are precipitated or **salted out**. Various proteins are precipitated at different salt concentrations so that salting out provides a convenient method for protein separation.

Proteins can be further purified by column chromatography on various media such as Diethylaminoethyl (DEAE) cellulose or carboxy methyl (CM) cellulose which separate by ionic exchange or by exclusion chromatography on such media as Sephadex, in which case separation is based on molecular size.

CLASSIFICATION OF PLANT PROTEINS

Historically proteins were classified on the basis of their solubility in various solvents. This fractionation technique originally designed by Osborne[489] is still used today to characterize proteins. The groupings are as follows (Table 6.4).

Table 6.4 Protein classification on basis of solubility.

Protein	Solubility
Albumins	Soluble in water and dilute salt solution.
Globulins	Insoluble in water but soluble in dilute salt solutions.
Glutelins	Insoluble in water or dilute salt solutions but soluble in dilute acid and bases.
Prolamins	Insoluble in water, dilute salt or dilute acid but soluble in 70–80 % ethanol.

The classification based on solubility is a gross separation and each fraction may contain a heterogeneous mixture of polypeptides which can be subjected to further fractionation. For example, electrophoresis of wheat (*Triticum aestivum*) proteins soluble in acetic acid demonstrates the presence of at least nine fractions[225] and legume globulins[301] can be separated into several components by the same technique. The albumins contain the majority of the cellular enzymes and are thus extremely heterogeneous.

An alternate classification is based on function; thus proteins are described as **enzymic**, **structural** or **reserve**. Structural proteins may be located in the cell wall for example. The protein extensin located in the cell wall is considered to control extensibility but has not been assigned enzymic function.[352] Likewise, little enzymic function has been ascribed to histones or the many proteins associated with the ribosomes, so they may be termed structural. Seeds contain some proteins which have an overall different amino acid composition from that of other plant proteins. No enzymic activity has been associated with these proteins and they are considered to be reserves.

Proteins may be characterized as **soluble** or **particulate.** Proteins which are associated with an organelle are usually assigned a location in that organelle and, in fact, certain proteins are used as characteristic markers for organelle fractions. Cytochrome oxidase is located in the mitochondria and serves as a convenient marker for this organelle, isocitrate lyase is a convenient marker for glyoxysomes, catalase is located in the peroxisome, etc. Such assignment requires rigorous testing to ensure that the association of a protein with an organelle is not merely an artifact of isolation during which a soluble enzyme becomes associated with an organelle. Conversely enzymes located in an organelle may be released and become soluble during the isolation.

Conjugate proteins

Proteins may occur as free proteins or they may exist in chemical or physical association with other compounds. Such complexes are termed conjugate proteins.

NUCLEOPROTEINS Conjugates of protein and nucleic acid are nucleoproteins. These are encountered in chromatin where the DNA is associated with proteins, primarily but not exclusively histones, and in ribosomes where the ribosomal RNA is associated with ribosomal proteins. The proteins can be dissociated from the nucleic acid component by concentrated salt solutions,[275] weak acids or urea[309] suggesting that the association of the components is non-covalent.

The histones have been extensively studied; they are generally basic owing to the presence of the basic amino acids lysine and arginine. Fractionation of histones by gel electrophoresis or ion exchange chromatography demonstrates[181] the occurrence of five components differing in

Table 6.5 Amino acid compositions of pea (*Pisum sativum*) bud histone components. (From Fambrough, Fujimura and Bonner,[181] Table II, p. 579. Reprinted with permission from *Biochemistry,* (1968), **7**. Copyright by the American Chemical Society.)

	Histones*				
Amino acid	I	IIa	IIb	III†	IV
Lys	25.5	16.1	10.6	8.6 (1.6)‡	8.5
His	1.1	1.1	1.6	2.1	2.4
Arg	2.8	6.5	9.0	13.1	15.6
Asp	2.3	6.0	6.1	4.5	5.6
Thr	4.0	4.8	4.1	6.6	7.3
Ser	4.9	6.7	5.6	4.1	2.2
Glu	7.3	8.0	6.6	10.8	6.2
Pro	9.9	6.7	7.1	4.5	1.4
Gly	2.3	8.8	11.4	6.6	17.2
Ala	22.8	12.3	12.8	12.9	7.5
Val	5.3	6.7	7.9	5.2	6.6
Met	—	0.5	—	Trace	Trace
I-Leu	1.9	4.5	3.1	5.1	6.3
Leu	4.1	7.9	10.6	9.4	7.6
Try	0.4	1.7	1.9	1.0	3.0
Phe	0.4	1.9	1.6	3.9	2.7

*Each histone has been shown by analytical disc electrophoresis to be free of contamination by other fractions. Compositions are expressed as moles/100 moles of amino acids. No corrections have been made for hydrolytic losses.

†Also contains 1 mole of cysteine/23 000 g of protein.

‡Value in parenthesis is ϵ-methyllysine in addition to lysine.

amino acid composition (Table 6.5). Similar fractions are encountered in tissues of animal origin. There is some variation in the distribution of the various histones in different plant tissues[631] but there is clearly insufficient diversity in the histones for specific histones to regulate specific genes as was originally speculated.

LIPOPROTEINS Conjugates of protein and lipid occur as lipoproteins which are ubiquitous in cellular membranes and are found in such organelles as mitochondria, chloroplasts and membranous structures such as the endoplasmic reticulum and plasmalemma. The lipid components in chloroplast lipoprotein complexes are enriched in sulpho and galactolipids whereas mitochondria contain a greater proportion of phospholipids. At this stage it is not clear if this lipid diversity between the organelles is associated with differences in the protein component. The lipid can be removed from the protein by detergent indicating that the lipid–protein association is physical rather than chemical.

GLYCOPROTEINS Conjugates of proteins with carbohydrate are termed glycoproteins. Several glycoproteins have been described in plants; they occur in cell wall proteins,[352] in some enzymes[559] and in the reserve proteins of legume seeds.[385, 511] In animal and bacterial systems the carbohydrate components are heteropolysaccharides.[582] The limited analyses conducted on the glycoproteins from plants demonstrate the presence of some of the following sugars, *N*-acetyl glucosamine, glucosamine, glucose, mannose, arabinose and xylose, suggesting the occurrence of heteropolysaccharides. The amino acid composition of some glycoproteins has been established (Table 6.6).

From the table below it can be seen that the cell wall glycoproteins are enriched in hydroxyproline, an amino acid not normally encountered in other cytoplasmic proteins although it is present in some peroxidases which however may be located in the cell wall. It is considered that hydroxyproline arises by the oxidation of proline by the sequence described below. The oxidation involves atmospheric oxygen and occurs after the incorporation of proline into the polypeptide chain.[544]

```
  H     H                                                HO         H
  |     |                                                  \        |
H-C-----C-H                                             H---C-------C-H
H\ |     | /H  + Ascorbate + O2  --α-Ketoglutarate-->       |       | /H
   C     C                            Fe2+              H---C       C
  / \   / \                                                /  \   /  \
 H   N    CO-                                             H    N     CO-
     H                                                         H
```

Peptidyl proline

Peptidyl-hydroxyproline
+ dehydroascorbate
+ H_2O

Table 6.6 Amino acid composition (moles percent of protein) of some glycoproteins.

Amino acid	Soybean (*Glycine max*) haemagglutinin[385]	*Nicotiana tabacum* cell wall[159]	Horseradish (*Armoracia rusticana*) peroxidase[559]	Bean (*Phaseolus aureus*) vicilin[178]
Hypro	—	18.1	—	—
Asp	9.7	4.9	14.4	13.4
Thre	5.0	5.4	12.5	3.0
Ser	7.6	9.2	12.4	7.3
Glu	5.2	3.8	7.4	19.9
Pro	4.7	6.1	3.3	2.9
Gly	4.0	4.8	3.7	5.4
Ala	6.1	4.0	10.7	5.6
Val	4.5	6.5	5.8	6.6
Met	0.6	0.4	0.4	0.3
I-Leu	3.8	2.0	4.1	4.5
Leu	6.3	3.3	9.5	9.4
Tyr	1.6	2.9	0.9	1.9
Phe	3.7	1.4	7.4	6.1
Lys	4.0	9.1	1.9	6.0
His	1.4	2.3	1.1	2.0
Arg	1.8	2.1	3.9	5.5
Free sugars	2.5	—	1.62	1.8 %
Amino sugars	3.5	—	0.19	0.2 %

Lamport[352] indicates that the hydroxyl group of the hydroxyproline is the point of covalent attachment of arabinose residues. Apart from this study of the cell wall glycoprotein, few investigations have been conducted to determine the site of attachment of the carbohydrate to protein in other plant glycoproteins. Investigations with bacterial and animal glycoproteins indicate that the association of carbohydrate and protein usually involves a covalent linkage of the carbon 1 of the carbohydrate and a functional group of the amino acid within the polypeptide chain. Amino acids which have been characterized as being bound to the carbohydrate residues are asparagine, serine and threonine.[582]

PHOSPHOPROTEINS Amino acids in the peptide chain may be phosphorylated and produce phosphoproteins. A great deal of interest has centred around these proteins recently since it has been shown that the secondary messenger effect of cyclic 3′,5′-AMP in animal tissues may involve the stimulation of the phosphorylation of protein.[534] An enzyme protein kinase has been characterized which utilizes ATP. Phosphoproteins have been demonstrated in plants; the phosphate is attached covalently to the serine residues.[639] However, cyclic AMP does not influence the activity of the protein kinase from plants[321] and the physiological role of phosphoproteins remains unresolved.

FLAVOPROTEINS AND HAEMOPROTEINS Flavoproteins and haemoproteins are conjugate proteins in which flavin nucleotides or haem residues are attached to proteins. Many of the enzymes involved in oxidation–reduction reactions are conjugates of this type. In the respiratory enzymes the dominant haem group is cytochrome.

PROTEIN SYNTHESIS

Studies of protein synthesis in plants have followed the investigations originally conducted in microbial and animal systems. It is now clear that the **Central Dogma** is followed in which the information in the DNA molecule is transcribed into RNA which is then translated into the sequence of amino acids of the polypeptide chain.

$$\text{DNA} \xrightarrow{\text{Transcription}} \text{RNA} \xrightarrow{\text{Translation}} \text{PROTEIN}$$

Transcription and the genetic code

The dogma implies that the amino acid sequence of the protein will correspond to base sequences in the RNA molecule. The base sequence in the RNA molecule is a transcribed complementary copy of the sequence originally contained in the DNA molecule.

The classical experiments of Nirenberg,[461] using a protein synthesizing system from *Escherichia coli*, demonstrated that certain synthetic polynucleotides could direct the incorporation of specific amino acids. By suitably manipulating the base composition of the polynucleotides it was possible to assign certain base combinations for the incorporation of specific amino acids. This work led to the establishment of the **genetic code** in which sequences of three bases are assigned to particular amino acids (Table 6.7). The genetic code is believed to be universal in that in all organisms a particular triplet of bases directs the incorporation of the same amino acid. Extensive comparative surveys have not been conducted in plants but experiments demonstrating the polyuridylic acid directed incorporation of phenylalanine,[43] the polyadenylic acid stimulation of lysine incorporation[244] and the AUG mediated incorporation of methionine[713] are all consistent with the predicted coding assignments. In this genetic code it is seen that some amino acids are assigned to more than one nucleotide triplet or codon. The reason for this redundancy or degeneracy is not clear; however, as will be apparent later, it may be of significance during the regulation of protein synthesis.

The transcription of the information of the DNA into the informational

Table 6.7 The genetic code.

Alanine	GCU, GCC, GCA, GCG
Arginine	AGA, AGG, CGU, CGC, CGA, CGG
Asparagine	AAU, AAC
Aspartic acid	GAU, GAC
Cysteine	UGU, UGC
Glutamic acid	GAA, GAG
Glutamine	CAA, CAG
Glycine	GGU, GGC, GGA, GGG
Histidine	CAU, CAC
Isoleucine	AUU, AUC, AUA
Leucine	CUU, CUC, CUA, CUG, UUA, UUG
Lysine	AAA, AAG
Methionine	AUG
Phenylalanine	UUU, UUC
Proline	CCA, CCC, CGU, CCG
Serine	AGU, AGC, UCU, UCC, UCA, UCG
Threonine	ACA, ACU, ACC, ACG
Tryptophan	UGG
Tyrosine	UAU, UAC
Valine	GUU, GUC, GUA, GUG
Chain termination	UAA, UAG, UGA

or mRNA will involve the enzyme RNA polymerase and an available DNA template. Protein synthesis occurs in the cytoplasm or in organelles. If the mRNA is translated in the cytoplasm then there is a requirement that the mRNA, or the transcription product synthesized in the nucleus, traverses the nuclear membrane. It is not clear how this process is achieved. In animal systems it has been suggested[134, 231] that polyadenylic acid residues are added to the transcriptional product at the time it traverses the nuclear membrane with the result that mRNA in the cytoplasm is enriched in adenylic acid. The findings that some plant mRNA's or D-RNA's[397, 679, 544a, 327] are enriched in adenine may indicate that a similar post-transcriptional modification takes place in plant mRNA as it crosses the nuclear membrane.

Translation—Ribosomes

Protein biosynthesis occurs on the subcellular organelles, the ribosomes. These are nucleoprotein structures and, depending on their location in the cell, have different size. The cytoplasmic ribosomes are characterized as having a sedimentation value of 80s and contain, as was shown earlier, 1.3×10^6 and 0.7×10^6, 5s and 5.8s RNA and proteins. The proteins are complex mixtures containing up to 85 polypeptide chains as indicated by two-dimensional electrophoresis.[447] Plant ribosomes show more diversity in their ribosomal proteins than do those from animals and bacterial systems. Ribosomes from different species contain different ribosomal proteins.[447] The chloroplast ribosomes have a sedimentation value of 70s and contain 1.1×10^6 and 0.56×10^6 and 5s RNA and proteins. Again there are numerous proteins associated with the RNA which can be separated from the ribosomes by acetic acid and urea treatment. The proteins of the chloroplast ribosomes differ from those of the cytoplasmic ribosomes;[309, 447] see Fig. 6.2. As in the case of proteins from cytoplasmic ribosomes there are also interspecies differences in proteins from chloroplast ribosomes.

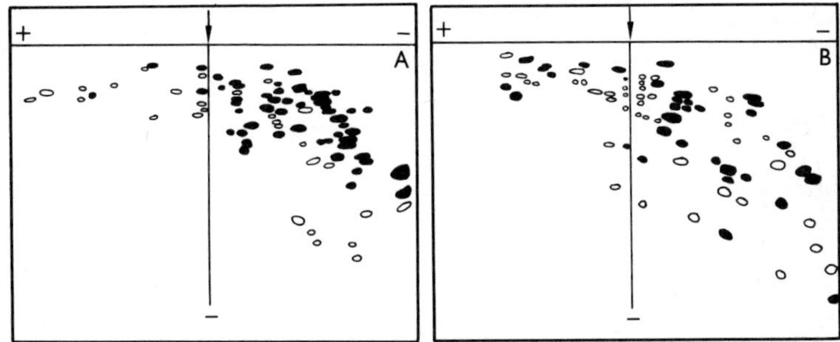

Fig. 6.2 Composite drawing of two dimensional acrylamide gel electrophoresis of wheat (*Triticum aestivum*) leaf cytoplasmic (A) and chloroplast (B) ribosomal proteins. (From Jones *et al.*,[309] Fig. 3, p. 169.)

Ribosomes have been detected in plant mitochondria and according to Leaver and Harmey[355] have a sedimentation value of 80s with RNA of molecular weight 1.2×10^6 and 0.7×10^6 and 5s. To date only preliminary characterizations have been made of the proteins associated with mitochondrial ribosomes.[655]

Ribosomes can be dissociated into subunits under appropriate conditions and in fact ribosome dissociation is an integral part of protein synthesis. Treatment of 80s ribosomes with concentrated ammonium or potassium chloride or with ethylene diamine tetra acetic acid, which chelates magnesium, dissociates the ribosomes into 60s and 40s subunits, whereas dissociation of chloroplast 70s ribosomes produces 50s and 30s subunits. Specific proteins are associated with each subunit. Various attempts have been made to ascribe a function to the ribosomal proteins. Some of the proteins removed from ribosomes by salt washing are required for the initiation of protein synthesis. However, no enzymatic functions have been delineated for the majority of the ribosomal proteins and they are currently considered as structural proteins.

In addition to their association with organelles, ribosomes may occur in the cytoplasm in the free state or bound to membranes. Although there is no discernible chemical or structural difference between free and membrane-bound ribosomes it is considered, on the basis of studies with animal systems, that the membrane-bound ribosomes synthesize proteins which are exported from the cell. Although the majority of plant cells are not immediately recognized as secretory or capable of export it should be realized that the cell wall is exocytoplasmic and the vacoule is also separated from the cytoplasm; thus proteins accumulating in the vacuole or in the cell wall may be considered as exported or secreted and such proteins may be synthesized in membrane-bound ribosomes.

Translation—Amino-acylation of transfer RNA

The first stage in the biosynthesis of proteins is the formation of amino-acyl-tRNA's. This is done in a two stage reaction catalysed by the amino-acyl-tRNA synthetases. The reaction may be represented as:

$$\text{Amino acid} + \text{ATP} \xrightarrow{\text{Enzyme}} \text{Enzyme–Amino-acyl-AMP} + \text{Pyrophosphate} \quad (1)$$

$$\text{Enzyme–Amino-acyl-AMP} + \text{tRNA} \longrightarrow \text{Enzyme} + \text{Amino-acyl-tRNA} + \text{AMP} \quad (2)$$

During the reaction the amino acid is attached to the terminal adenosine in the tRNA molecule. Most recent evidence indicates that the amino acid is initially attached to the 2′ carbon of the ribose moiety of the adenosine and is subsequently transferred to the 3′ carbon.[584]

The amino-acylation shows great specificity and there are specific synthetases for each amino acid.[354] The assay for amino-acyl synthetase activity can be followed by following pyrophosphate exchanges as shown in reaction (1) but caution must be used to eliminate ATP hydrolysis by ATP-ase and to ensure that only the specific acid for which the cognate synthetase is being studied is supplied. Such assays cannot discriminate between different synthetases which may be capable of acylating the iso-accepting tRNA.

Since, as indicated earlier (p. 157), there are iso-accepting species of tRNA the question is raised whether or not the iso-accepting tRNA's for a particular amino acid are acylated by the same or different synthetases. It is not clear to what extent formation of the iso-accepting tRNA's is due to differences in the anti codon region, as would be predicted on the basis of the degeneracy of the genetic code, or due to such differences as side chain substitution on the bases or due to different cellular origins. Some of the multiplicity of iso-accepting tRNA species can be related to the association of specific iso-accepting tRNA's with specific organelles.

The specific organelles have been shown to possess specific synthetases. These different synthetases show preferential acylation or charging of certain iso-accepting tRNA species. The feature has been particularly well demonstrated by Guderian *et al.*[234] Tobacco (*Nicotiana tabacum*) leaves contain $tRNA^{leu}$ which, following charging, can be fractionated by reverse phase chromatography. Six leucyl-tRNA peaks can be identified, as indicated in Fig. 6.3.

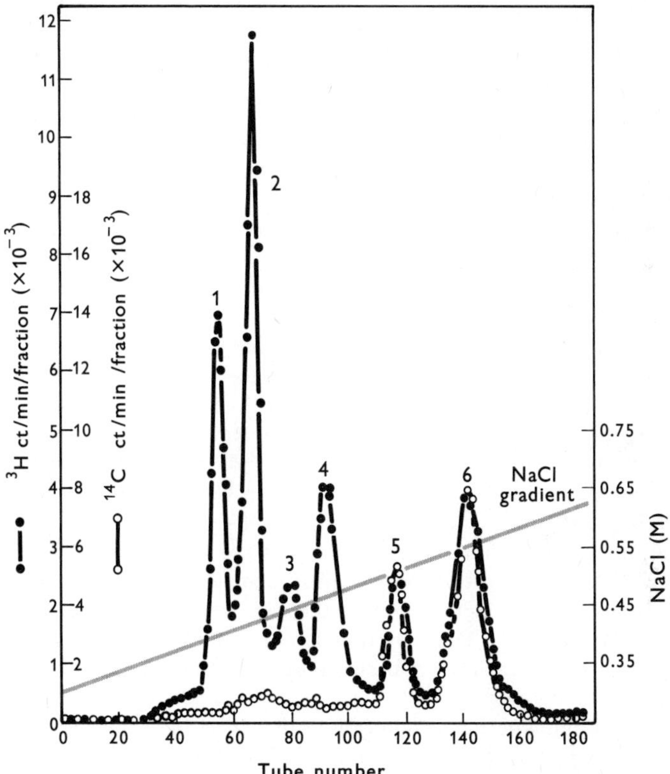

Fig. 6.3 Reverse-phase chromatography of total leucyl-tRNA from tobacco leaves amino-acylated by total and chloroplast amino-acyl-tRNA synthetases. Co-chromatography of ^{3}H-leucyl-tRNA (●——●) (2.65 $A_{260\ nm}$ units of total tRNA amino-acylated by total amino-acyl-tRNA synthetases) and ^{14}C-leucyl-tRNA (○——○) (5.12 $A_{260\ nm}$ units of total tRNA amino-acylated by chloroplast amino-acyl-tRNA synthetases) on a reverse-phase Freon column.

Other experiments indicate that iso-accepting species 5 and 6 are located in the chloroplast. 3 and 4 are located in the mitochondria and are amino-acylated by mitochondrial synthetase which can also acylate 1 and 2 which are cytoplasmic. (From Guderian, Pulliam, and Gordon.[234])

Two tyrosyl synthetases have been characterized from pea roots which selectively charge the iso-accepting $tRNA^{tyr}$ species.[129] However, as yet the different synthetases have not been assigned to specific organelles.

Translation—Initiation

Studies with bacterial systems originally indicated that polypeptide chain formation depended initially on the incorporation of *N*-formyl methionyl-tRNA (f-met-tRNA). This same initiator tRNA has been detected in chloroplasts from wheat,[369] cotton[418] and beans[86] and thus protein chain initiation in chloroplast ribosomes is probably similar to that occurring in 70s bacterial ribosomes.

In the cytoplasm, protein chain initiation on the 80s ribosomes involves a specific non-formylatable initiator methionyl-tRNA which has been isolated from wheat (*Triticum aestivum*),[369] peas (*Pisum sativum*),[679] and beans (*Vicia faba*).[713]

The mechanism of protein chain initiation has been extensively studied in bacterial and animal systems and more recently the process has been studied in plant systems. The process is complex and the steps involved are not fully resolved.

Studies with bacterial systems indicated that the initiation step involved the binding of f-met-tRNA to the 70s ribosome. The binding was dependent upon the presence of certain proteinaceous factors which could be removed from the ribosome by treatment with 1.0 molar solutions of ammonium chloride or potassium chloride.[392] The binding of f-met-tRNA catalysed by these proteins was dependent on GTP and AUG, the triplet codon for methionine. Natural messenger RNA could substitute for AUG in the binding reaction. It was shown that the salt washing of the ribosomes released at least three proteinaceous factors all of which were essential for the binding reaction and these are designated initiation factors IF-1, IF-2, and IF-3. The currently accepted mode of operation of these factors in protein synthesis in bacterial systems and probably chloroplasts is depicted in Fig. 6.4.

The 30s subunit of the ribosome binds to mRNA in the presence of IF-3. The binding is near the 5′ of the mRNA. Binding of synthetic messenger RNA such as polyuridylic acid does not require IF-3. Subsequently, the initiating f-met-tRNA binds to the 30s-subunit-mRNA complex in a step catalysed by IF-1 and IF-2 and requiring GTP. The initiating f-met-tRNA is considered to bind at the P site of the ribosomal subunit.

The 30s subunit initiator complex binds to the 50s subunit with the hydrolysis of GTP to GDP and a release of initiation factors IF-1, IF-2, and IF-3.[370, 392]

Similar proteinaceous factors have been isolated from animal cytoplasmic 80s ribosomes and the initiation process has been extensively studied

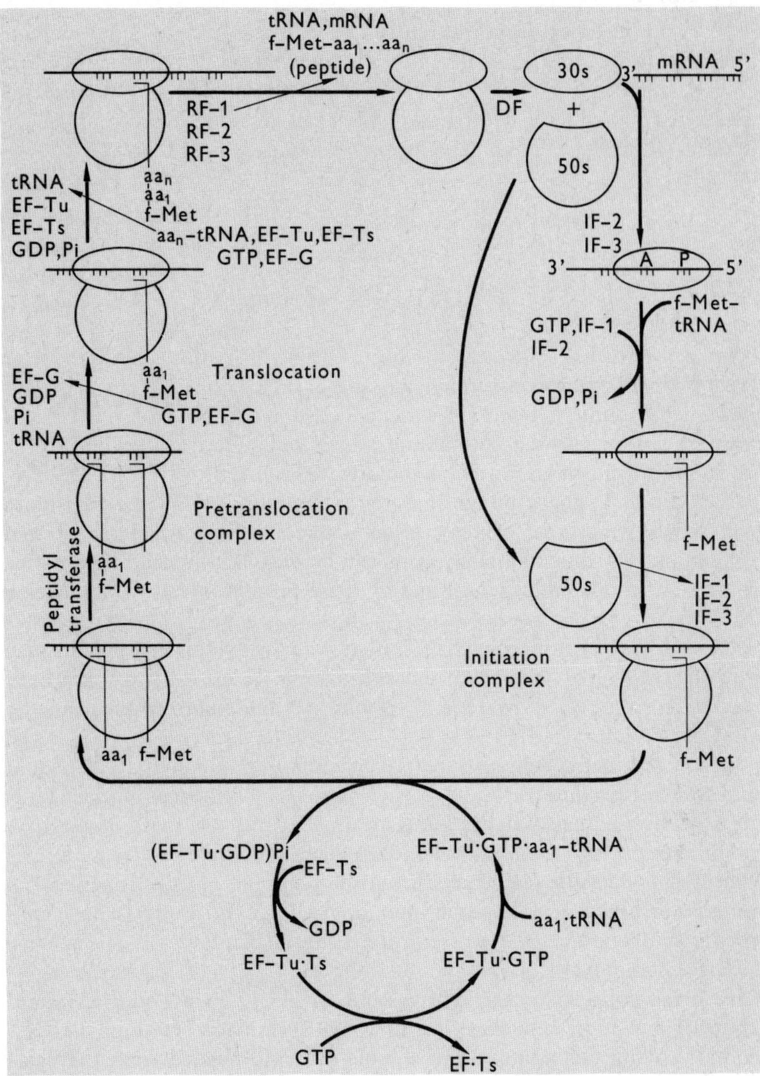

Fig. 6.4 Proposed scheme for protein synthesis by bacterial ribosomes.

mRNA, messenger RNA ; IF-1, IF-2, IF-3, initiation factors ; f-Met-tRNA, formyl methionyl-transfer RNA ; GTP, guanosine triphosphate ; A, acceptor site on 30s ribosomal subunit ; P, peptidyl site on 30s ribosomal subunit ; EF-Tu, EF-Ts, EF-G, elongation factors ; RF-1, RF-2, RF-3, release factors ; DF, ribosome dissociation factors.

The operation of these components in protein synthesis is described on pp. 190–195.

by using ribosomes from rabbit reticulocytes. To date four proteins functional in initiation complex formation have been demonstrated.[510, 555] The binding of initiator methionyl-tRNA to ribosomes using the proteins extracted from reticulocyte ribosomes by salt washing has recently been investigated in detail[372] and the proposed sequence is as follows

$$\text{Ribosome factor C} + \text{met-tRNA}_f + \text{GTP} \longrightarrow [\text{C-met-tRNA}_f\text{-GTP}]$$

$$[\text{C-met-tRNA}_f\text{-GTP}] + 40\text{s} + \text{Ribosomal factors} + \text{GTP} \longrightarrow 48\text{s complex}$$

$$48\text{s complex} + 60\text{s} + \text{factors} + \text{GTP} \longrightarrow 80\text{s complex}$$

Noteworthy in this scheme is the observation that a three component or ternary complex can be formed in the absence of mRNA or AUG or ribosomal subunits. There is a dependence on GTP and the ternary complex can bind to the 40s subunit which will subsequently bind to the 60s subunit to form an initiation complex deficient in mRNA.

In addition to the binding of initiator methionyl-tRNA to ribosomes catalysed by ribosomal proteins other workers[207, 719] have demonstrated that binding in some animal systems can be catalysed by soluble proteins from the cytosol. Binding catalysed by these proteins is not dependent on GTP.

The evidence that protein chain initiation in the cytoplasm of plants may proceed in a similar manner to those initiations described above has been provided primarily by Marcus,[400] Boulter,[713] and more recently in the author's laboratory.[679, 680]

It was originally demonstrated that initiator methionyl-tRNA would bind to 80s ribosomes in the presence of AUG or tobacco mosaic virus RNA.[553] Subsequently it has been shown that initiator methionyl-tRNA will bind to polysomes in which endogenous mRNA provides the codon.[679] A messenger-like RNA extracted from the polysomes is also functional in stimulating binding of initiator methionyl-tRNA. So there is now convincing evidence that *in vivo* chain initiation in plants involves a specific initiator methionyl-tRNA.

By using tobacco mosaic virus RNA or AUG as a codon Marcus' group[553] has indicated that the binding of initiator methionyl-tRNA requires the participation of two soluble proteins isolated from the post-ribosomal supernatant from wheat (*Triticum aestivum*) embryos. The binding was dependent on ATP and GTP. During the formation of the initiation complex the initiator methionyl-tRNA was bound originally to the 40s subunit.[676]

Wells and Beevers[681] demonstrated that proteins which are functional in catalysing the binding of initiator methionyl-tRNA can be removed from the 80s ribosomes prepared from dry pea (*Pisum sativum*) seeds. Although these proteins have not been fully characterized their properties are similar to those described above from mammalian systems and the proposed sequence of events is as follows.

Ribosomal factor + initiator Methionyl-tRNA + GTP ⟶
Ternary complex (Ribosomal factor-initiator Met-tRNA-GTP).
(Ribosomal factor-initiator Met-tRNA-GTP) + 40s subunit ⟶
(40s-Ribosomal factor-initiator Met-tRNA-GTP).
(40s-Ribosomal factor-initiator Met-tRNA-GTP) + 60s subunit ⟶
(80s-initiator Met-tRNA) + GDP + Ribosomal factor.

The time of release of the ribosomal proteins and their composition have not been definitely established but the obvious point to be made from this research and the investigation from mammalian systems is that the initial binding of initiator methionyl-tRNA does not require mRNA. The binding of mRNA may be a secondary independent process. Formation of the 80s complex involves hydrolysis of GTP.

A further feature concerning the initiation of protein synthesis is that it involves ribosomal subunits; thus some mechanism is required for ribosome dissociation. It has been suggested that the dissociation function was performed by the same initiation factor IF-3 which catalysed the binding of mRNA. However, the most recent studies with mammalian systems indicate that dissociation and mRNA binding are catalysed by different proteins.[419] Dissociation factor activity has recently been demonstrated in extracts from the ribosomes of beans[693] (*Phaseolus vulgaris*); however it has not been established whether or not this factor also catalyses mRNA binding.

Chain elongation

After the formation of the ribosomal initiation complex and association of the mRNA, amino acids are incorporated into a growing peptide chain in the sequence coded for by the base sequence in the mRNA. The process is extremely complex as indicated in Fig. 6.4 and in bacterial systems involves three proteins usually found in the cytoplasm. In mammalian and plant systems elongation requires two elongation factors EF-1 and EF-2 or Amino-acyl transferase I and II.[368, 381]

During chain elongation the amino-acyl-tRNA is bound to a ribosomal site A adjacent to the P site occupied originally by initiator methionyl-tRNA (Fig. 6.4). Peptide bond formation then occurs between the newly bound amino-acyl-tRNA at the A site and the initiator methionyl-tRNA at the P site. The P site may also be called the donor site since it donates its amino group of the amino-acyl-tRNA to the amino-acyl-tRNA at the acceptor site. The next step in elongation involves the translocation of newly formed peptidyl-tRNA from the A site back to the P site. Simultaneously the ribosome advances along the mRNA by one codon in the $5' \rightarrow 3'$ direction. The acceptor site A is now available for the next amino-acyl-tRNA and subsequent elongation step.

In bacterial systems the elongation factors are believed to function as follows. Elongation factor EF-Tu binds with GTP and amino-acyl-tRNA in a stepwise manner to form a complex EF-Tu-GTP-aa_1-tRNA which reacts with ribosomes transferring the amino-acyl-tRNA to the A site. An EF-Tu-GDP complex and inorganic phosphate are released. The second elongation factor EF-Ts then displaces the GDP with the formation of a complex EF-Tu-Ts which is dissociated by GTP to regenerate EF-Tu-GTP. The peptide bond formed between the terminal carboxyl of the peptidyl-tRNA at the P site and the α-amino group of the amino-acyl-tRNA at the A site is formed by the enzyme peptidyl transferase located in the 50s ribosomal subunit. The final bacterial elongation factor EF-G is required along with GTP to translocate the newly formed peptidyl-tRNA to the P site and move the ribosome along the mRNA.[370, 392]

Studies with elongation factors isolated from wheat embryos indicate that EF-1 stimulates the GTP-dependent binding of amino-acyl-tRNA to the ribosome whereas EF-2 functions in the translocation step.[223, 647] Peptide bond formation is catalysed by peptidyl transferase located in the ribosome; this enzyme is somewhat remarkable in that its activity appears to be stimulated by ethanol.[208]

During the elongation process many ribosomes may be simultaneously attached to a specific mRNA. This feature results in the formation of polysomes.

The elongation process is susceptible to inhibition by various compounds. Puromycin is important in this respect. Structurally it resembles the terminal portion of an amino-acylated tRNA (Fig. 6.5). Puromycin

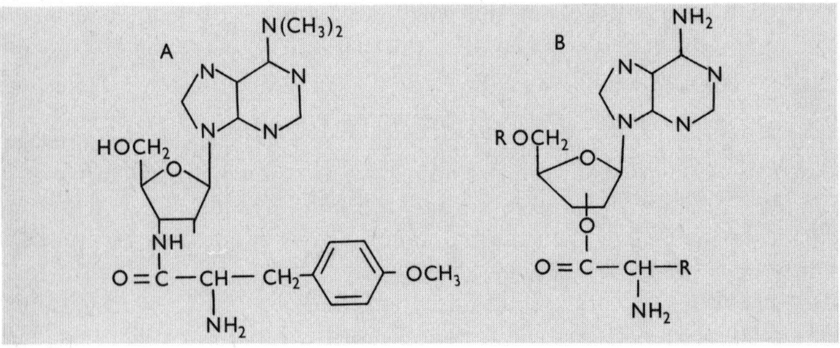

Fig. 6.5 Structure of puromycin (A) and of adenosine end of amino-acyl-tRNA (B).

terminates peptide chain elongation by reacting with peptidyl-tRNA at the P site with the resultant formation of puromycin peptides which are released. Since the P site is thus vacated further protein synthesis is arrested.

Chloramphenicol binds to the 50s subunit of 70s ribosomes and inhibits peptide chain elongation in an as yet unspecified manner. Following the demonstration that the inhibitor bound to bacterial ribosomes, it was subsequently shown[16] that chloramphenicol also bound to chloroplast but not 80s cytoplasmic ribosomes.

In contrast to chloramphenicol, cycloheximide specifically binds to 80s ribosomes and again interferes with the process of chain elongation.

These two inhibitors have been used to determine whether a particular protein is synthesized in chloroplast or cytoplasmic ribosomes.

Chain termination

The mechanism of chain termination in plants has not been investigated. Studies with bacterial systems indicate that the triplets UAG, UAA and UGA serve as termination codons. Release factors have been characterized from bacterial systems and these appear to show specificity to a particular termination codon. It is believed that during termination the peptidyl-tRNA, with a now complete peptide chain, is translocated to the P site and the ribosome advances along the mRNA and the termination codon becomes aligned with the A site. In response to one of the termination codons the appropriate termination or release factor binds to the ribosomal complex and the peptidyl-tRNA is hydrolysed releasing the polypeptide.

Although protein chain initiation involves the amino acid methionine very few proteins contain an N-terminal methionine. Thus following completion of the peptide chain, the methionine must be released from the newly synthesized polypeptide to expose the usual N-terminal amino acid. This presumably involves a specific amino peptidase; however to date no studies have been made of such enzymes in plants.

The ribosome cycle

It is not clear at this stage in what form the ribosomes are released from the mRNA at the termination of polypeptide synthesis. The process of initiation involves dissociated ribosomes, with the small subunit being required for the initial events. The observation that monosomes rather than subunits accumulate during periods of reduced protein synthesis suggests that at the termination of peptide synthesis the ribosomes are released as 80s or 70s monosomes and dissociation immediately precedes initiation. Reassociation of the subunit takes place during the formation of the initiation complex.

EVIDENCE THAT PROTEIN SYNTHESIS OCCURS ACCORDING TO THE DESCRIBED SEQUENCE

In addition to the demonstration of many of the partial reactions involved in protein synthesis, described above, evidence that protein synthesis probably occurs according to the described sequence can be provided by various *in vitro* and *in vivo* studies. The *in vivo* demonstrations utilize various inhibitors. As shown in Chapter 5, Actinomycin D and 6-methyl purine are potent inhibitors of RNA synthesis. Since transcription is an essential requirement for protein synthesis it is to be expected that an inhibition of protein synthesis would follow a curtailment of RNA synthesis. Various experiments indicate that this prediction is satisfied (Table 6.8). In contrast 5-fluorouracil, which inhibits the synthesis of the majority of RNA but does not impede synthesis of mRNA, does not restrict protein synthesis, suggesting that sustained protein synthesis requires the production of mRNA. Translation can be inhibited by puromycin and cycloheximide in cytoplasmic systems and by puromycin and chloramphenicol in the chloroplast. Application of puromycin and cycloheximide results in a decline in the capacity of tissue to incorporate amino acids into protein. A similar inhibition has also been shown by using chloramphenicol. However in these reports of the apparent inhibition of protein synthesis by chloramphenicol high concentrations of the inhibitor were used. It appears that in these instances inhibition of protein synthesis was probably due to a reduction of oxidative phosphorylation by the high levels of chloramphenicol.[603] Ellis[175] has discussed the precautions necessary in interpreting protein synthesis studies in which chloramphenicol is used.

Table 6.8 The influence of inhibitors of RNA and protein synthesis on the amino acid incorporating capacity of soybean (*Glycine max*) hypocotyl. (From Key.[323])

Addition	Incorporation (c.p.m. into proteins/g fresh wt.)
None	14 040
10 μg/ml Actinomycin D	9 480
2.5×10^{-3} M azaguanine	7 770
5×10^{-4} M puromycin	6 840

In vitro studies have demonstrated the capacity of cell-free extracts from plants to support the incorporation of labelled amino acids into material precipitable by hot trichloracetic acid.[66] Demonstration of amino acid incorporation in such cell-free extracts requires the presence of ribosomes, a post-ribosomal supernatant fraction, ATP and GTP, a

Table 6.9 Requirements for cell-free amino acid incorporation by ribosomal preparations from developing pea (*Pisum sativum*) cotyledons. (From Beevers and Poulson,[43] Table 1, p. 478.)

Treatment	Amino acid incorporation c.p.m./mg rRNA	% of control
Complete system	21 903	100
−KCl	10 420	47
$-MgCl_2$	7 345	33
−β-mercaptoethanol	7 155	32
−ATP	8 085	36
−GTP	4 890	22
−ATP, GTP, P-enolpyruvate, pyruvate-kinase	880	4
−supernatant	815	3
−ribosomes	795	2
+0.5 mg RNase	890	4
+0.5 μg cycloheximide	12 485	57
+5.0 μg cycloheximide	1 800	8
+5.0 μg puromycin	2 630	12
+50.0 μg chloramphenicol	19 955	95

sulphydryl reagent and potassium and magnesium ions (Table 6.9). The post-ribosomal supernatant fraction provides amino acids, tRNA, amino-acyl-tRNA synthetases, amino-acyl transferases and perhaps initiation factors. The amino acid incorporating capacity of the cell-free system depends to a great extent on the ribosomal component; ribosomal preparations with abundant polysomes, and thus having a high endogenous mRNA, show good incorporating capacity. In contrast ribosomal preparations with predominantly monosomes have a reduced incorporating capacity but usually respond to exogenous synthetic mRNA such as polyuridylic acid (Fig. 6.6).

ATP is required for the formation of amino-acyl-tRNA. Usually ATP is produced by an ATP-generating system which uses phospho-enol-pyruvate and pyruvic kinase to convert ADP to ATP. GTP is required, as was shown earlier in the initiation step and elongation phase.

The *in vitro* incorporation is inhibited by ribonuclease which destroys the endogenous mRNA and by puromycin and cycloheximide. This latter observation indicates that the bulk of the ribosomes isolated from plant material are of the 80s type.

No studies have been reported of *in vitro* studies using 70s ribosomes isolated from chloroplasts; generally amino acid incorporation is determined using intact chloroplasts.[67] In such systems light stimulates amino acid incorporation presumably by generating ATP required for protein synthesis. Incorporation is inhibited by chloramphenicol and also by

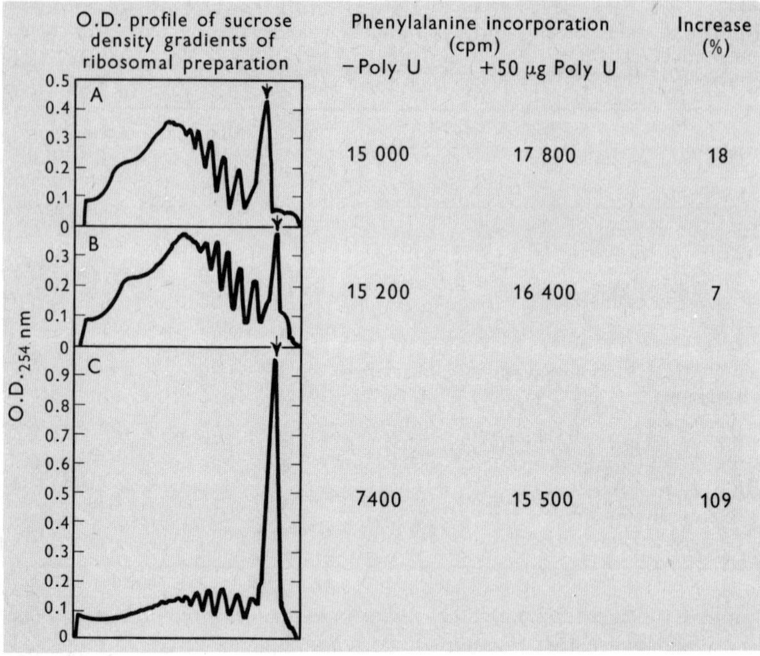

Fig. 6.6 The influence of polysome content of ribosomal preparation on amino acid incorporation and response to synthetic mRNA, polyuridylic acid. Ribosomal preparations A, B, C, containing different levels of polysomes as demonstrated by sucrose density gradient centrifugation, were incubated in an incorporating system with ^{14}C-phenylalanine in the presence or absence of polyuridylic acid. Ribosomal preparations (C) with few polysomes but a preponderance of monomeric ribosomes (arrow in Fig. A, B, C) respond to exogenous synthetic mRNA.[43]

spectinomycin, lincomycin and erythromycin which are specific inhibitors of protein synthesis by chloroplast ribosomes.[176]

Despite early claims to the contrary, it has not been possible to demonstrate net protein synthesis by using cell-free amino acid incorporating systems. Some evidence has been provided for the synthesis of specific proteins by cell-free systems. Cellulase has been reported[141] to be produced by ribosomal preparations from pea (*Pisum sativum*) stems and chloroplasts have been reported to produce ribulose diphosphate carboxylase in an *in vitro* system.[402]

REGULATION OF PROTEIN SYNTHESIS

From the foregoing description of the reactions involved in protein synthesis it is apparent that there are several loci at which polypeptide production may be regulated. Some of these are summarized in Table 6.10.

Table 6.10 Possible sites for regulation of protein synthesis.

The synthesis of mRNA
- Initiation
- Accessibility of the template
- Availability of RNA polymerase and its affinity to accessible template
- Number of gene copies per cell
- Number of cells capable of synthesis
- Rate of transcription
- Termination
- Post transcriptional processing and rate of release from nucleus to cytoplasm

The functioning of mRNA
- Complexing of mRNA with ribosomes. Are specific initiation factors IF-3 required for each mRNA?
- Movement of ribosomes along mRNA
- Number of translations of specific mRNA
- Rate of degradation of mRNA

tRNA synthesis and functioning
- Specification of kinds of protein synthesized by the rate of synthesis of iso-accepting tRNA species
- Multiple forms of amino-acyl synthetases
- Modification of bases in tRNA
- Availability of tRNA with appropriate anticodon for appropriate codon for amino acids with degenerate codes

Polypeptide synthesis
- Availability of ribosomes, free or membrane-bound
- Availability of initiation factors
- Dissociation of ribosomes
- Level of elongation factors
- Activity of peptidyl transferase
- Availability of termination and release factors

The formation of an active enzyme may be dependent upon the addition of prosthetic groups or combination of subunits, etc.

ENZYME INDUCTION

In many instances the level of a particular enzyme can be increased by appropriate treatments. Illumination, application of growth regulators or various compounds may bring about an increase in enzyme level. If the compound causing an increase in the tissue level of a particular enzyme is the substrate for that enzyme then the increased enzyme level may be due to substrate induction.

Experiments demonstrating an increased enzyme level in tissues following a particular treatment are relatively easy to perform but often are difficult to interpret.

The usual procedure is to treat a plant tissue in some manner, illumination, growth regulator application, etc., and then extracts are prepared from treated and untreated plants and the activity of the enzyme in question determined. If it is found that the extract from the treated plant has a greater enzyme activity than that from the control the question is raised as to whether the different enzyme activity of the extracts is due to the synthesis of more enzyme or due to activation of existing enzymes. Another possibility not usually considered is that an increased enzyme level following treatment may be caused by the treatment preventing the normal breakdown of the particular enzyme.

Various tests can be conducted to test whether the production of an enzyme in response to a treatment is due to synthesis or activation. Preliminary tests may be conducted by treating the tissue in the presence or absence of inhibitors of RNA or protein synthesis (Table 6.11). If it is

Table 6.11 Some enzymes which increase in response to a particular treatment. Note that the increase is prevented by inhibitors of RNA or protein synthesis. For a more complete listing see Filner, Wray and Varner.[187]

Enzyme	Inductive treatment	Inhibitors
Invertase	GA	Actinomycin D, Puromycin
Nitrate reductase	NO_3^-	Cycloheximide, Puromycin, Actinomycin D, 8-Azaguanine, Amino acid analogues
Phenylalanine deaminase	Light	Cycloheximide
Cellulase	IAA	Puromycin, Actinomycin D, Azaguanine
Amylase	GA	Cycloheximide, Actinomycin D, 6-Methyl Purine
ATP phosphatase	Age	Dinitrophenol, *p*-Fluorophenylalanine, Chloramphenicol

found that there is no increment in activity following treatment in the presence of the inhibitors, then it is unlikely that the increased activity is due to activation and it can be concluded that the increased enzyme level depends on an unimpaired protein synthesis. Of course this does not conclusively indicate that the particular treatment stimulates the synthesis of a specific enzyme. In order to prove that a particular treatment induces the synthesis of a specific protein it is necessary to show that labelled amino acids are built into the enzyme protein only during the inductive treatment and that they are not incorporated into the assigned protein in the absence of induction. Such unequivocal proof is difficult to obtain owing to the fact that very few enzymes can be purified to homogeneity. However, an alternative approach has been developed by Filner and Varner.[186] This method capitalizes on the fact that proteins synthesized from labelled amino acids have a higher molecular weight than those produced from the non-labelled precursors. The proteins which differ in molecular weight can be separated by isopycnic equilibrium sedimentation (Fig. 6.7). This technique has been used effectively to demonstrate the *de novo*

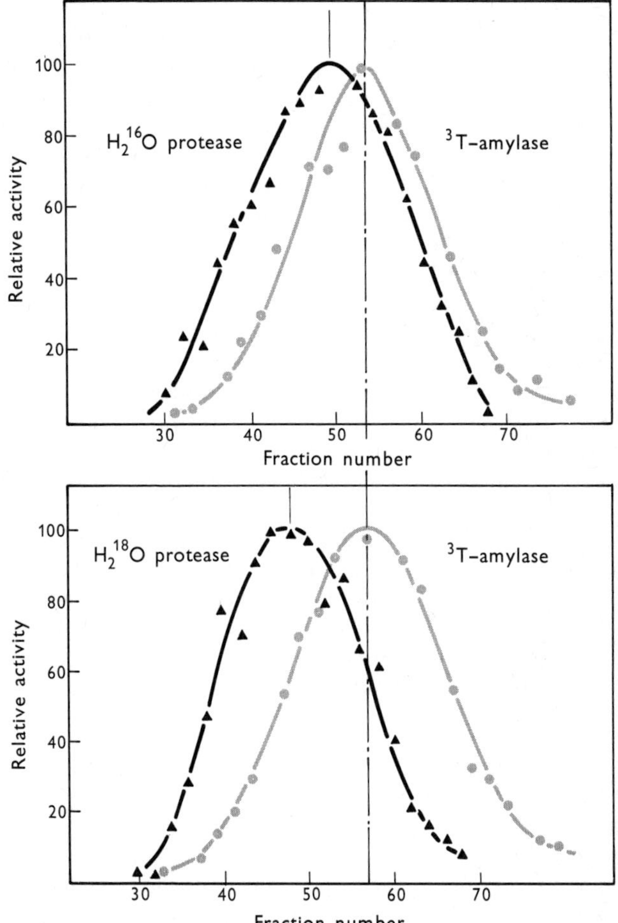

Fig. 6.7 The distribution of protease, released by barley (*Hordeum vulgare*) aleurone layers, in the presence of $H_2{}^{16}O$ and $H_2{}^{18}O$, after equilibrium centrifugation in caesium chloride. Density increases to the left. The increase in density in protease produced in $H_2{}^{18}O$ is explained as follows.

$$\text{Reserve proteins} \xrightarrow{H_2{}^{18}O} \text{R.CH(NH}_2\text{)CO}^{18}\text{OH} \dashrightarrow \text{Protease } {}^{18}\text{O.}$$

(From Jacobsen and Varner.[304])

synthesis of various hydrolytic enzymes induced by gibberellic acid and the induction of the enzyme nitrate reductase by its substrate nitrate.

Since it can now be conclusively demonstrated that various treatments can induce the synthesis of specific enzymes the question is then raised, at what point is the synthesis controlled? It can be demonstrated that the induction of the synthesis of many enzymes can be prevented by inhibitors of RNA synthesis such as Actinomycin D or 6-methyl purine. Although such observations suggest that the inducers exert their effect through regulating RNA synthesis, in reality such experiments only demonstrate the requirement for a sustained RNA synthesis for the expression of the inductive effect. While it is attractive to interpret results of this sort as supporting the contention that inducers function in the manner predicted by Jacob and Monod,[303] there is as yet no evidence that this is the case in higher plants. It has been possible to demonstrate[180, 506] an apparent increased synthesis of mRNA in response to hormones as indicated by an increased level of polysomes (Fig. 6.8) or more label in heterodisperse RNA separated by gel electrophoresis (Fig. 6.9).[723] This heterodisperse RNA contains poly (A).[304a]

However, it has not been possible to associate the induction of the synthesis of a specific enzyme with the induced production of a cognate mRNA. The state of knowledge on the induction of enzyme synthesis can perhaps best be summarized by paraphrasing statements of Chrispeels and Varner.[114] 'Much of the data obtained by using inhibitors is consistent with the hypothesis that inducers exert their control at the level of the gene to bring about the synthesis of an RNA fraction specific for the proteins being synthesized. However, much caution must be used in basing a model on the mode of action of inducers on a study of the effects of inhibitors of RNA synthesis alone. The data are equally consistent with a control mechanism at the level of translation with a requirement for continued RNA synthesis.'

PROTEIN TURNOVER

Studies in the 1940's which involved feeding plants with ammonium salts labelled with the non-radioactive isotope of nitrogen ^{15}N demonstrated that the isotope could be incorporated into proteins without a concomitant increase in the total protein content of the tissue.[260, 657] These observations have been confirmed using other protein precursors and it can frequently be demonstrated that amino acid incorporation into proteins can occur even though the tissue is undergoing a decline in protein content. Such findings support the original hypothesis of Gregory and Sen[232] that a protein cycle exists in plants in which proteins are undergoing continuous synthesis and breakdown. In conditions of steady

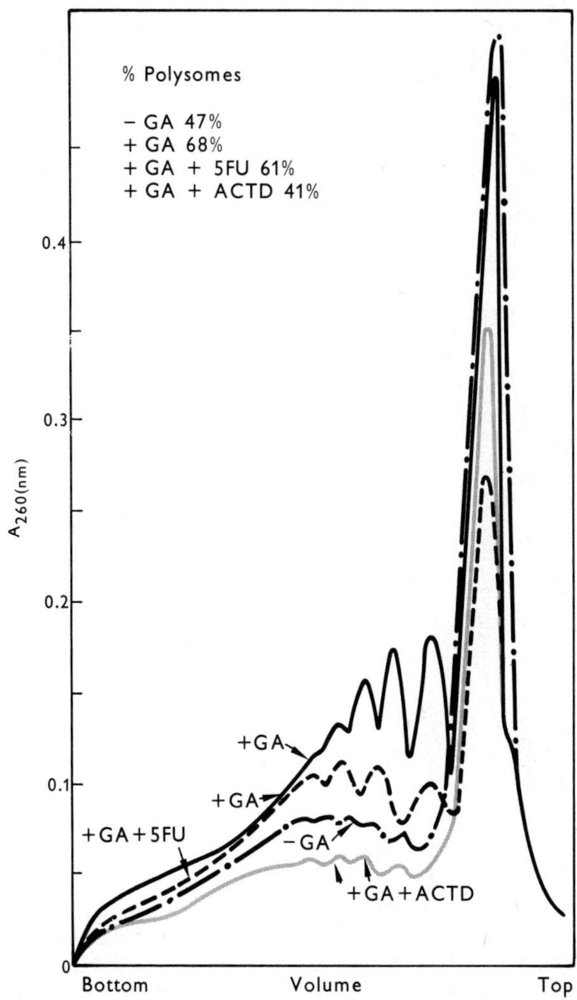

Fig. 6.8 Effect of GA, Actinomycin D and 5-fluorouracil on polyribosome formation. Polysomes were isolated from 40 barley (*Hordeum vulgare*) aleurone layers that were incubated for 8 h in the presence of 1 μM GA_3(+GA), or GA_3 and 2.5 mM fluorouracil (+GA+FU), or GA_3 and 100 μg/ml Actinomycin D (+GA+ACTD), or in the absence of the hormone (–GA) at 25°C on a Dubnoff metabolic shaker. Polysome profiles were determined in 0.3 to 1.0 M isokinetic sucrose gradients. (From Evins and Varner,[180] Fig. 3, p. 351.)

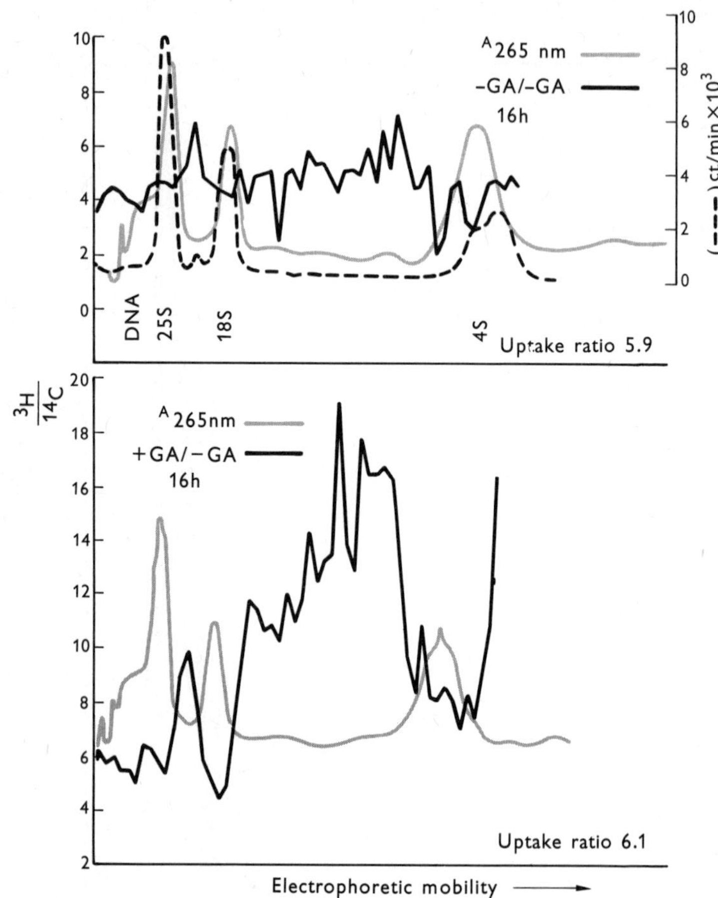

Fig. 6.9 $A_{265\ nm}$ traces of RNA fractionated on acrylamide gels and $^3H/^{14}C$ d.p.m. ratios of labelled RNA contained in the gel slices. The RNA had been prepared from isolated barley (*Hordeum vulgare*) aleurone layers incubated for 6 h. In −GA/−GA neither the ^3H nor the ^{14}C-labelled layers received GA, in +GA/−GA only the ^3H-labelled layers received GA. Note increase in heterodisperse ^3H-labelled RNA in gibberellic acid treated material. (From Zwar and Jacobsen.[723])

state protein levels synthesis and breakdown are equal; during protein accumulation synthesis exceeds protein breakdown and during protein depletion breakdown is greater than synthesis.

The situation can be represented as follows:

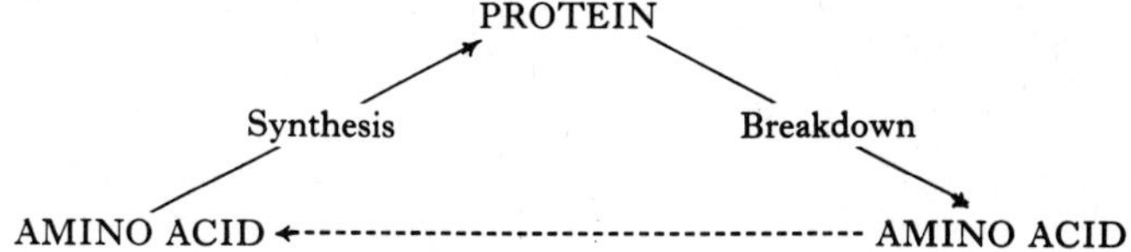

Superficially it appears that it would be a simple task to measure protein turnover. The rate of incorporation of supplied labelled amino acid into protein could be measured and the rate of release of the labelled amino acid from the protein determined after removal of the supplied labelled precursor. Unfortunately this is an over-simplification of the true situation. Complications arise because it is not clear to what extent the supplied label has access to the sites of protein synthesis. Even during the period of incorporation of label into the protein the amino acid may be released as a result of protein breakdown. It is also difficult to remove the applied label to establish the rate of release of label from the synthesized protein. In spite of these difficulties studies have been made of the rate of turnover and although different techniques have been used most authors indicate a turnover rate of 0.1–2% per hour.[259, 514]

The most recent data on turnover provided by Trewavas[637, 638] utilized amino-acylated tRNA as the initial amino acid source. This approach to some extent circumvents the problem of dilution of the supplied labelled exogenous amino acids by the endogenous amino acid so that there is a more reliable estimate of the immediate precursor of protein synthesis. In *Lemna minor* Trewavas measured turnover rates of about 7% per day (half life of 7 days). The turnover rate is affected by growth conditions, hormones, nutrient supply, etc., and as would be expected the altered rate of turnover can be related to either a changed rate of synthesis or a changed rate of degradation.

In the late 1950's and early 60's it was observed that labelled carbohydrates were more efficient precursors for protein synthesis than exogenously supplied amino acids. Such findings suggested that amino acids newly synthesized from the supplied carbohydrate had a greater access to protein synthesizing sites than exogenously supplied amino acids. Such observations are interpreted as demonstrating the occurrence of two amino acid pools, one of which functions as a precursor for protein synthesis and one which originates from the breakdown of protein. Interchange between the two pools appears to be limited and thus recycling of amino acids during protein turnover may be minimal (Fig. 6.10).

Estimates of protein turnover measure the rate of total protein turnover; however, within the protein population of the cell individual proteins

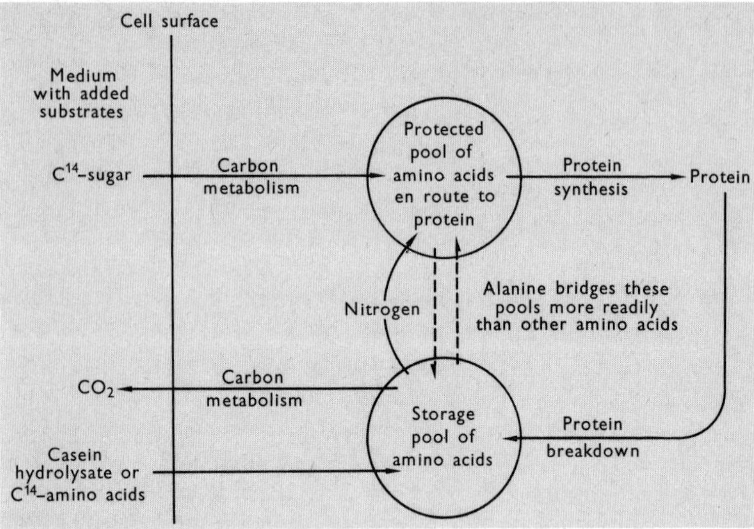

Fig. 6.10 The relationship of exogenous sugars and amino acids to soluble pools in the cell and their involvement in protein turnover. (From Bidwell, Barr and Steward,[51] Fig. 1, p. 370.)

may turnover at different rates. Studies of the turnover of specific enzymes have been made by measuring the decline in a particular enzyme following application of an inhibitor of protein synthesis. These measurements assume that the inhibitors of protein synthesis do not alter the rate of protein degradation. Schrader *et al.*[550] inhibited protein synthesis by cycloheximide in maize (*Zea mays*) seedlings receiving inducer nitrate. At intervals extracts were prepared from the seedlings and assayed for nitrate reductase activity. They found that although inducer nitrate was present the activity of nitrate reductase in the tissue treated with cycloheximide declined and it was estimated that nitrate reductase has a half life of 4.2 hours. This technique is not however universally applicable. In barley (*Hordeum vulgare*) nitrate reductase level does not decline in the presence of cycloheximide and it is postulated that protein synthesis is necessary for the decrease in nitrate reductase level.[632] Zucker[722] has similarly reported that protein synthesis may be required during the degradation of phenylalanine ammonia lyase.

In measurements of this type it is not determined whether the lowered enzyme activity in the extracts following cycloheximide treatment is due to actual degradation of the enzyme molecule or due to some transient inactivation of the enzyme. Definitive measurements of the rate of degradation of a particular enzyme should assess the disappearance of a specific enzyme protein and to date such experiments have not been conducted.

PROTEIN DEGRADATION

The observations that proteins are in a dynamic state and turnover can be measured indicate that the plant must possess mechanisms for protein hydrolysis. Very few investigations have been conducted in this area.

Protein hydrolysis is achieved through the activity of the proteolytic enzymes—**proteases**. These can be divided into four principal groups depending on their active site and catalytic mechanisms. The categories are serine proteinases, sulphydryl proteinases, metalloproteinases and acid proteinases. Within this broad classification the enzymes can be divided into endopeptidases, carboxypeptidases or aminopeptidases depending on whether they attack internally or cleave amino acids sequentially from the carboxyl or amino terminal ends of the peptide chain.[251] Although this classification is commonly used in bacterial and mammalian systems, few plant proteinases have been characterized sufficiently to allow for their assignment to a specific class.

The most detailed investigations of plant proteinases have been conducted on the enzymes papain, ficin, and bromelain which are derived from the latex of papaya (*Carica papaya*), fig (*Ficus glabrata*) and pineapple (*Ananas comosus*) respectively. Many tropical fruits contain active proteases and this characteristic has been utilized by the inhabitants of some tropical areas who use papaya leaves, pineapples, and chinese gooseberries (*Actinidia chinensis*) as meat tenderizers. The proteolytic enzymes of papain and bromelain are both sulphydryl proteinases and function as endopeptidases in hydrolysing a variety of peptide bonds. It is considered that peptide bond cleavage involves the formation of an enzyme substrate complex in which an acyl thioester linkage is formed on the free SH group of the cysteine residue in the papain polypeptide.[221] In spite of the detailed studies of these sulphydryl proteinases little is known of their physiological or metabolic function. It is speculated that they may play a role in resisting attack by insect or fungal pathogens.[221]

Protein hydrolysis has been extensively studied in germinating seeds as discussed in Chapter 7. Some proteolytic enzymes have been partially characterized from germinating seeds and they are classified as peptidases and proteases. The proteinases frequently show maximal activity at low pH and thus may be categorized as acid proteases. Ihle and Dure[280, 281] have extensively purified a carboxypeptidase which has a pH optimum of 6.7 and hydrolysed polypeptides sequentially from the carboxy terminal amino acid. An enzyme with similar properties has been isolated from bean[683] (*Phaseolus vulgaris*) leaves and citrus fruits.[721] Significantly these plant carboxypeptidases show little specificity and can release a range of amino acids from the terminal carboxy position, a situation in marked contrast to carboxypeptidases of animal origin.

In addition to the carboxypeptidases, bean (*Phaseolus vulgaris*) leaves also contain other proteinases capable of hydrolysing casein.[683] Caseolytic activity has further been demonstrated in extracts of tobacco[14] (*Nicotiana tabacum*) and nasturtium (*Tropaeolum majus*) leaves.[39] The physiological role of these proteinases has not been established.

Carboxypeptidases and other acid hydrolases have been located in the plant vacuole. As discussed in Chapter 9 it has been suggested that these vacuolar enzymes play an important role in senescence.[406]

7

Nitrogen Metabolism in Seeds

SEED FORMATION

Except in the cases of apomixis, which occurs for example in dandelion (*Taraxacum officinale*) and annual meadow-grass (*Poa annua*), seed development in most instances commences following fertilization of the ovule. In Angiosperms a double fertilization occurs in which one of the male nuclei (or sperm) from the pollen grain fuses with the egg cell and the other nucleus (sperm) combines with the polar nucleus to give rise to the triple fusion nucleus. The fertilized egg cell or zygote produces the embryo and the triple fusion nucleus develops into the endosperm (Fig. 7.1).

As the seed develops the ovule undergoes characteristic changes. The nucellus is usually degraded and absorbed as the endosperm develops; however in some instances it may persist and form the perisperm of the mature seed. The integuments surrounding the ovule become thickened to form the testa or seed coat. In most instances endosperm development precedes extensive embryo development so that initially there is a preponderance of endosperm tissue. The proliferation of the endosperm is achieved by divisions of the triploid fusion nucleus which may or may not be accompanied by cellular divisions. In some species a cellular endosperm is formed whereas in other cases the endosperm is liquid. The endosperm may persist and be present in the mature seed as occurs in many monocotyledons or alternatively the endosperm, which initially develops, is absorbed by the developing embryo.

Embryo development has been studied in detail. Initially the zygote divides to produce a basal suspensor. The function of this structure is not

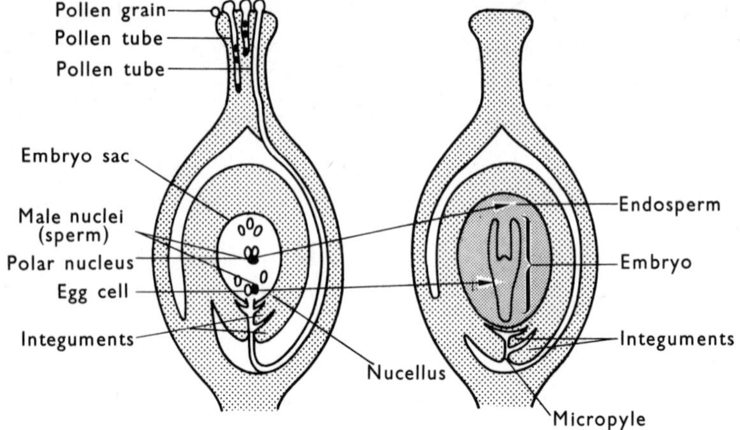

Fig. 7.1 Diagrammatic representation of fertilization and embryo development in Angiosperms.

certain but it is characterized by having a high cellular DNA content brought about by endoduplication of the chromosomes.[661] The embryo proper is differentiated by activities of the cells distal to the micropyle which undergo extensive division and differentiation. The embryo may develop extensively, as in the case of peas and beans for example, to occupy the bulk of the seed. In other cases extensive development of the embryo may be restricted until germination. In those cases where embryo development is extensive the majority of the structure consists of large cotyledons, which function as food reserve containing structures, and a relatively small embryonic axis.

As a result of the differential development of the embryo or endosperm, the seed at maturity may have food reserves deposited principally in the endosperm or the cotyledons of the embryo.

There is considerable variation between seeds in their chemical composition which reflects to a great extent the nature of the main food reserves (Table 7.1).

It should be noted that the composition of the seed reserves differs markedly from that encountered in the vegetative parts of the plant. Thus in cereals, for example, the principal carbohydrate reserve in the seed is starch; this polysaccharide is often absent in vegetative tissues. The main lipid reserves in seeds are triglycerides, and in contrast in vegetative tissues mixed glycerides such as phospholipids, sulpholipids and galactolipids predominate. The proteins encountered in seeds have a different amino acid composition from those usually found in the vegetative tissue.

Table 7.1 Chemical composition of seeds.

	Percentage (by wt.) of air-dry seed		
	Carbohydrates	Proteins	Fats
Zea mays	50–70	10	5
Pisum sativum	30–40	20	2
Arachis hypogaea	10–30	20–30	40–50
Ricinus communis	0	18	64
Acer saccharinum	60	28	4
Brassica rapa	25	20	34

Clearly the deposition of food reserves in the seed must involve the expression of metabolic sequences which are not utilized extensively during vegetative growth of the plant.

SEED PROTEINS

The principal nitrogenous reserve component in seeds is protein. The classical investigations of Osborne[489] fractionated proteins on the basis of their solubility and this method is still used extensively today. Seeds vary considerably in the nature of their protein reserves (Table 7.2).

Table 7.2 The protein content of various seeds.

	Fractions as % of total protein			
	Albumin	Globulin	Glutelin	Prolamin
Wheat (*Triticum aestivum*)	5	10	40	45
Maize (*Zea mays* (IND 260))	14	—	31	48
(*Zea mays* (Opaque 2))	25	—	39	24
Pea (*Pisum sativum*)	40	60	—	—
Oat (*Avena sativa*)	Trace	80	5	15
Squash (*Cucurbita pepo*)	Very little	92	Small amount	Very little
Rice (*Oryza sativa*)	5	10	80	5

Dicotyledons contain principally globulin protein reserves whereas cereals are enriched in alcohol-soluble prolamins and alkali- or acid-soluble glutelins. The prolamins are almost exclusively located in the seeds of members of the Gramineae; however in some grasses, oats (*Avena sativa*) and rice (*Oryza sativa*) for example, the main protein reserves are globulins and glutelins respectively.

The fractionation of proteins on the basis of their solubility is dependent upon differences in amino acid composition of the protein fractions.

The globulins are characterized by being enriched in glutamic acid, aspartic acid, and their amides, and arginine. The prolamins contain proline and glutamic acid (mainly as glutamine) as the principal amino acids and are low in lysine (Table 7.3).

Table 7.3 The amino acid composition of four protein fractions from barley grain. Based on data of Folkes and Yemm[193] (g amino acid per 16 g N).

	Globulin	Albumin	Prolamin (Hordein)	Glutelin (Hordenin)
Arg	11	6.5	3	6
His	1.8	2.5	1.3	2.5
Lys	5.3	6.7	0.7	4.0
Try	0.8	1.5	0.8	1.3
Phe	2.8	5.1	3.0	3.6
Cys	3.6	2.1	2.1	1.2
Met	1.5	2.4	1.3	1.9
Ser	4.7	4.9	3.8	5.0
Thr	3.3	4.6	2.6	4.2
Leu	6.8	8.6	6.9	8.7
I-Leu	3.3	6.2	5.4	5.2
Val	5.5	7.8	4.7	6.6
Glu	11.9	12.9	39.6	19.8
Asp	8.5	12.2	1.8	7.1
Gly	9.2	5.7	1.5	4.5
Ala	0.7	7.3	2.2	6.7
Pro	3.6	5.5	20.1	8.7
Amide N	5.1	5.9	23.0	10.3

The relative proportion of the different proteins in the seed grossly affects its overall amino acid composition. In this regard it is interesting to note, in passing, that corn (*Zea mays*) meal normally is deficient in the amino acid lysine and is thus of restricted nutritive value. However mutants opaque-2 and floury-2 are exceptional in containing higher levels of the essential amino acid lysine. Analyses have indicated that enrichment in this amino acid was not due to an increased level of lysine in the prolamin fraction but was related to the fact that opaque-2 and floury-2 contained greater proportions of albumins and globulins and lower levels of the lysine deficient prolamins (see Table 7.2).

The protein fractions separated on the basis of their solubility are heterogeneous and vary from species to species. The albumins are composed primarily of the enzymatic proteins and obviously contain many polypeptides. The globulin component may also contain a range of proteins. In peas (*Pisum sativum*) and beans (*Vicia faba*) for example the globulin can be fractionated by isoelectric precipitation at pH 4.5 into insoluble legumin and soluble vicilin. These proteins can be further dissociated and fractionated by gel electrophoresis. Vicilin from *Vicia faba*[21] is dissociated into four subunits whereas that from peas produces five subunits. The globulins from peanuts (*Arachis hypogaea*)[138] have been shown to contain several polypeptides and soybean (*Glycine max*) globulins can be fractionated into several discrete components.[100] Differences, although minor, also exist in the amino acid composition of the vicilin and legumin fractions from various species. Even greater differences in amino acid composition are apparent when globulins from cereals are compared with those from legumes or other dicotyledons (Table 7.4).

Table 7.4 Amino acid composition of some reserve proteins from different species (g amino acid/100 g protein).

	GLOBULIN			GLUTELIN		PROLAMIN	
	Soybean [100] (*Glycine max*)	Barley [193] (*Hordeum vulgare*)	Pumpkin [113] (*Cucurbita moschata*)	Barley[193] Hordenin	Maize [116] (*Zea mays*) Glutelin	Barley [193] Hordein	Maize [445] Zein
Arg	7.8	11	17.5	6	5.0	3	2.1
His	2.4	1.8	2.1	2.5	4.4	1.3	1.4
Lys	6.0	5.3	1.7	4.0	2.1	0.7	0.1
Try	3.8	0.8	2.05	1.3	5.6	0.8	5.6
Phe	5.5	2.8	5.7	3.6	5.1	3.0	7.6
Cys	1.2	3.6	.7	1.2	1.4	2.1	1.0
Met	1.2	1.5	1.6	1.9	2.4	1.3	1.7
Ser	5.7	4.7	5.9	5.0	5.5	3.8	6.2
Thr	3.6	3.3	3.06	4.2	4.2	2.6	3.2
Leu	7.9	6.8	8.3	8.7	12.9	6.9	22.4
I-Leu	4.8	3.3	3.8	5.2	3.3	5.4	4.2
Val	4.7	5.5	5.1	6.6	4.3	4.7	4.5
Glu	22.4	11.9	23.01	19.8	22.0	39.6	27.4
Asp	12.6	8.5	10.06	7.1	5.4	1.8	5.9
Gly	4.1	9.2	4.5	4.5	4.5	1.5	1.6
Ala	3.9	0.7	4.5	6.7	7.2	2.2	10.9
Pro	5.4	3.6	0.4	8.7	13.9	20.1	10.7

Glutelins are complex proteins of high molecular weight and appear to be composed of several polypeptide subunits which can be separated by electrophoresis in urea. The urea apparently cleaves –S–S– disulphide bridges between component polypeptides.[151] Again differences exist in the amino acid composition of glutelins derived from different sources.

The prolamin zein, from maize, can be fractionated into several components by electrophoresis in starch gel following urea treatment, again suggesting the occurrence of several component polypeptides.[445] The amino acid composition of prolamins extracted from various cereals varies slightly.[662]

The principal reserve proteins in some cases are located in subcellular structures called protein bodies or aleurins. Histological and immunological techniques have demonstrated that both the reserve globulins, legumin and vicilin, are located in protein bodies in the cotyledons of beans.[227] Protein bodies from maize (*Zea mays*)[116] and barley (*Hordeum vulgare*)[644] contain only prolamins with glutelins occurring in the endosperm matrix. Australian workers,[226] however, have indicated that the protein bodies from wheat endosperm contain two classes of protein, one soluble in 0.05 N acetic acid and the other extractable in 0.1 N sodium hydroxide.

In addition to the principal protein reserves it is becoming evident that seeds also contain small amounts of other proteins. Among these are various low molecular weight soluble proteins which *in vitro* inhibit the activity of the proteolytic enzyme trypsin. The original trypsin inhibitor was characterized from soybean;[348] however, since then there has been an accumulation of evidence for the occurrence of other protease inhibitors in a variety of seeds.[542] Additionally recent interest has centred around a group of proteins isolated from seeds and other plant parts which have the ability to cause agglutination of red blood cells. These proteins are characteristically but not invariably glycoproteins of molecular weights in the vicinity of 120 000. Their physiological role in plants has not been established; however they are used extensively in animal and biochemical studies owing to their characteristic properties; in addition to causing agglutination of erythrocytes these proteins preferentially agglutinate malignant cells, they bind sugars specifically and precipitate polysaccharides and glycoproteins.[384] Some of these phytohaemagglutinins or lectins, as they are now called, also stimulate mitosis in certain cell types.

NUCLEIC ACIDS IN SEEDS

Seeds vary in their nucleic acid content and, although few comparative studies have been performed, it appears that nucleic acid content may be dependent upon whether the predominant reserve tissue is endosperm or cotyledon. Corn (*Zea mays*)[289] and rye (*Secale cereale*)[270] endosperms contain relatively low levels of DNA and RNA whereas the embryonic axis contains high levels. In contrast the cotyledonary tissue of peas (*Pisum sativum*),[507] beans (*Vicia faba*),[687] and peanuts (*Arachis hypogaea*)[117] contain much greater levels of nucleic acid.

SOLUBLE AMINO ACIDS

At maturity, seeds contain only a small percentage of their total nitrogen in the form of soluble amino acids. In wheat (*Triticum aestivum*)[308] they represent only 2% of the total nitrogenous material. In peas (*Pisum sativum*) less than 10% of the nitrogen is soluble in alcohol.[415] The bulk of the alcohol-soluble nitrogen is derived from the soluble amino acids. The amino acid composition of the soluble fraction differs from that present in the proteins.

NITROGEN METABOLISM DURING SEED DEVELOPMENT

The changes which occur in the principal nitrogenous components are shown in Figs. 7.2 and 7.3.

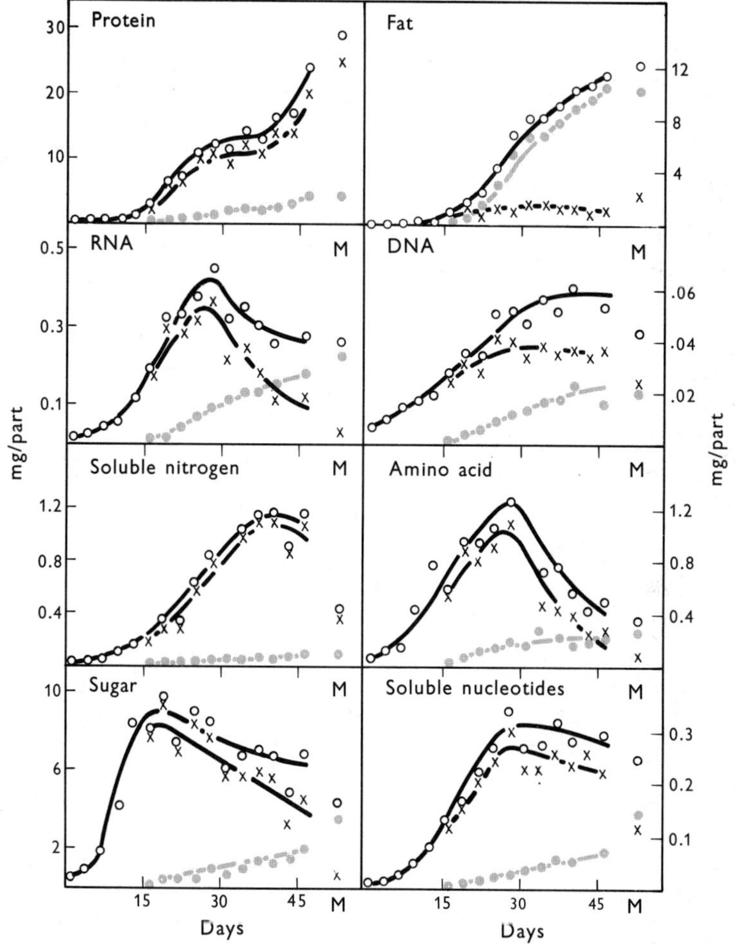

Fig. 7.2 Changes in various components of the maize (*Zea mays*) grain (○), embryo (●), and endosperm (X), over a 46-day developmental period after pollination; M represents mature grain at harvest. (From Ingle *et al.*,[289] Fig. 4.11, p. 837.)

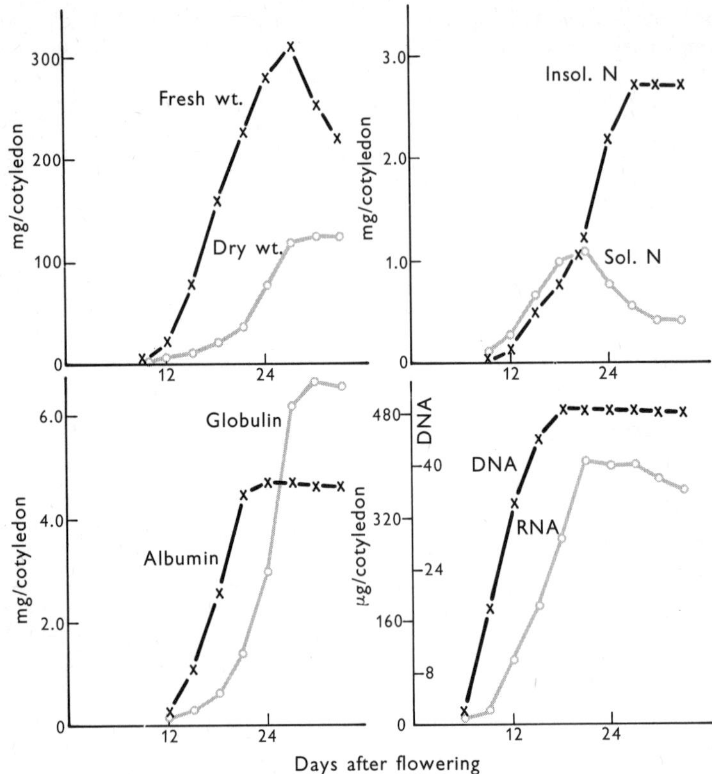

Fig. 7.3 Changes in nitrogenous components of pea (*Pisum sativum*) cotyledons during development.

In developing maize (*Zea mays*) the initial accumulation of materials occurs in the endosperm with a less extensive deposition of nitrogenous components in the embryo.[289] In peas (*Pisum sativum*) there is an initial deposition of materials in the endosperm following anthesis but these components are depleted and materials accumulate extensively in the developing embryo.[22]

In maize the protein accumulation in the endosperm appears to be biphasic with an initial accumulation occurring between days 15–30 after anthesis and a subsequent second deposition occurring between 30 days and maturity. Although no analyses have been conducted in this regard the different phases may represent the deposition of different reserve proteins.

In developing peas the proteins accumulate rapidly between day 12 and 25 following anthesis.[43] Bean (*Vicia faba*)[426] and field peas (*Pisum arvense*)[570] show similar sequences although on a somewhat different time scale. In peas the initial proteins accumulated in the cotyledons are the albumins whereas at later stages of seed development the globulins vicilin and legumin are deposited (Fig. 7.3).

The initial phase of protein accumulation in the maize endosperm is accompanied by a rapid increase in RNA and DNA. The attainment of maximal DNA content has been assumed to coincide with the termination of nuclear and cell division in the endosperm and DNA content remains constant during the remaining period of grain maturation. The RNA content of the endosperm declines during the second phase of protein accumulation and seed maturation.[289]

The nucleic acid metabolism in the pea cotyledon contrasts markedly with that of the maize endosperm. During early embryogenesis there is an early increase in DNA coinciding with cell division. However following the termination of cell division there is a sustained increase in DNA content which slightly precedes the onset of rapid RNA and protein synthesis.[507, 548] This increased DNA content without accompanying cell division which results in cells with up to 64C in comparison to the usual diploid 2C levels of DNA also occurs in other legumes (*Pisum arvense*),[570] (*Vicia faba*).[427] According to Millerd and Whitfeld[427] it represents an endoreduplication of the whole genome. Such endoreduplication provides a mechanism for producing extra gene copies which may be required for the synthesis of the large amounts of RNA which accumulate during cotyledonary development. During seed development ribosomal RNA and the soluble RNA components are produced in constant proportions and there is no accumulation of any one specific RNA component.[507] Of course there may be an enhanced transcription of globulin mRNA's at this time, but such a synthesis has not been detected. As the seeds mature and dehydrate the level of RNA and DNA remains essentially constant and during this time there is only a limited incorporation of exogenously supplied labelled precursors so turnover of RNA in the maturing cotyledon is minimal.[507] The constant RNA and DNA level during dehydration is clearly in marked contrast to the situation in the maize endosperm.

In vitro studies of protein synthesis

The changes in the rate of deposition of protein during seed development have been related to changes in components required for protein synthesis.[43] Ribosomal preparations extracted from developing peas up until the commencement of dehydration contain a preponderance of polysomes (Fig. 7.4). The number of ribosomes per cell must increase during early seed development as indicated by an enhancement of cotyledonary RNA. The majority of the ribosomes produced during the period of rapid RNA and protein synthesis are membrane bound.[498] It has been suggested that the membrane-bound ribosomes are responsible for the

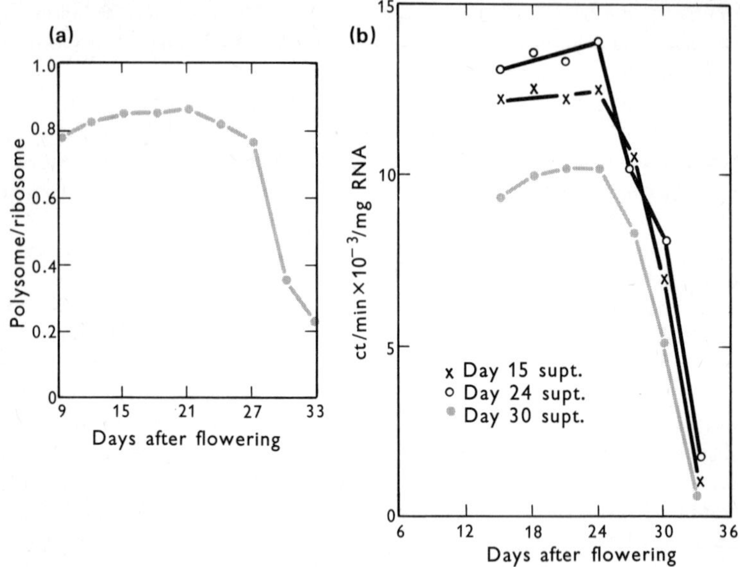

Fig. 7.4a The distribution of polysomes in the ribosomal preparations from pea cotyledons. Results are expressed as the proportion of the total ribosome profile occurring in polysomic configuration after sucrose density centrifugation. (From Beevers and Poulson,[43] Fig. 3, p. 478.)

Fig. 7.4b The influence of the supernatant fraction from 15-, 24-, and 30-day pea cotyledons on the capacity of ribosomal preparations obtained from cotyledons of varying age to incorporate ^{14}C-leucine. (From Beevers and Poulson,[43] Fig. 6, p. 479.)

synthesis of the reserve proteins ultimately deposited in the protein bodies. As the seed matures the level of polysomes decreases and there is a corresponding increase in monosomes in the ribosomal preparations. The monosomes appear to be 'free' and are derived from originally membrane-bound ribosomes.

The ribosomal preparations from cotyledons of different developmental stages show varying amino acid incorporating capacities related to the relative distribution of polysomes and monosomes.[43] Reciprocal mixing experiments in which ribosomal preparations from one developmental age are mixed with supernatant fractions from cotyledons of another developmental age suggest that there is some overall change in tRNA, amino-acyl transferase and amino-acyl synthetases during development (Fig. 7.4b). There may be changes in specific tRNA species and amino-acyl synthetases which may facilitate the incorporation of those amino acids which occur in greater abundance in reserve proteins. In a preliminary investigation of this aspect Fowden's group[464] has shown that in wheat endosperm amino acid activating enzymes show a peak of activity six weeks after anthesis, followed by a rapid decline. Individual amino-acyl synthetases generally followed this overall pattern but there were exceptions which might indicate that the availability of appropriate amino-acyl tRNA's could vary during endosperm development.

In early investigations of protein synthesis in maize endosperm[513] it was found that the capacity of ribosomal preparations to support *in vitro* amino acid incorporation was initially low but increased to a peak 25 days after pollination and then declined as the grain matured. This is in contrast to the situation in peas where ribosomal preparations from young cotyledons were active in *in vitro* amino acid incorporation. A detailed analysis of the level of polysomes in the ribosomal preparations of maize grains of different developmental ages has not been made so it is not possible to account for the low incorporating capacity of ribosomes extracted from the young endosperm. The decline in incorporating capacity with increasing endosperm maturity may be attributable to a decrease in polysome level similar to that reported in castor bean (*Ricinus communis*) endosperm[403] and pea cotyledons.[43] However, it appears that the monosomes are degraded in endosperm tissue since there is a decline in total RNA during development.[289]

Morton and Raison[442, 443] indicated that in wheat endosperm the reserve proteins may be synthesized by semi-autonomous discrete organelles called proteoplasts. However subsequent investigators[698] have suggested that the amino acid incorporation observed in the particulate fraction may have been due to contaminating bacteria or amyloplasts which in similarity to chloroplasts do contain some of the components involved in protein synthesis.

Formation of protein bodies

Since the current evidence indicates that reserve proteins are synthesized by cytoplasmic ribosomes the problem of formation of discrete bodies housing the reserves must be resolved. Ultrastructural studies[22, 73, 485] and those illustrated in Fig. 7.5 indicate that the protein bodies are formed in the vacuoles of cotyledonary cells.

At the time corresponding to the onset of maximal protein deposition the tonoplast shows considerable invaginations with long cytoplasmic strands extending into the vacuole (Fig. 7.5a). Associated with the invagination of the tonoplast there is an extensive proliferation of the endoplasmic reticulum. As the cotyledonary cells mature the number of cytoplasmic strands increase and the vacuolar ground substance increases in electron density (Fig. 5.7b). This increase in vacuolar density may be due to the reserve proteins accumulating in the vacuolar space. The cytoplasmic

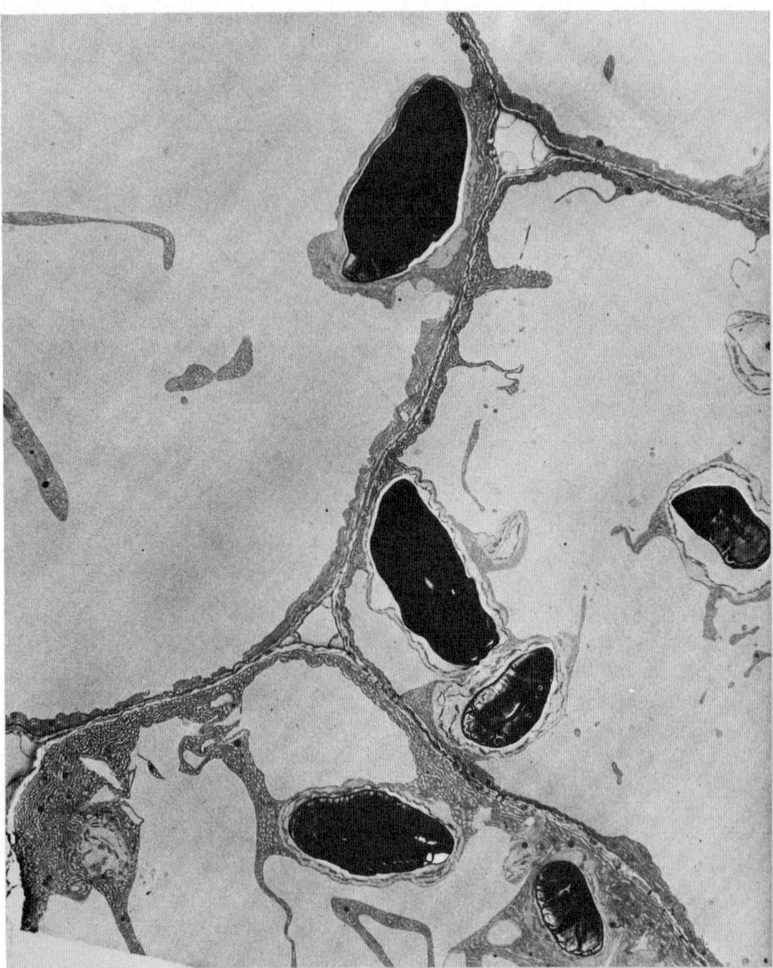

Fig. 7.5a Early stages of protein body synthesis in cotyledonary cells of *Pisum sativum*. Note extensive invaginations of the tonoplast with abundant endoplasmic reticulum. Electron micrograph (×2340) by courtesy of H. H. Mollenhauer.

strands ultimately divide the vacuole into segments (Fig. 7.5c). The segments correspond to the pre-aleurone grains and contain dense matrices of reserve proteins. As the seed approaches maturity the pre-aleurone grains become spherical and, apart from cytoplasmic strands extending into the matrix, resemble the protein bodies of mature seeds. At maturity the protein bodies are characteristically spherical with no visible cytoplasmic strands in them. The endoplasmic reticulum is no longer a conspicuous feature of the cytoplasm and the ribosomes appear to be dispersed randomly (Fig. 7.5d).

It is suggested[149a] that in oil bearing seeds such as cotton (*Gossypium hirsutum*), peanuts (*Arachis hypogaea*) and shepherds purse (*Capsella*

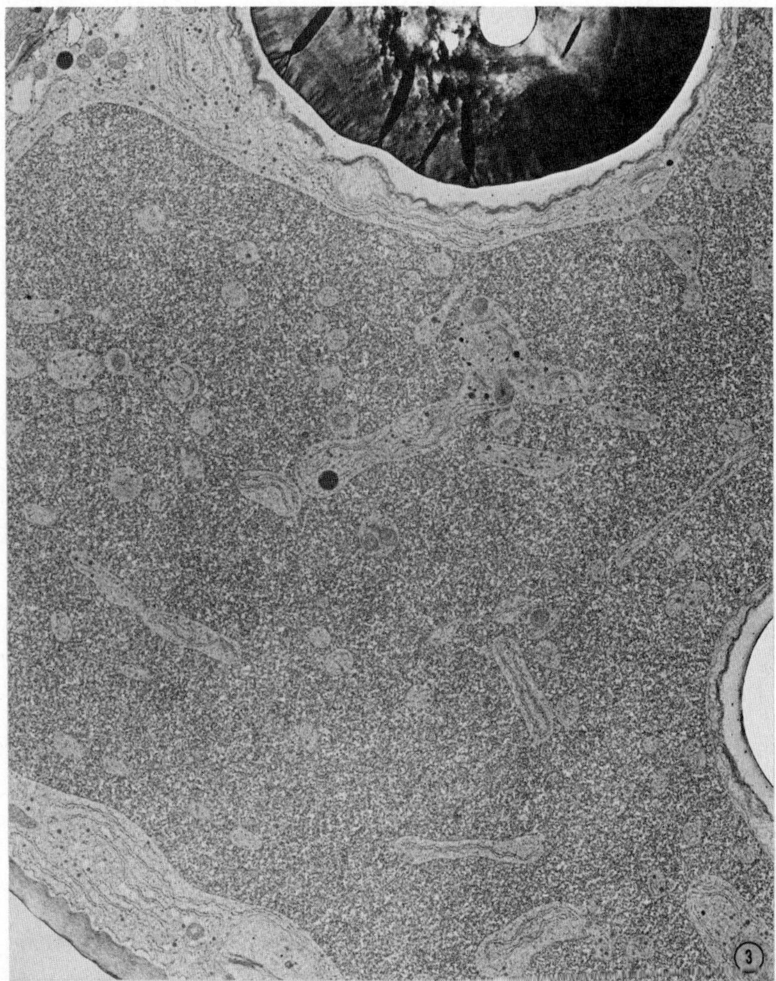

Fig. 7.5b Later stage in protein body formation. Note change in electron density of vacuolar contents and cytoplasmic strands containing endoplasmic reticulum protruding into the vacuole. Electron micrograph ($\times$5850) by courtesy of H. H. Mollenhauer.

bursa-pastoris), the reserve proteins which accumulate in the vacuolar space and form the protein bodies are synthesized on the polysomes of the rough endoplasmic reticulum. The proteins are believed to be passed through the membranes to the lumen of the endoplasmic reticulum and are transported to the dictyosome (Golgi apparatus) where they are concentrated into droplets. These membrane-bound protein droplets then

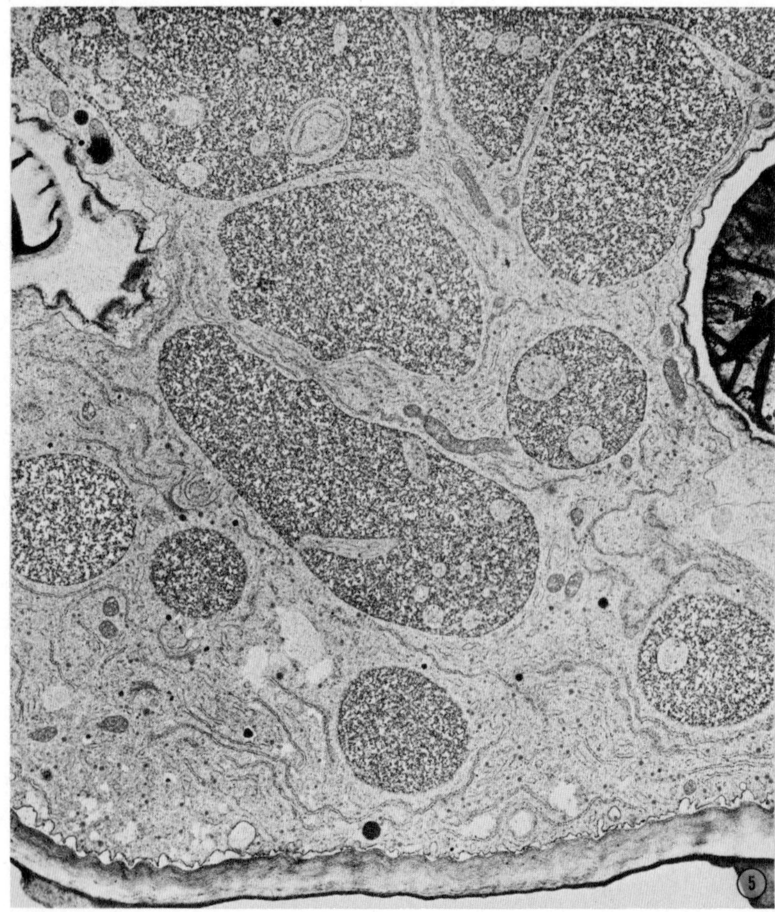

Fig. 7.5c Intermediate stage in protein body development. Cytoplasmic strands have completely divided the vacuole into segments which give rise to protein bodies. Note at this stage cytoplasmic strands extend into the matrix of the protein bodies. Electron micrograph (×5265) by courtesy of H. H. Mollenhauer.

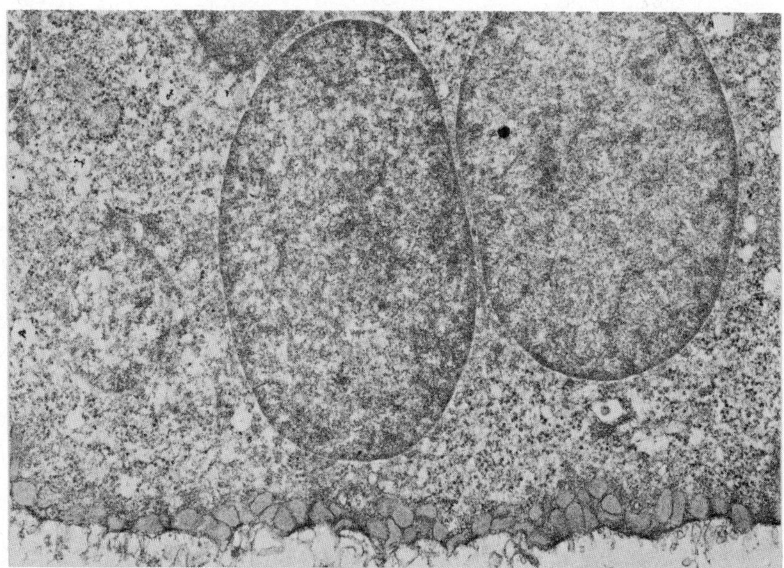

Fig. 7.5d A cotyledonary cell of *Pisum sativum* at maturity with characteristic protein bodies. Note absence of endoplasmic reticulum in the surrounding cytoplasm and randomly dispersed ribosomes. Electron micrograph (× 24 640) by courtesy of H. H. Mollenhauer.

migrate through the cytoplasm to the vacuolar space and the protein is transferred into the vacuole by a process of membrane fusion. However, the data of Fig. 7.5 fail to demonstrate a similar mechanism of deposition of protein into the protein body of peas.

It seems that protein body formation in endosperm tissue may involve different mechanisms from that described in cotyledons. Morton *et al.*[441] have indicated that some of the protein bodies from wheat (*Triticum aestivum*) have endoplasmic reticulum associated with them. Buttrose[92] has reported on the occurrence of protein deposits lying free in the cytoplasm without a surrounding membrane. Other protein bodies in the endosperm from wheat possessed membranes which it was suggested originated from the Golgi apparatus. More recently Khoo and Wolf[328] have indicated that in the maize (*Zea mays*) endosperm the protein bodies develop from vesicles produced from the endoplasmic reticulum or Golgi apparatus (Fig. 7.6a and 7.6b). The protein accumulated in the vesicle appears to be synthesized outside the membranes; the occurrence of ribosomes and polyribosomes on the surface of the protein granule supports this proposition. At maturity the protein bodies are embedded in an amorphous protein matrix.

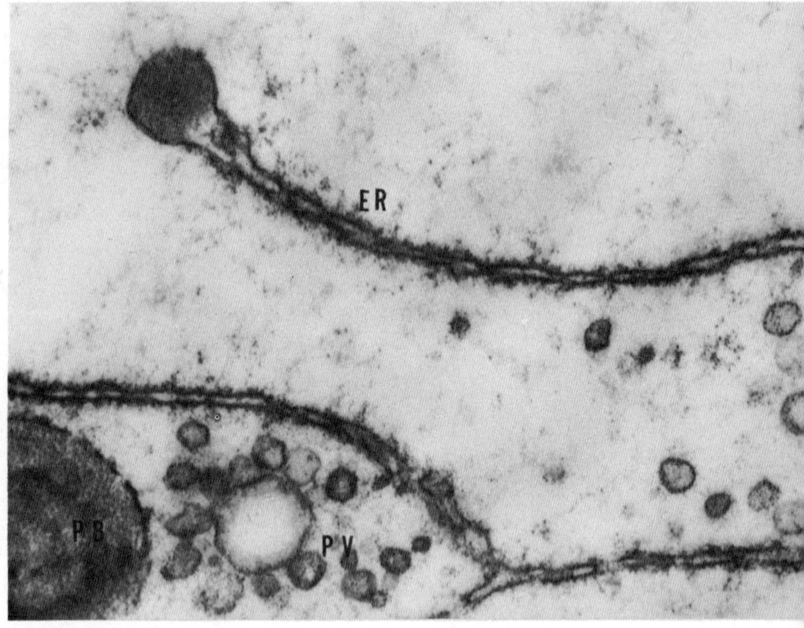

Fig. 7.6a Protein body formation in maize (*Zea mays*) endosperm (× 54 000). The figure shows an enlarged end of endoplasmic reticulum (ER) with protein deposits. Also several protein vesicles (PV) clustered near a larger protein body (PB). (From Khoo and Wolf,[328] Fig. 6, p. 1046.)

Synthesis of non-reserve proteins

In addition to the protein reserves, many seeds accumulate major amounts of other food materials, the synthesis of which requires the presence of appropriate enzymes. In the endosperm of cereals the onset of starch synthesis is characterized by an increase in the levels of UDP-G and ADP-G phosphorylase which subsequently declines as the rate of starch deposition decreases.[646]

The synthesis of reserve materials involves the expenditure of a great deal of energy derived from ATP. Thus it is to be expected that tissues in which reserves are being deposited exhibit high rates of respiration. Only limited studies have been made of this aspect. On the basis of CO_2 evolution the respiratory metabolism of the developing pea (*Pisum sativum*) cotyledon is lower than that of the surrounding hull and pod tissue.[416] However this low CO_2 release may be due to the fact that carbon is retained in the developing cotyledon to provide the skeletons for amino

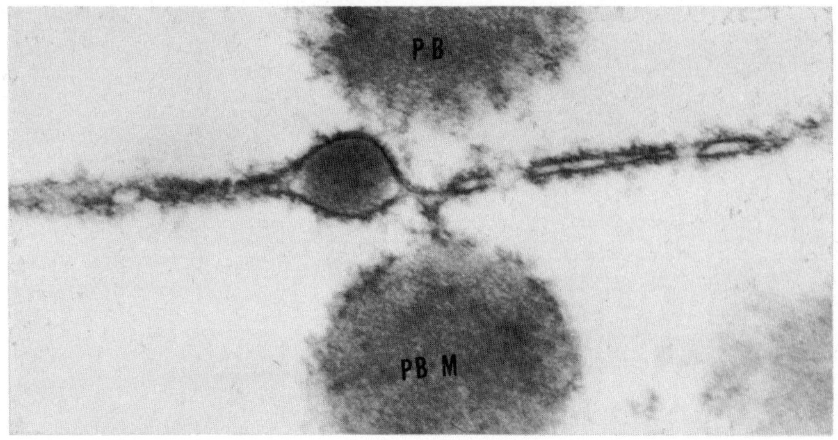

Fig. 7.6b Small localized dilation of cisterna with protein deposit (×57 000). Note some protein bodies have discrete membranes (PBM) whereas it is absent in others (PB). (From Khoo and Wolf,[328] Fig. 7, p. 1046.)

acid biosynthesis. In castor bean (*Ricinus communis*) endosperm respiration as indicated by O_2 uptake shows a maximum level at about the half-way point of seed maturity with an R.Q. above unity. At later stages of maturation the R.Q. decreases progressively. The increase in O_2 uptake at the early stages of development is associated with increases in the levels (on a per seed basis) of many glycolytic and mitochondrial enzymes. As the seed matures the enzyme activity declines to a very low level.[403]

As discussed in the subsequent section, amino acids undergo considerable interconversion during the deposition of reserve proteins and thus reserve tissues ought to be rich sources of enzymes involved in amino acid metabolism. Few studies have exploited this possibility. However it has recently been demonstrated[341] that the cotyledons of *Vicia faba* contain the enzyme ornithine carbamyl transferase required for the biosynthesis of arginine. The enzyme activity was found to be high in developing cotyledons but declined during maturation.

Amino acid metabolism in developing seeds

In the maize endosperm and in pea cotyledons amino acids accumulate during the time of rapid protein synthesis and then as the rate of protein accumulation decreases there is a decline in the soluble amino acid level (Figs. 7.2, 7.3). The changes in soluble amino acid level are accompanied by alterations in the content of the component soluble amino acid.[192, 415]

Different amino acids reach maximal levels in the developing seed at different times (Fig. 7.7).

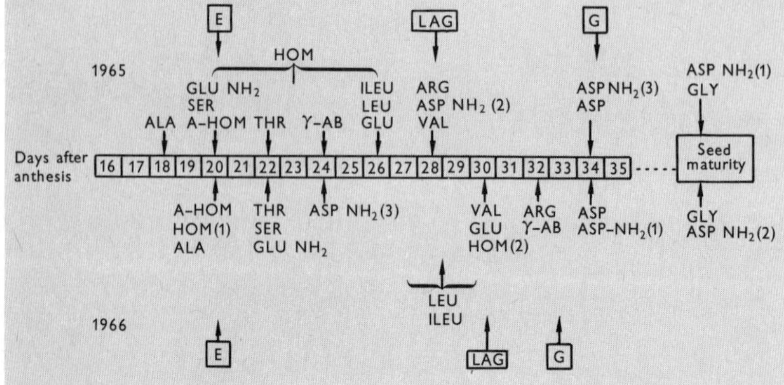

Fig. 7.7 The timing of events during maturation of seeds of field pea (*Pisum arvense*) during two seasons of growth. Arrows marking specific amino compounds refer to the times at which those compounds reach a maximum amount in the seed. Where two or more peaks are recorded for the content of a particular amino acid these are numbered in decreasing order of magnitude.
E—the time when there is a maximum amount of endospermic fluid in the seed.
LAG—the period showing a temporary cessation of fresh weight increase by the seed.
G—the time when globulin-type protein is first detected in the cotyledons of the embryo.
(From Flinn and Pate,[192] Fig. 7, p. 493.)

These soluble amino acids in the seed to some extent function as precursors for the amino acids incorporated into the reserve proteins. However there are major differences between the amino acid composition of the soluble nitrogen fraction and the protein component.[192, 308]

The amino acids in seeds can be derived from a variety of sources. In monocarpic annual plants the onset of fruit and seed development is frequently accompanied by the correlated senescence of leaves on the plant. During the senescence of these leaves, proteins may be hydrolysed and the products translocated to the developing seed and fruit. However the amino acid composition of the leaf protein differs from that of the soluble pool in the seed so that extensive interconversion must occur.

In some plants the carpellary tissues become extensively developed into large ovaries and these may function in producing amino acids to be translocated to the developing seed. In peas and beans for example the pod develops extensively prior to the rapid phase of seed growth. In the initial phase of seed development the endosperm develops more rapidly than the embryo. Thus amino acids accumulating in the pod and endosperm could serve as precursors for the synthesis of soluble and protein amino acids in the later developing embryo. Flinn and Pate[192] estimate that about 20% of the nitrogen of the seed of *Pisum arvense* could be furnished by metabolites stored earlier in the endosperm and pod.

In many perennials, fruit and seed development occur without leaf senescence and thus amino acids in the seed must be derived from components translocated from the photosynthetically active leaves. In other instances such as some spring-flowering trees, e.g. elm (*Ulmus*), and silver maple (*Acer saccharinum*), fruit and seed development precede leaf production. In these cases the soluble and protein amino acids in the seed must be derived from reserve components stored within the plant. The reserves may be in the form of proteins or amino acids; in this regard it is noteworthy that arginine plays an important role as a storage product in the bark of many tree species.[267]

Few studies have been made to establish to what extent the amino acids derived from the different sources contribute to the nitrogen budget of the developing seed. Observations that amino acids accumulate in seeds in ratios different from those encountered in vegetative tissues suggest that the seed possesses synthetic capabilities. Some interconversion of amino acids can occur as indicated by the experiments of Sodek and Wilson[577] which demonstrated the transfer of carbon from lysine into glutamic acid and proline in the developing maize (*Zea mays*) grain. In contrast leucine did not undergo extensive interconversion.

It appears that developing seeds synthesize many amino acids *de novo*. Lewis and Pate[375] indicate that the principal nitrogenous compounds translocated to the field pea (*Pisum arvense*) seed are the amides, glutamine and asparagine, and the photosynthetically generated amino acids serine, glycine and alanine. Many of the amino acids required in large amounts for synthesis of the reserve proteins were found to be poorly represented in the phloem sap which did, however, contain large amounts of sugars. It was concluded that the synthesis of tyrosine, leucine, phenylalanine, histidine, lysine and arginine was a major function of the developing seed. The nitrogenous component of these amino acids can be derived from the amido group of the amides in the phloem sap and the carbon skeletons are furnished by the translocated carbohydrates. Similar synthetic capabilities have been demonstrated in the seeds of *Datura stramonium* which can manufacture a range of amino acids from sucrose and nitrate.[374]

GERMINATION

The exposure of a mature seed to appropriate conditions of temperature, moisture and aeration usually results in the resumed metabolism of the previously quiescent seed. Usually the first visible change accompanying the exposure of the seed to moisture is an increase in seed size. This initial increase in volume is due to imbibition and is primarily a physical process related to the chemical composition of the seed. Protein-rich seeds imbibe extensively whereas seeds containing lipid or insoluble starch residues take up less water. Imbibition occurs in both viable and non-viable seeds.

In viable seeds imbibition is accompanied by an increased respiratory activity of the seed.[486] Respiration in dry seeds is minimal and those rates which have been measured have been criticized as being due to contaminating micro-organisms and thus not representative of seed respiration. The initial respiratory increase (A) during imbibition is directly related to the degree of hydration. With an increasing period of exposure to moisture the respiratory rate reaches a transient plateau (B) or declines (Fig. 7.8), at which time the rate of water uptake also declines. This phase

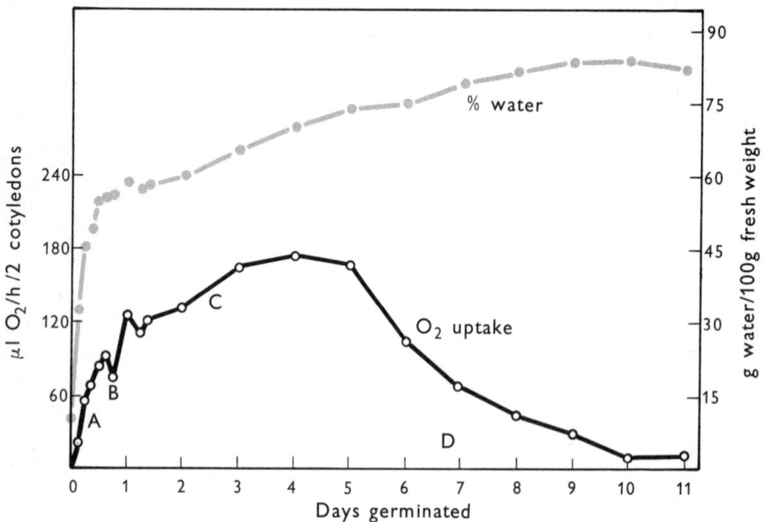

Fig. 7.8 Changes in respiration rate and water content in the cotyledons of *Phaseolus vulgaris* during germination at 25°C. (Data from Öpik and Simon,[486] Fig. 2, p. 301.)

is followed by an increased respiratory activity (C) and a resumption of water uptake. This has been designated the metabolic phase of water uptake.

Various studies have been made to account for the resumption of respiratory activity.[403] Levels of respiratory enzymes in the dry seed are low, and it appears that the onset of respiration during imbibition may be due to activation by hydration of enzymes already existing in the mature seed. However, subsequently the increasing respiratory activity of the seed appears to be associated with the synthesis of respiratory enzymes. Marré[403] has indicated that the normal increase in several glycolytic enzymes and the mitochondrial enzymes isocitrate dehydrogenase, cytochrome oxidase and malate dehydrogenase does not occur when seeds are germinated at low temperature, or in the lack of oxygen or in the presence of cycloheximide. These observations suggest that the increase in respiration is associated with an increase in the level of respiratory enzymes requiring protein synthesis. There is an increase in mitochondrial nitrogen during germination.[6,72,107] This increase may be partially associated with an increase in the enzymatic and phosphorylative efficiency of existing mitochondria which undergo ultrastructural changes.[72,454] However during germination the enhanced respiratory activity is also associated with an increase in the number of mitochondria.[72]

The respiratory activity of the whole seed is a reflection of the respiration occurring in the reserve tissue and the growing embryonic axis. During germination the respiratory capability of the axis continues to increase whereas that of the storage tissues, endosperm or cotyledon usually increases for a few days and is followed by a decline. Most studies of the respiratory activity of endosperm tissue have been confined to the lipid-rich castor bean.

CHANGES IN NITROGENOUS COMPONENTS DURING GERMINATION

Seed germination is characterized by a decline in the nitrogen content of the storage tissue and an increase in this component in the growing embryonic axis (Fig. 7.9a, b).

Protein

The major decline in nitrogen content of the endosperm or the cotyledons can be attributed to a decrease in protein level in the reserve tissues. During the decrease in protein there is an accumulation of soluble nitrogen, primarily α-amino nitrogen, in the embryonic axis and to a lesser extent in the reserve tissue (Fig. 7.9). These findings suggest that during germination the proteins in the reserve tissue are hydrolysed to amino acids which are translocated to the growing axis.

There are several reports of proteolytic enzymes in germinating seeds which may function in hydrolysing the reserve proteins. However, there are conflicting reports on the levels of proteolytic enzymes in reserve tissues and the role of these enzymes in protein hydrolysis in many instances has not been established.[143,542]

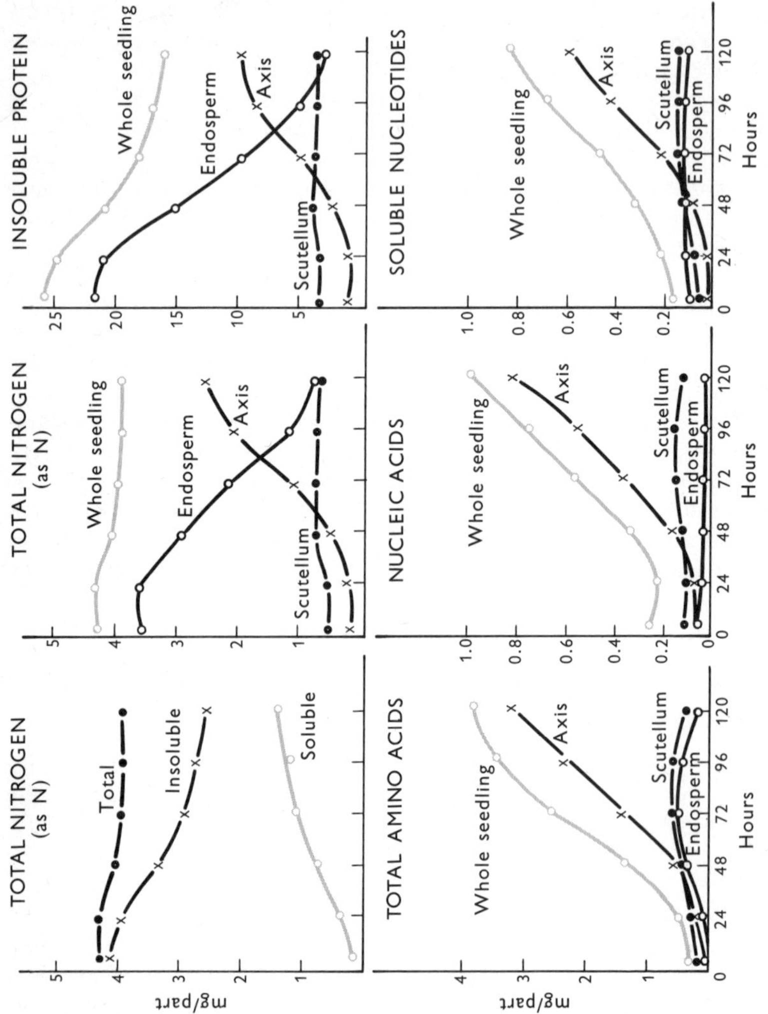

Fig. 7.9a Changes in nitrogenous components during the germination of maize (*Zea mays*). (From Ingle, Beevers and Hageman,[288] Fig. 1.12, p. 737.)

Part of the confusion and difficulty in interpreting the reports of proteolytic activity in germinating seeds can be attributed to the method of enzyme assay. The usual assay for proteolytic activity involves the incubation of enzyme extracts with casein or haemoglobin, obviously not normal substrates encountered by plant proteases, and measuring the

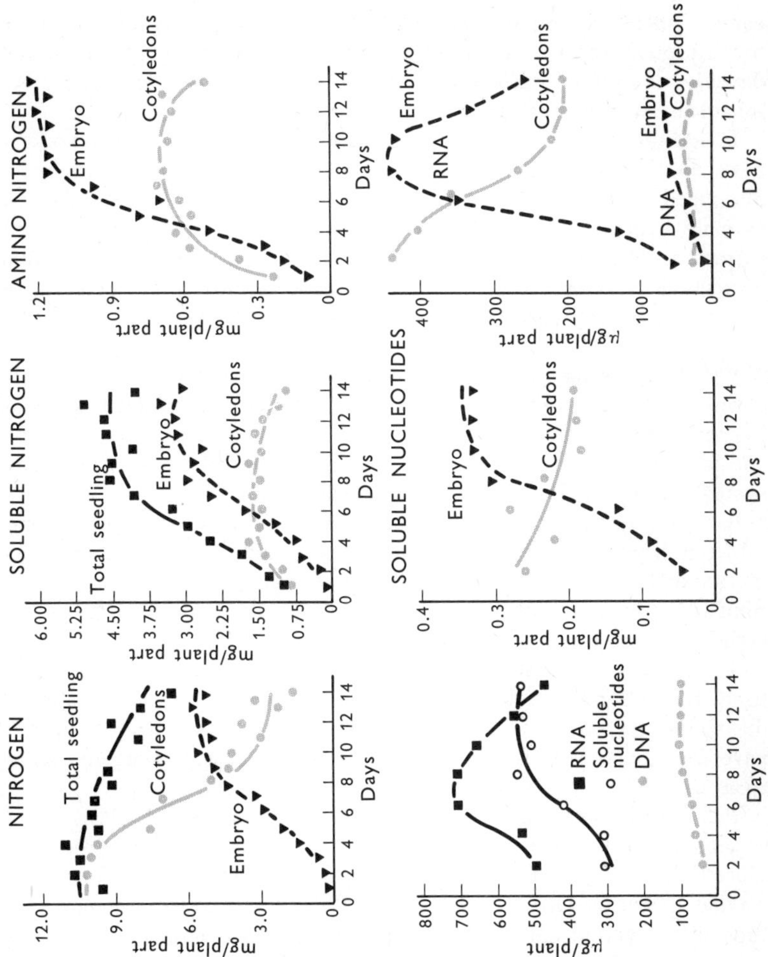

Fig. 7.9b The changes of various nitrogenous components in the cotyledons and embryo-axis of peas during germination. (From Beevers and Guernsey,[40] Figs. 2–7, p. 1457.)

release of trichloracetic acid soluble peptides and amino acids. The trichloracetic acid soluble components ideally can be detected by measuring the absorption at 280 nm which detects tyrosine residues according to the method originally developed by Kunitz[348] for determining trypsin-like activity in extracts of animal tissues. Alternatively, the trichloracetic acid soluble α-amino nitrogen can be determined. Since crude plant extracts contain phenolic compounds, nucleotides etc., which may all contribute to absorption at 280 nm unrelated to the hydrolysis of proteins, the detection of α-amino nitrogen by a colorimetric method provides a more reliable assay of proteolytic activity.

In addition to using non-physiological protein substrates, other workers have used various synthetic peptides, some of which were originally developed to measure trypsin-like enzymes in extracts of animal tissues. In many instances there is little relationship between the capacity of plant extracts to hydrolyse the synthetic peptides and the ability to hydrolyse proteins.[38, 96] Thus assays based on synthetic peptides may not measure proteolytic activity and should more correctly be termed peptide hydrolases.

Very few investigators have studied the capacity of extracts from germinating seeds to hydrolyse reserve proteins from plants. Jacobsen and Varner[304] used gliadin to assess proteolytic activity in barley aleurone cells and an azo derivative of pea globulins has recently been used to study proteolytic activity in extracts from pea cotyledons.[268]

Based on the measurement of the capacity of extracts to release α-amino nitrogen from casein substrates, it appears that the protease level increases in the cotyledons of peas during germination[38, 45, 233] (Fig. 7.10). Similar increases in proteolytic activity have been reported during the germination of beans (*Phaseolus vulgaris*).[512, 717] Palmiano & Juliano[492] have also reported that proteolytic activity, as measured by release of trichloracetic acid soluble material absorbing at 280 nm, also increases during the germination of rice (*Oryza sativa*). Jacobsen and Varner[304] have demonstrated an increase in gliadin hydrolysing capacity following treatment of barley aleurone cells with gibberellic acid.

The proteolysis occurring during germination may be catalysed by several enzymes, few of which have been characterized. Ihle and Dure[280, 281, 282] indicate that carboxypeptidase level increased in cotton cotyledons during germination. Eight peptidases (three carboxypeptidases, three amino peptidases and two dipeptidases) have been demonstrated in extracts of barley grains.[424] The carboxypeptidase activity was located in the aleurone layer of the endosperm, in the starchy endosperm and in the scutellum and in these tissues activity of the enzymes increased during germination. The two dipeptidases and aminopeptidase were found in the scutellum and there was little change in activity of these enzymes during germination. Harvey and Oaks[253] have recently characterized an acid protease from maize (*Zea mays*) which increases during the depletion of the endosperm protein.

In peas (*Pisum sativum*) peptide hydrolase activity (measured by hydrolysis of *N*-benzoyl-L-arginine-*p*-nitroanilide (BAPA) and leucyl-*p*-nitroanilide (LPA)) is initially high in the cotyledons and declines as

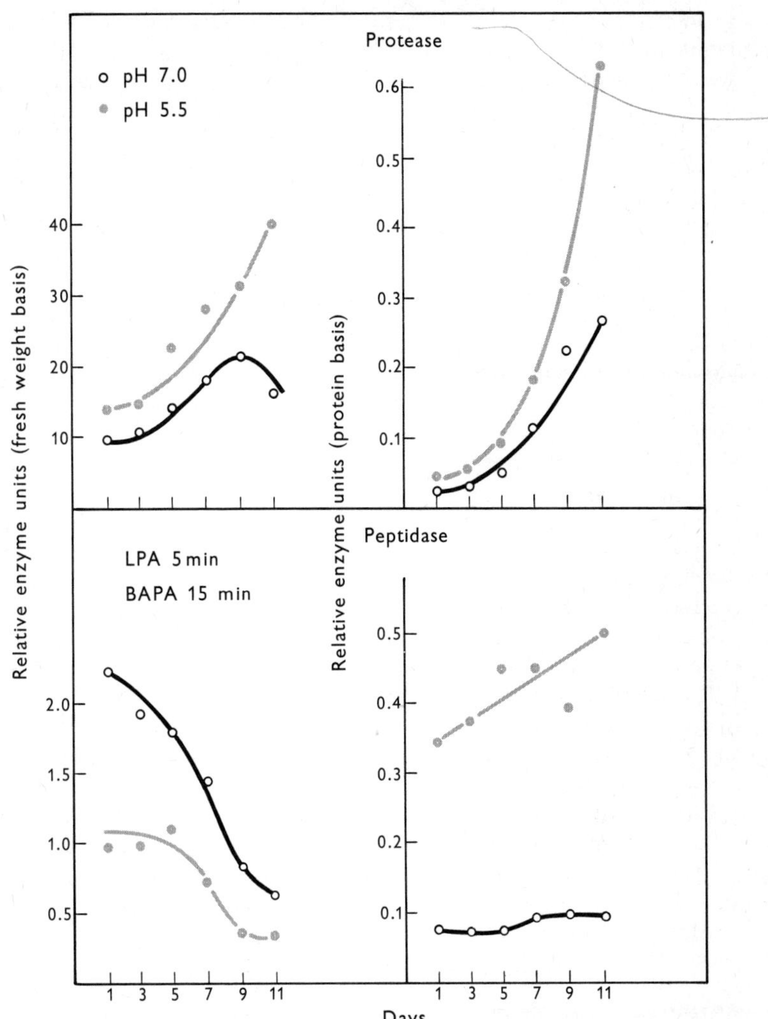

Fig. 7.10 Changes in protease and peptidase activity in extracts of pea cotyledons during gerimnation. (From Beevers,[38] Figs. 5 and 8, pp. 1841–1842). In the protease assay extracts were prepared from pea cotyledons after various periods of germination. The capacity of the extracts to release trichloracetic acid soluble ninhydrin reactive material from casein was determined at pH 7.0 and 5.5. Peptidase activity was measured on the same extracts by colorimetrically determining the release of *p*-nitroanilide from the synthetic peptides leucine-*p*-nitroanilide (LPA) and benzoyl-arginine-*p*-nitroanilide (BAPA). Enzyme activity is expressed per unit fresh weight or per unit protein.

germination progresses[38] (Fig. 7.10). In beans (*Phaseolus vulgaris*) peptide hydrolase activity remained constant during initial phases of germination.[512] Similar constant levels of peptide hydrolases have been reported in barley grains during germination; however, the capacity to hydrolyse LPA increased.[542]

Clearly the physiological role of these peptide hydrolases remains to be established; likewise it has yet to be determined whether the measured increases in proteolytic activity represents a build up of the enzymes responsible for degrading the reserve proteins.

Associated with our limited knowledge of the enzymes involved in reserve protein degradation is a lack of information on the mechanisms involved. Catsimpoolas and coworkers[99] have observed that the principal globulin components of soybeans (*Glycine max*) are degraded at different rates. The electrophoretic mobility of the reserve polypeptides was altered early during germination suggesting that deamidation of asparagine and glutamine residues of the proteins produced an increase in negative charges by exposure of carboxyl groups. The major globulin α-arachin in peanuts (*Arachis hypogaea*) shows a similar anodic shift in electrophoretic mobility early during germination, again suggesting the loss of basic residues, which may involve deamidation.[139] Significantly these early changes in electrophoretic mobility are not accompanied by change in antigenic properties of the reserve proteins so that the initial degradation must occur at antigenically unimportant sites.

Ultrastructural studies[23,74] indicate that during germination protein hydrolysis occurs uniformly throughout the protein body. Such findings suggest that the hydrolysis may be achieved by proteinases included in or associated with the protein bodies. In this regard protein bodies from cotton (*Gossypium hirsutum*),[714] barley (*Hordeum vulgare*),[487] and hemp (*Cannabis sativa*)[588] seed have been reported to contain acid proteases supporting the notion that protein bodies are lysosomal-like organelles in which intracellular digestion occurs. During the early stages of germination protein degradation in the protein body may be achieved by these acid proteases; however this possibility remains to be proven.

Amino acid metabolism during germination

Reference to Fig. 7.9a, b indicates that during germination soluble amino acids accumulate in the reserve tissue and to a greater extent in the developing embryonic axis. Some of these amino acids could originate directly from the hydrolysis of the reserve proteins. However, analyses indicate that the amino acid composition of the reserve protein is considerably different from the soluble amino acid pool in either the reserve tissue or the axis (see also Table 7.5).

Studies using labelled amino acids indicate that the products of reserve

Table 7.5 Amino acid composition (moles percent) of the reserve proteins and the soluble amino acids found in cotyledons, exudate and axis hypocotyl of germinating pumpkin (*Cucurbita moschata*) seeds. (From Chou and Splittstoesser,[113] Table 2, p. 112.)

Amino acid	Globulin	5-day			7-day		
		Cotyledon	Exudate	Hypocotyl	Cotyledon	Exudate	Hypocotyl
Aspartate	10.0*	3.2	6.4	1.1	4.2	5.4	1.6
Asparagine	—	4.0	3.2	3.4	4.2	1.9	4.4
Glutamate	20.7*	8.9	4.1	1.2	10.0	7.9	0.9
Glutamine	—	15.4	18.0	24.3	10.0	19.7	25.8
Threonine	3.4	1.5	2.5	2.3	2.2	2.5	2.4
Serine	7.5	3.8	11.5	11.8	2.4	10.4	12.4
Glycine	8.0	2.0	10.0	6.5	1.9	9.2	7.8
Alanine	7.0	2.3	10.0	22.6	2.4	9.4	15.5
Cysteine	0.7	Trace	Trace	Trace	Trace	Trace	Trace
Methionine	1.5	0.8	Trace	0.6	0.5	Trace	0.7
Isoleucine	3.8	1.1	1.2	2.2	1.8	1.2	3.1
Valine	5.8	2.7	2.7	3.3	5.5	2.5	5.6
Leucine	8.4	8.1	1.8	3.0	9.2	1.5	3.8
Proline	0.4	3.2	1.9	1.0	3.0	1.0	1.2
γ-Aminobutyrate	None	2.6	5.1	0.9	2.8	3.6	0.8
Tyrosine	1.5	3.9	Trace	1.5	4.6	1.0	1.5
Phenylalanine	4.6	4.5	1.7	8.8	4.8	2.1	4.6
Histidine	1.8	3.0	2.2	1.7	2.1	2.1	2.0
Ornithine	None	0.9	14.4	1.2	1.5	15.7	2.1
Lysine	1.5	1.4	2.5	0.4	1.8	2.1	1.4
Arginine	13.3	26.7	0.7	2.1	24.9	0.6	2.4

*Includes its corresponding amide.

protein digestion may be utilized in several ways.[45] The amino acids may be incorporated into enzymic proteins in the reserve tissues. They may be translocated to the axis and serve as a source of precursor for protein synthesis in the developing axis. Folkes and Yemm[194] have estimated that 75 per cent of the amino acid required for protein synthesis in the axis can be satisfied by amino acids derived directly from the hydrolysis of endosperm protein reserves in barley. However embryos may be cultured on simple organic media and nitrate nitrogen source; so they apparently contain the metabolic capability of synthesizing many of their amino acids.[422]

The amino acids derived from reserve protein hydrolysis provide carbon skeletons for other syntheses in the developing axis,[45,353] and they may also serve as nitrogen source for the synthesis of other nitrogenous components.

The mechanisms of amino acid metabolism to provide carbon skeletons have not been established. However, as indicated in Chapter 3, for the most part it appears that amino acids undergo transamination and the resulting keto acids are eventually metabolized through the Krebs' tricarboxylic acid cycle.

The nitrogenous component of the amino acids appears to be converted to the amido nitrogen of glutamine and this is the principal nitrogen-containing compound exported from the cotyledons and endosperm to the developing axis. Although the reserve proteins are enriched in asparagine and arginine and cotyledons may contain soluble nitrogen-rich amino acids such as canavanine, it appears that these compounds are not extensively translocated to the embryonic axis; instead they are metabolized in the reserve tissue.[311, 537, 583, 688] Initially arginine and canavanine are hydrolysed by arginase which accumulates in the cotyledons during germination. The products from arginine hydrolysis are ornithine and urea. The occurrence of ornithine in the exudate of pumpkin (*Cucurbita moschata*) cotyledons (see Table 7.5) suggests that this amino acid may be transported to the growing axis; additionally it can be metabolized to glutamate, γ-amino butyrate and proline. The urea produced as a result of arginase activity does not accumulate but is hydrolysed by urease to ammonia and carbon dioxide. The ammonia produced in this hydrolysis may be incorporated as the amido nitrogen of glutamine through the activities of the enzyme glutamine synthetase which accumulates in cotyledons during germination.[380] The glutamine may then be translocated to the growing axis; high levels of glutamine in the cotyledonary exudate are consistent with this suggestion.

Studies of amino acid metabolism during germination, which utilize labelled compounds, may be criticized in that the exogenously supplied metabolite may artificially expand precursor pools and thus not be metabolized in the 'normal' fashion. This criticism has partially been overcome in the recent investigations of Sodek and Wilson[578] who studied the metabolism of amino acids from reserve proteins of maize (*Zea mays*) which had been labelled during seed development. These studies demonstrate that leucine, derived from the storage protein, was incorporated into proteins in the developing axis; additionally some of the amino acid was oxidized resulting in the production of labelled carbon dioxide.

Nucleic acids

During germination of peas (*Pisum sativum*) the RNA content of the cotyledon declines and the DNA content increases slightly and then decreases (Fig. 7.9b). Similar changes occur in beans (*Phaseolus vulgaris*).[660] In the peanut (*Arachis hypogaea*)[107] RNA content increases initially in the cotyledon and then declines during the senescence of the cotyledon. DNA content increases slightly. The small increases in DNA content of these reserve tissues occur in the absence of any detectable cell division and to a great extent can be attributed to the production of mitochondria, with their associated DNA, which occurs during the early stages of germination.[72]

In species with epigeal germination the RNA content of the cotyledons may decline initially; however following exposure to light and accompanying the greening of the cotyledon RNA levels may increase. The enhanced RNA levels are associated with a production of chloroplast RNA fractions and an increased synthesis of cytoplasmic RNA.[286]

In cereal endosperms an initially low RNA content is further decreased during the course of germination (Fig. 7.9a).

The decline in RNA content of the storage tissue is accompanied by an increase in RNA levels in the embryonic axis (Fig. 7.9b). Also there is an accumulation of soluble nucleotides in the reserve tissues and embryonic axis during the initial phases of germination (Fig. 7.9b).

These observations that soluble nucleotides accumulate during RNA depletion in the reserve tissue suggest that they may arise as a result of hydrolysis of RNA. However the total nucleic acid and soluble nucleotides accumulating in the axis exceeds the RNA level present in the reserve tissue so that during germination there must be extensive *de novo* synthesis of nucleotides and nucleic acids.

Various ideas have been formulated to account for the decrease in RNA content. It has been suggested that the RNA may be transported from the reserve tissue to the growing axis as a macromolecule.[367,484] Others contend that the RNA is hydrolysed to its component nucleotides which are then transported. This later contention is partially supported by the frequent observation[28,290,409] that ribonuclease levels increase in storage tissue during germination; however similar increases in ribonuclease level occur in axis tissue in which RNA levels are increasing. These conflicting observations may be accounted for by the fact that in very few instances have the enzymes responsible for the measured ribonuclease activity been characterized. However Ingle and Hageman[290] demonstrated that the enhanced ribonuclease level in maize (*Zea mays*) endosperm was associated with ribonuclease I type activity, whereas in the axis tissue both ribonuclease I and ribonuclease II increased during germination with a greater production of the latter enzyme. Somewhat ironically this most detailed study involves a system which is relatively low in reserve RNA.

If, as the accumulating evidence tends to indicate, the RNA of the reserve tissue is hydrolysed by ribonucleases then there is a need to understand the mechanisms involved. Foreseeably if the hydrolysis involved endonuclease activity intermediate size polynucleotide fragments would be produced. However gel electrophoresis of RNA extracted from germinating seeds has, so far, failed to demonstrate the accumulation of such fragments.

Although RNA content of the principal reserve tissue may decline during germination, labelling studies[117,201] indicate that simultaneous RNA synthesis may be occurring. Studies which demonstrate the failure of germination or the production of a specific enzyme, in the presence of inhibitors of RNA synthesis, demonstrate the essentiality of the production of RNA. In castor bean (*Ricinus communis*) endosperm it appears that part of the requirement for RNA synthesis represents a need for ribosomal RNA synthesis; ribosome content increases initially in this tissue during germination.[403] In peas and beans, however, abundant ribosomes are present in the dry seed and it has been demonstrated[660] that in the bean (*Phaseolus vulgaris*) the polysomes formed during germination utilize preformed ribosomal RNA. Thus it may be that the requirement for RNA synthesis in this case may involve production of mRNA.

CHANGES IN NON-NITROGENOUS COMPONENTS

In addition to the breakdown of the major nitrogenous components other materials are depleted from the reserve tissues during germination. These depletions involve the hydrolysis of the reserve material and a subsequent translocation and/or metabolism of the products. Starch, for example, is hydrolysed to glucose by the combined activities of the enzymes α-amylase, β-amylase, and 1,6-glucosidase, and maltase. An alternative hydrolysis could occur by the phosphorolytic pathway involving the enzyme phosphorylase. However studies of the changes in enzyme levels during the germination of peas (*Pisum sativum*)[616] and barley (*Hordeum vulgare*)[314] demonstrate that the onset of starch degradation coincides with the build up of α-amylase indicating that the amylolytic pathway is favoured.

In lipid-rich seeds such as castor bean (*Ricinus communis*) it is observed that the decrease in fat content is associated with a stoichiometric accumulation of sucrose.[37] This interconversion involves an initial hydrolysis of the reserve triglycerides to fatty acids and glycerol by the enzyme(s) lipase and subsequent oxidation of the fatty acids to acetyl-CoA by the β-oxidation pathway. The acetyl-CoA is metabolized in the glyoxylate cycle, which involves participation of the key enzymes malate synthetase and isocitrate lyase, and ultimately forms oxaloacetate. This compound can be converted to phospho-enol-pyruvate and by a reversal of glycolysis produce glucose and fructose phosphates which are utilized for the production of sucrose. The enzymes involved in lipid hydrolysis, fatty acid oxidation and the glyoxylate cycle are undetectable or present in low levels in dry seeds but their activities, along with those of some of the glycolytic enzymes, increase as germination progresses.

The majority of the metabolism of the fatty acids is carried out in subcellular organelles, the glyoxysomes, which are synthesized extensively in the reserve endosperm during germination.[209] Thus in addition to the production of the enzymes necessary for the degradation of fats there is a commensurate requirement, during germination, for the synthesis of the structural proteins and lipids necessary for the formation of the organelles in which the enzymic machinery is housed.

SOURCE OF ENZYMES IN GERMINATING SEEDS

The resurgence of metabolic activity following imbibition of non-dormant viable seeds is associated with an increase in enzymic activity. This enhanced activity can arise in various ways (Table 7.6).

Table 7.6 Possible sources of increased enzyme activity during germination.

Method	Enzyme
Activation by hydration of existing enzymes	Some transaminases ; glutamic decarboxylase ; ribonuclease
Modification of inert precursor into active enzyme	α-1,6-glucosidase trypsin-like enzyme
De novo synthesis using stored mRNA	Carboxypeptidase
De novo synthesis requiring mRNA synthesis	Various hydrolases ; some respiratory enzymes

In some instances the enzymes may be present in the dry seed and are activated by the hydration which occurs during imbibition. That increase in enzyme activity is due simply to activation can be demonstrated by the fact that an elevation of enzyme level occurs in cell-free homogenates incubated in the cold. This increase in enzyme level is insensitive to inhibitors of protein or RNA synthesis. It has been indicated that at least part of the increase in ribonuclease level which occurs in maize (*Zea mays*) endosperm during germination is due to activation of existing enzymes.[291] In wheat (*Triticum aestivum*) grains the activity of glutamic acid decarboxylase and glutamic acid–alanine transaminase are related to moisture content of the grain, suggesting that the enzymes are activated by hydration.[383]

In other cases the activation and increase in enzyme activity may involve more complex reactions than simple hydration of existing proteins. The increase in α-1,6-glucosidase in germinating peas (*Pisum sativum*) is a result of its release from an inactive precursor form.[557] Similarly, it is suggested that the increase in trypsin-like activity which occurs in lettuce (*Lactuca sativa*) seeds is due to the liberation of the enzyme from some inactive precursor form. Since the level of trypsin inhibitor also decreased during the initial phases of germination it is speculated that the increase in trypsin-like enzyme may involve its release from a complex of enzyme and trypsin inhibitor; the inhibitor may be degraded by other proteolytic enzymes.[558] In other systems, however, there is no convincing evidence that the frequently encountered trypsin inhibitors inhibit endogenous proteolytic enzymes.[542]

In many instances the normal increase in enzyme activity following imbibition can be prevented by inhibitors of RNA and protein synthesis indicating that the *de novo* synthesis of the enzymic proteins involves transcription and subsequent translation. In addition to the previously referred to respiratory enzymes which appear to be synthesized *de novo* during germination, the earlier investigations of Young and Varner[718] clearly indicated that the production of various hydrolases in the cotyledons of germinating peas (*Pisum sativum*) required protein synthesis. This synthesis occurs at a time when storage proteins are being degraded. The many investigations of the production of hydrolytic enzymes by the aleurone cells of barley indicate that the build up of α-amylase, protease and ribonuclease involves protein synthesis dependent upon a normal transcription and translation.[304]

The increase in carboxypeptidase which occurs in cotton (*Gossypium hirsutum*) cotyledons during germination can be prevented by cycloheximide but not by concentrations of Actinomycin D which inhibit the bulk of RNA synthesis. These findings, indicating that the accumulation of an enzyme depends on protein but not RNA synthesis, suggest that enzyme production may involve the translation of pre-existing mRNA coding for the carboxypeptidase. In a detailed analysis of this system it has been established that immature cotton seeds can be induced to germinate precociously and produce normal amounts of carboxypeptidase in the presence of Actinomycin D. However with progressively younger, more immature, seeds the capacity to produce carboxypeptidase during precocious germination in the presence of Actinomycin D is lost. These data are consistent with the concept that mRNA directing carboxypeptidase synthesis is produced at some time during seed development and maturation. The translation of this mRNA normally does not occur until germination.[282] It has been suggested[661a] that the mRNA, or its precursor, for carboxypeptidase which is transcribed during seed development is not translated until the mRNA undergoes post-transcriptional poly-adenylation (see p. 172) during germination. This concept may be supported by the preliminary observations that cordycepin (3-deoxyadenosine), an inhibitor of post-transcriptional poly (A) addition, prevents the build up of carboxypeptidase in the cotyledons of germinating cotton seeds.[661a]

The production of the hydrolytic enzymes in the reserve tissues is in some cases dependent upon the presence of axis tissue. This observation was originally made by Haberlandt.[240] It is now recognized that in cereal grains the gibberellins produced in the embryonic axis are necessary for stimulating the onset of the synthesis of hydrolases in the reserve tissue. In the early 1960's it was speculated that gibberellins may function as inducers in facilitating the production of the mRNA's for the various hydrolases. However, this idea has not been confirmed and the mechanism by which gibberellins enhance hydrolase synthesis in the cereal grain remains unresolved.

A similar mechanism operates in some dicotyledons; removal of the cotyledons at the initiation of germination prevents the build-up of various hydrolases.[24, 500, 653] Axis removal also prevents the accumulation of polysomes which normally occurs in the cotyledon during germination of peas (*Pisum sativum*).[112] However, the factor(s) from the axis which controls the production of hydrolases and regulates polysome formation has not been identified.

THE ONSET OF PROTEIN SYNTHESIS DURING GERMINATION

A great deal of research has been directed at elucidating the mechanism by which protein synthesis is triggered following the transfer of a quiescent seed to germinating conditions. Initial investigations[398, 455] indicated that during the early imbibition phase labelled amino acids became incorporated into proteins before nucleic acid precursors were combined into polynucleotides. More recent investigations, however, indicate that the onset of RNA and protein synthesis may be concurrent events.[414]

It has been established[398] using *in vitro* systems that ribosomal preparations from imbibed seeds are able to catalyse the incorporation of amino acids into trichloracetic acid insoluble products when fortified with appropriate cofactors and supernatant components from either dry or imbibed seed. In contrast, in reciprocal mixing experiments, ribosomal preparations from dry seeds were unable to direct amino acid incorporation (Table 7.7).

Table 7.7 Amino acid incorporation by ribosomal (microsomal) preparations from dry and imbibed peanut (*Arachis hypogaea*) cotyledons. (After Marcus and Feeley.[398])

Ribosomes	Supernatant	Leucine incorporation (c.p.m.)
0 day	—	15
0 day	0 day	12
0 day	1 day	0
1 day imbibed	—	167
1 day imbibed	0 day	214
1 day imbibed	1 day	276

During imbibition there appears to be an alteration of the ribosomal component. Ribosomal preparations from dry seeds were able to direct the incorporation of phenylalanine when supplied with synthetic mRNA polyuridylic acid. These findings indicate that ribosomes from dry seeds and the soluble factors required for protein synthesis are enzymatically functional prior to imbibition but the ribosomes are deficient in mRNA. Subsequent investigations have indicated that ribosomal preparations from imbibed seeds contain polysomes whereas those from dry seeds do not.[399]

Some of the mRNA utilized in polysome formation and protein synthesis in the early phases of imbibition is apparently stored in the dry seed and *in vitro* polysome formation has been demonstrated in preparations from wheat germ. Although a ribonucleoprotein fraction from dry seeds has been shown to function in stimulating amino acid incorporation in an *in vitro* system it has not been possible thus far to isolate the stored mRNA.[675]

In addition to the stored mRNA directing the initial protein synthesis following imbibition there is also evidence for an early synthesis of RNA during germination.[414] This synthesis is presumably catalysed by an RNA polymerase which is activated following hydration of the seed. Jendrisak and Becker[307] have recently indicated that commercial wheat germ is a good source of polymerase, chiefly of the polymerase II type. In addition, studies by Payne[497] have demonstrated *in vivo* polymerase activity during the early germination of onion (*Allium cepa*) seeds. The polymerase activity was distributed along the length of the chromosome and not confined to the nucleolus. This nucleoplasmic polymerase therefore may be directing the synthesis of mRNA during the early phases of germination. Attempts to identify the nature of the RNA synthesized during the initial phases of germination have been made by Chen *et al.*[104] They showed that labelled RNA synthesized during the early phases of wheat seed germination could hybridize with DNA from wheat. This hybridization could be prevented by inclusion of unlabelled RNA extracted from dry seed. In contrast the hybridization of labelled RNA from seeds at 72 hours after germination was not restricted to such a great extent by RNA from dry seeds. This evidence indicates that the RNA synthesized during the early phases of germination is similar to that already present in the dry seed. Imbibition triggers protein synthesis by allowing for the combination of the existing ribosomes with the stored mRNA. Subsequently RNA synthesis proceeds with the initial production of mRNA similar to that stored in the dry seed. However at later stages of germination the RNA population is different from that present in the dry seed, suggesting that as germination proceeds there is production of new mRNA species; this situation would allow for the progressive production of new enzymes required for the development of new metabolic pathways which become established during germination.

In addition to this change in mRNA population during the germination process it has been shown that there are concurrent changes in tRNA species.[464,659] Accompanying changes in amino-acyl synthetases also occur, so it is possible that some regulation of enzyme synthesis during germination may be exerted at the translational level.

NITROGEN METABOLISM DURING SEED STORAGE AND DORMANCY

Viability

The metabolism of the mature seed during dry storage is minimal and seeds can be stored for prolonged periods without deleterious effects. However with progressively increasing storage periods there is frequently a loss of viability of the seeds. The rate of loss of viability during storage is dependent upon such conditions as moisture, oxygen levels, and temperature. Various hypotheses have been advanced to account for the loss of viability.[527] During storage under conditions which result in a decreased viability there is frequently an associated increase in chromosomal aberrations and nuclear damage which would result in altered DNA metabolism. However irradiated seeds, in which DNA synthesis and cell division are suppressed, are still capable of germinating and producing seedlings (gamma plantlets)[238] so that the loss of seed viability cannot be attributed solely to a failure of DNA production.

Ultrastructural studies indicate that during the loss of viability there is a deterioration of plastid membranes, the endoplasmic reticulum becomes distorted and the plasmalemma and mitochondria become disorganized.[246] Non-viable embryos are incapable of incorporating amino acids into protein and this may be associated with a decreased capacity of the supernatant fraction from non-viable seeds to bind amino-acyl tRNA to ribosomes.[526] Clearly the events leading to these biochemical lesions associated with the loss of viability during storage require extensive further study. At present the loss of viability indicates that some metabolic, albeit deleterious, processes do occur in the dry seed.

Dormancy

Freshly harvested seeds frequently fail to germinate when placed under apparently optimal conditions of temperature, moisture and aeration. The mechanisms preventing the germination of the freshly harvested seeds are not understood but frequently the block to the germination process can be removed by a period of dry storage or **after-ripening**. Other seeds fail to germinate even after a period of after-ripening and germination can proceed only after various manipulations. Such viable seeds which fail to germinate under apparently optimal conditions are considered to be **dormant**. The dormancy may be due to a variety of causes which have been classified as follows:[130] (1) immaturity of the embryo; (2) impermeability of the seed coat to water; (3) mechanical resistance of the seed coat to embryo growth; (4) low permeability of the seed coat to gases; (5) dormancy resulting from a metabolic block of the embryo which can be overcome by light and/or chilling; (6) a combination of the above.

Most detailed studies of metabolism of dormant seeds have been conducted on those types in which dormancy can be broken by a chilling treatment of the imbibed aerated seeds (**stratification**). During the chilling treatment changes occur in the level of various growth regulators in the seed which it is suggested control metabolic events in the embryo.[12] This contention is to some extent supported by experiments which showed that the requirement for a cold treatment for the breaking of dormancy can be replaced by treatment of the seed with gibberellic acid or kinetin.

During the low temperature treatment there may be extensive changes in nitrogenous compounds. In *Heracleum sphondylium*, which is dormant due to an immature embryo and is stimulated to develop by low temperature treatment, there is a depletion of protein in the endosperm and an increase in soluble amino acids during stratification (Fig. 7.11). Stokes[602] suggests that the cold treatment stimulates protein hydrolysis in the endosperm with the resultant production of amino acids essential for the nutrition

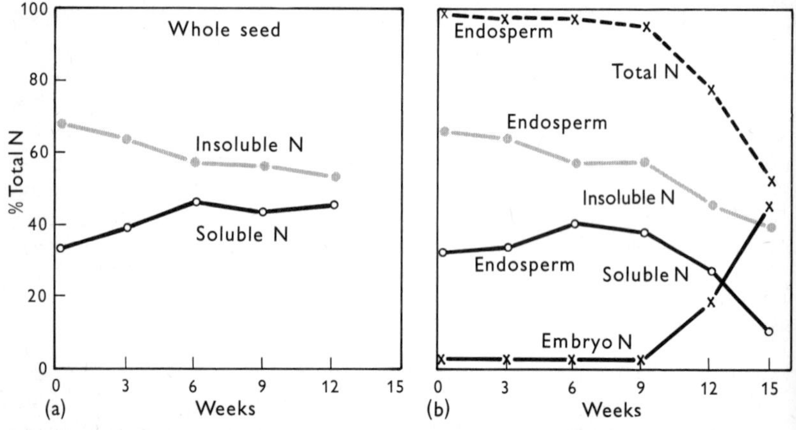

Fig. 7.11 Soluble and insoluble seed nitrogen contents of *Heracleum sphondylium* during treatment at 2°C.

(**a**) ○, soluble N ; ●, insoluble N (expressed as a percentage of total N).

(**b**) Endosperm and embryo nitrogen contents during treatment at 2°C expressed as a percentage of total seed nitrogen : ●, insoluble endosperm N ; ○, soluble endosperm N ; X-----X, endosperm total N (soluble + insoluble) ; X——X, embryo total N.

There was no change in the composition of the nitrogen fractions at 15°C. (From Stokes,[602] Fig. 3A, p. 164.)

of the immature embryo which increases and grows during the cold period. The mechanism of protein hydrolysis in the endosperm has not been established but it was indicated that it was unlikely that the embryo excreted the enzymes involved in protein breakdown; rather it seemed that some product of embryo metabolism at low temperature diffused into

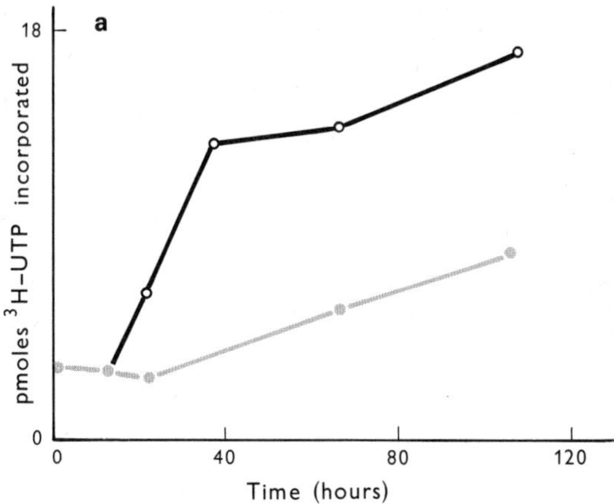

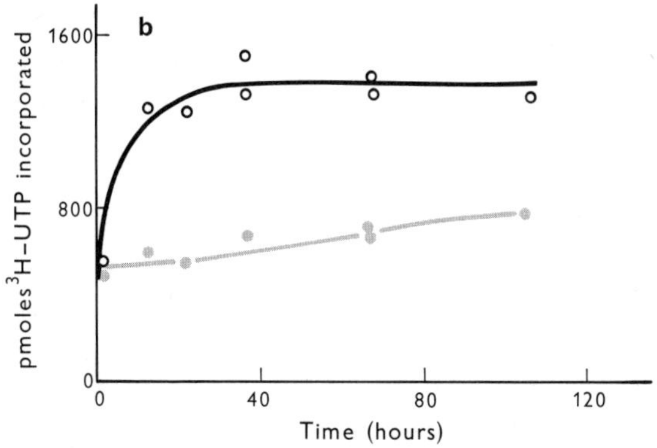

Fig. 7.12a The RNA polymerase activity associated with chromatin prepared from embryonic axes of seeds of *Corylus avellana* sown in gibberellic acid (○) or water (●). Results expressed as pmoles ^{3}H-UTP incorporated/100 μg DNA.
Fig. 7.12b DNA template availability of chromatin prepared from embryonic axes of seeds sown in gibberellic acid (○) or water (●). Results expressed as pmoles ^{3}H-UTP incorporated into RNA/100 μg DNA. (From Jarvis, Frankland and Cherry,[306] Fig. 2 and 3, p. 1735.)

the endosperm and induced the breakdown of storage protein. Olney and Pollock[479] have similarly shown that in cherry (*Prunus cerasus*) seeds there is a mobilization of nitrogenous compounds from the cotyledons into the axis during moist chilling; no such transfer occurs in moist seeds maintained at room temperature. Chilling of seeds of *Corylus avellana*[69] enhanced their capacity to convert adenine into adenosine, AMP and ADP and the chilling of pear (*Pyrus communis*) embryos is associated with an increased capacity for nucleic acid synthesis.[315]

In hazel (*Corylus avellana*) seeds the chilling requirement can be replaced by treatment with gibberellic acid. Jarvis *et al.*[306] indicated that the gibberellic acid treatment enhanced RNA synthesis. This was related to the fact that chromatin preparations from gibberellic acid treated embryos had a greater capacity for RNA synthesis than similar extracts from untreated embryos (Fig. 7.12a, b). Also the chromatin from gibberellic acid treated embryos was a more effective template for RNA synthesis catalysed by *Escherichia coli* RNA polymerase indicating that hormone treatment increased the template availability of the chromatin. This increase in template availability of the chromatin following gibberellic acid treatment preceded the increase in endogenous polymerase activity.

Amen[12] suggests that the breaking of dormancy may be related to a change in endogenous hormones which regulate (presumably at the transcriptional level) synthesis of various hydrolytic enzymes required for the germination process. However Chen[105] cautions that although application of growth hormones to dormant seeds may induce changes in the level of various hydrolytic enzymes germination may not be necessarily stimulated.

8

Nitrogen Metabolism in the Whole Plant

EMBRYO DEVELOPMENT

The growth of the embryo, which occurs during germination, is achieved primarily by the processes of cell enlargement and cell division. The capacity to form seedlings (gamma plantlets) from irradiated seeds in which DNA synthesis is curtailed indicates that extensive development can occur without cell division.[239] Normally, however, during seedling growth cell enlargement and cell division occur simultaneously.

Cell division in the growing embryos is usually confined to certain meristematic regions. In dicotyledons active meristems are found in the root apex and in the apical and sub-apical region of the shoot. In cereals the elongation of the root and coleoptile are in part attributable to the production of new cells from apical meristems; however, the meristems producing the leaves occur at the base of the leaf and are called intercalary. The cells of the meristematic regions presumably contain the requisite enzymic machinery necessary for the synthesis of nucleotide precursors, and for DNA synthesis as well as being capable of synthesizing those proteins involved in nuclear division and development.

At points distal to the meristems the cells undergo elongation and differentiation. During elongation extensive changes occur in cell composition (Fig. 8.1). Detailed analysis[81] of the changes indicate that as the cells increase in volume they increase in dry weight, respiratory activity and protein content so that during this period there must be major changes in nitrogen metabolism.

The increase in volume during elongation is accompanied by a vacuolation of the cell. Various hypotheses have been advanced to account for the origin of vacuoles;[122] on the one hand they may arise by enlargement of

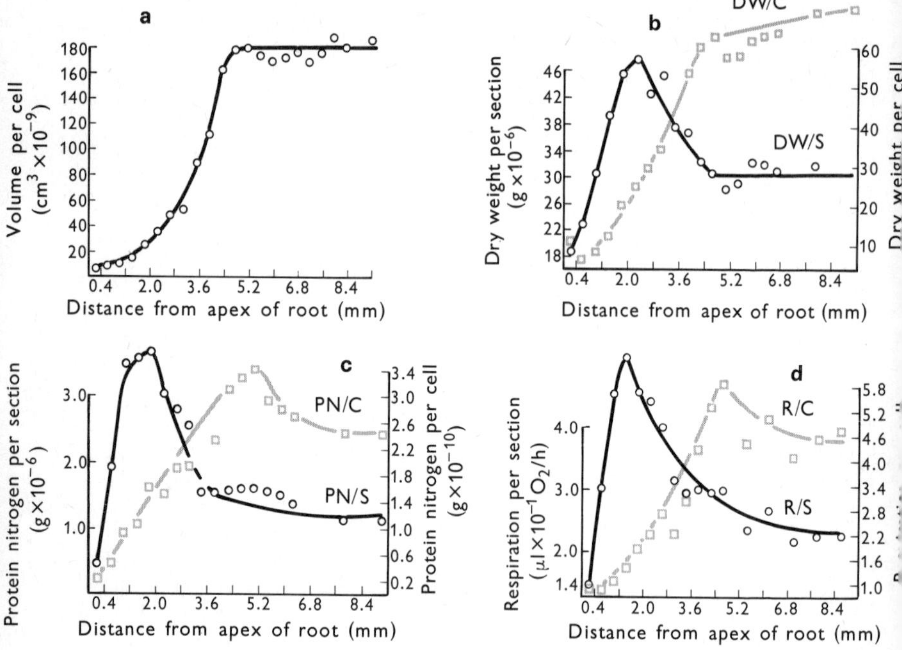

Fig. 8.1 Changes in cell composition and respiration during elongation. (From Brown and Broadbent,[81] Figs. 1, 3, 4 and 5, pp. 254–257.)

(a) Average volume per cell at increasing distances from the apex of the root.

(b) Dry weight of sections (DW/S) and of average cells in the sections (DW/C) at increasing distances from the apex of the root.

(c) Protein nitrogen content of sections (PN/S) and average cells in sections (PN/C) at increasing distances from the apex of the root.

(d) Respiration of sections (R/S) and of average cells in sections (R/C) at increasing distances from the apex of the root.

existing prevacuoles or they may be formed from vesicles pinched off from the endoplasmic reticulum or from the Golgi cisternae. Regardless of the origin of the vacuole, it is clear that its production must require extensive synthesis of the membranous component (tonoplast) surrounding the vacuole. Consequently in the transition from the meristematic apices to the elongating cell there is probably a termination of the production of those enzymes required for nuclear replication and the synthesis of enzymes involved in membrane synthesis. The increase in cell volume is usually associated with an enhancement of cell dry weight. Part of the accumulating dry weight is due to an increased production of cell wall during the elongation phase. Since the protein extensin seems to be an important component of the cell wall[352] it is to be expected that synthesis of this protein is active at this time. Additionally, as well as increase in dry weight there is a change in carbohydrate composition of the cell wall during elongation.[148] Thus there appears to be a development of new metabolic sequences for carbohydrate metabolism.

The increased respiratory activity of cells during elongation has been attributed to an increased number of the mitochondria per cell and an associated enhanced respiratory efficiency of this organelle. The increased efficiency of the mitochondria has been related to changes in their ultrastructure.[394] In meristematic cells two types of mitochondria are apparent, a mature form which has a heterogeneous internal matrix and prominent cristae and a second type which has a dense membrane and a nearly homogeneous matrix. During elongation this latter mitochondrial type develops internal vesicular structures similar to cristae. It is believed that it is the development of these cristae in the 'immature' mitochondria which leads to the increased respiratory efficiency of mitochondrial preparations from elongating as opposed to meristematic root segments. It should be recalled from Chapter 5 on nucleic acids that mitochondria possess their own complement of nucleic acids and may to some extent be autonomous. However, this above evidence of development and maturation associated with cell elongation indicates that events in the cell nucleus may control mitochondrial ontogeny. The number of mitochondria per cell varies with cell type; transfer cells, which seem to be specialized in solute transport, are characterized by having numerous well developed mitochondria.[496] It seems that the growth and development of mitochondria are regulated by the cells in which they are contained.

The increases in cell dry weight and respiratory activity of the elongating cell are accompanied by an elevation in cellular protein content. The preliminary research of Brown's group[531] indicated that this increased protein content was associated with changes in the enzymic complement of the cells, suggesting that there must be protein turnover. The demonstration that the activity of various proteolytic enzymes changed during cell elongation is consistent with this concept. More recent investigations[592] using gel electrophoresis confirm the suggestion that the complement of cellular proteins changes during the period of elongation.

This changed protein complement which occurs during the transition from a meristematic to an elongated cell is accompanied by an increase in cellular RNA. Preliminary studies indicated that in addition to an increase in RNA there was a change in its distribution. In meristematic cells the RNA was extracted readily with perchloric acid whereas in expanding cells the extraction was more difficult.[264] More recent information[115] suggests that this observed difference in extractability might be associated with the fact that during elongation there is a proliferation of endoplasmic reticulum and an increase in membrane-bound ribosomes. These bound ribosomes may be responsible for the production of enzymes accumulating in the enlarging vacuole. The endoplasmic reticulum has additional non-ribosomal RNA associated with it. This membrane RNA has not been characterized with regard to composition or function but obviously the proliferation of membranes during cellular elongation is associated with the production of this nucleic acid fraction.

If the changing enzyme complement during cell elongation is due to changes in transcription it is to be expected that during cell expansion there will be alteration in mRNA population; however, so far such differences have not been established. It is also possible that a regulation of enzyme synthesis could occur at the translational level. Some preliminary evidence of such control is provided by information which demonstrates a difference in the ratios of iso-accepting $tRNA^{tyr}$ species in dividing and non-dividing cells of pea roots.[650]

The observed changes in protein and RNA content which occur during cellular expansion are not merely a consequence of elongation but are essential requirements for the sustained growth. This fact is demonstrated by the observation that continued cellular expansion is prevented by treatment of tissues with inhibitors of RNA and protein synthesis.[326, 463] Detailed analyses, however, indicate that synthesis of all RNA components is not required for growth. Thus while growth is inhibited by Actinomycin D which prevents all RNA synthesis, it is not restricted by 5-fluorouracil. In the presence of 5-fluorouracil the production of the D-RNA component is continued and it appears that synthesis of this fraction is a necessary prerequisite for sustained growth.[326]

As the seed germinates in the soil it has access to the mineral nutrients contained in the soil solution. During germination the nitrate and/or ammonia in the soil solution are presumably increasingly utilized to fulfil the nitrogen demands of the growing embryo. Very few studies have been conducted to establish the relative contribution to the growing embryo of exogenously supplied nitrogenous compounds in comparison with reserve nitrogenous constituents. The demonstration of the induction of nitrate reductase in cotyledons[44, 292] and the scutellum[276] of germinating seeds indicates the potential for an early utilization of soil nitrate as a nitrogen source for the growing embryo. However, the contribution of nitrate to the nitrogen budget of the young growing embryo may initially be limited due to a lack of carbohydrate to provide the reductants required in the conversion of nitrate to ammonia and the low levels of nitrate reductase induced in dark-grown tissue in comparison with the activity produced in illuminated leaves and cotyledons.

DE-ETIOLATION AND GREENING

Seedlings may be grown for periods of several days in the dark resulting in the production of etiolated seedlings. In dicotyledons such seedlings are characterized by having elongated hypocotyls or epicotyls. In instances where the hypocotyl elongates (epigeal germination) the cotyledons remain small and surround the apical regions of the stem. In those cases where the epicotyl elongates (hypogeal germination) the etiolated shoot is characterized by having small, poorly developed leaves devoid of chlorophyll and the shoot apex remains bent in the plumular hook. In cereal seedlings maintained in the dark, growth is characterized by an extensive elongation of the mesocotyl. In grasses the leaves elongate in the dark and break through the surrounding coleoptile. This enlargement of the leaves in the dark is in marked contrast to the situation in dicotyledons; the cereal leaves, however, usually remain rolled.

Illumination of the etiolated seedlings produces marked changes in growth habit (Fig. 8.2 and Table 8.1). In dicotyledons, hypocotyl or epicotyl growth is retarded by illumination and the plumular hook unfolds. In those species with epigeal germination, the cotyledons expand and become green, the exposed apical region (plumule) begins to develop and leaf production proceeds. In seedlings with hypogeal germination, illumination causes the etiolated leaves on the epicotyl to expand and become green and stem growth is retarded. Illumination of etiolated cereal seedlings results in a cessation of mesocotyl elongation and the leaves unroll and become green.

These light-induced changes in morphology are mediated by the pigment phytochrome. However, although it is recognized that phytochrome absorbs light, the mechanism by which the photomorphogenic changes are manifest is not resolved. If it is assumed that the growth and development of the seedling are the result of the expression of metabolic activities then it is to be expected that changes in ontogeny and morphology will be associated with an altered metabolism. This altered metabolism in part will be the result of a changing enzyme complement.

The demonstration that sustained growth is dependent upon a continued synthesis of protein and D-RNA or mRNA discussed earlier suggests that during the retardation of growth, such as occurs during illumination of etiolated stems, there may be a reduction in the synthesis of protein and

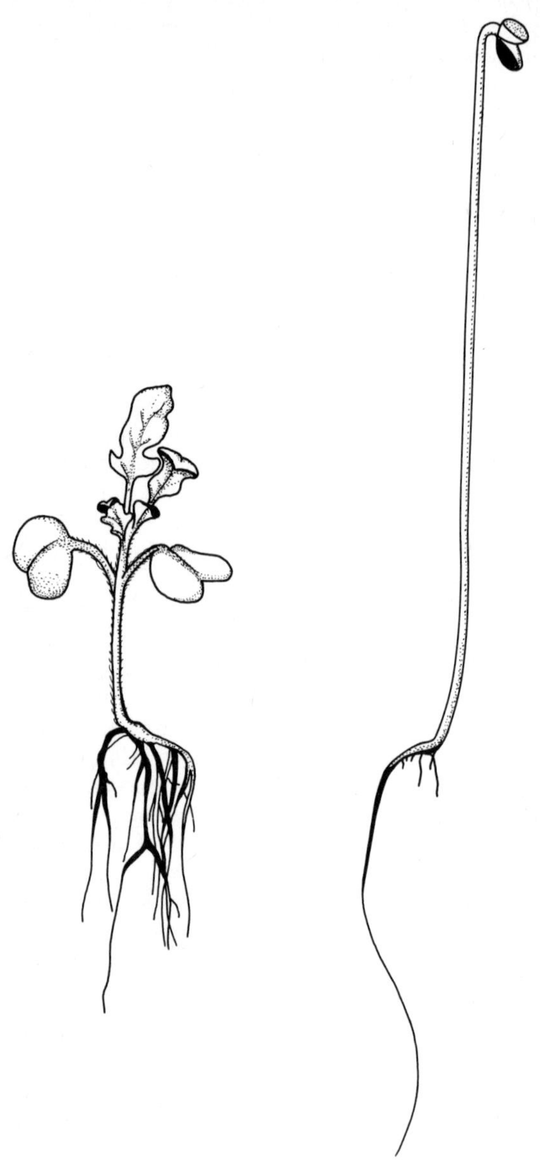

Fig. 8.2 Influence of light on the growth of mustard seedlings (*Sinapis alba*). (From Mohr,[434] Fig. 15.1, p. 510.)

Table 8.1 Photoresponses of the mustard seedling (*Sinapis alba*). (From Mohr,[434] Table 15.1, p. 524.)

Inhibition of hypocotyl lengthening
Inhibition of translocation from the cotyledons
Enlargement of cotyledons
Unfolding of the lamina of the cotyledons
Hair formation along the hypocotyl
Opening of the hypocotylar ('plumular') hook
Formation of leaf primordia
Differentiation of primary leaves
Increase of negative geotropic reactivity of the hypocotyl
Formation of tracheary elements
Differentiation of stomata in the epidermis of the cotyledons
Formation of plastids in the mesophyll of the cotyledons
Changes in the rate of cell respiration
Synthesis of anthocyanin
Increase in the rate of ascorbic acid synthesis
Increase in the rate of chlorophyll accumulation
Increase of RNA synthesis in the cotyledons
Increase of protein synthesis in the cotyledons
Changes in the rate of degradation of storage fat
Changes in the rate of degradation of storage protein

RNA. However studies which compare the capacity of illuminated or dark-grown stems to incorporate amino acids or nucleosides into proteins or nucleic acid demonstrate little differences in the two systems.[435] Thus although growth is arrested nucleic acid and protein synthesis are sustained. This continued synthesis will allow for the production of other enzymes and development of the new metabolic processes which occur after illumination. For example in stem tissues, anthocyanin and lignin production are enhanced following illumination. These increased syntheses may be related to an enhanced supply of the common precursor cinnamic acid: this compound is produced from phenylalanine through the activities of the enzyme phenylalanine ammonia lyase. The level of this enzyme is enhanced after illumination of etiolated seedlings.[571]

Many of these photomorphogenic effects can be prevented by treatment of dark-grown tissue with inhibitors of protein or RNA synthesis prior to or during illumination.[435] Such findings were initially interpreted as supporting the concept that light functioned in some manner as an inducer causing a de-repression of an operator gene and thus allowing for the transcription of the structural gene which is translated into the enzyme protein. In spite of the attractiveness of this scheme, it has not been possible to detect a light-triggered production of specific mRNA specific for a particular enzyme and all that can be safely concluded is that the enzyme production or photomorphogenic response requires the sustained synthesis of protein and RNA.

Greening

Extensive studies have been conducted of the metabolic events associated with leaf expansion and greening which occur following illumination of etiolated plants.

In dicotyledons the expansion of the leaf after illumination is due to an increase in cell size and in young etiolated leaves may also be accompanied by cell division. Illumination of etiolated leaves of cereals causes an unrolling of the leaf. The expansion or unrolling of leaves is accompanied by a rapid increase in protein and RNA content. In etiolated dicotyledons where leaf expansion is accompanied by cell division there is an increase in DNA content per leaf. During expansion and unrolling chlorophyll is produced in the leaf (Fig. 8.3). Since chlorophyll is located in the chloroplasts it might be expected that following illumination a great deal of the cellular metabolism would be directed towards the synthesis of these subcellular organelles. However, following illumination of leaves the chlorophyll is synthesized in existing structures, the etioplasts.[68] Illumination may increase the number of plastids per leaf; however, the newly developed plastids seem to be confined to recently divided cells so that the plastids per cell remains constant.[236]

During the greening process the etioplasts undergo characteristic ultrastructural changes. The etioplast develops from the proplastid in leaves maintained in the dark and are characterized by being about 2 μm in diameter surrounded by a double membrane enclosing a lattice crystalline-like prolamellar body and some lamellar sheets. After illumination, this prolamellar body is converted into sheets, thylakoids begin to develop and eventually the characteristic granal formations are produced (Fig. 8.4).

The changes in ultrastructure are accompanied by increases in various photosynthetic enzymes and ultimately the development of photosynthetically competent chloroplasts.[68] In view of the facts that extensive chloroplast development follows illumination and the suggestion that 70–80% of the leaf protein is confined to this organelle it seems that the increase in protein content of leaves is due primarily to the light-stimulated production of chloroplast proteins. However, investigations[184] indicate that there is also an extensive increase in the activities of mitochondrial and soluble cytoplasmic enzymes. Of course not all enzyme activities increase to the same extent so that on a specific activity basis (enzyme activity/unit

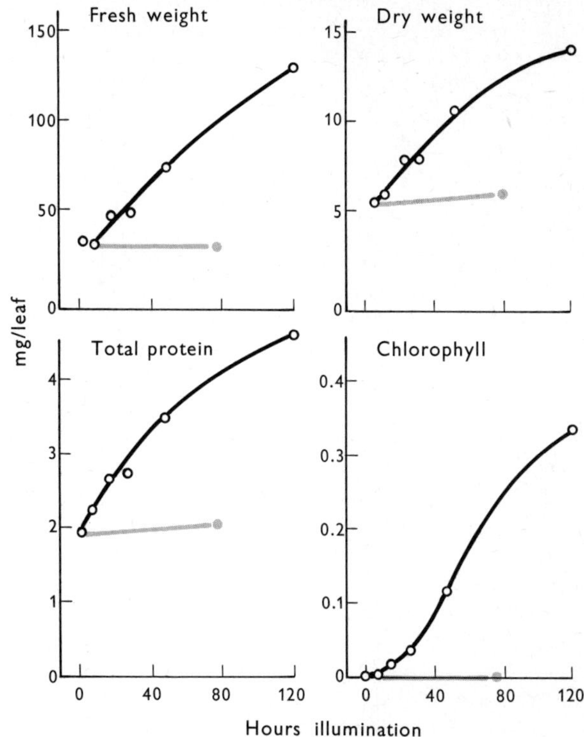

Fig. 8.3 The effects of illumination on the primary leaf development of 14-day-old dark-grown beans (*Phaseolus vulgaris*) (○), in comparison with controls maintained in darkness (●). (From Bradbeer.[68])

protein) the level of some enzymes declines following illumination (Fig. 8.5).

Since, as was indicated in Chapter 5, chloroplasts possess their own machinery for polypeptide synthesis it is to be expected that they may synthesize some of their own proteins. The light-stimulated production of chloroplast protein has provided an experimental tool by which to measure these capabilities. Essentially leaves from etiolated plants are illuminated in the presence of either cycloheximide or chloramphenicol. Those chloroplast proteins whose synthesis is prevented by chloramphenicol are considered to be produced on chloroplast ribosomes. If the production of a chloroplast enzyme is inhibited by cycloheximide it appears that the protein is synthesized on cytoplasmic ribosomes and must be incorporated into the organelle during its ontogeny. As indicated in Table 8.2, very few of the proteins usually associated with chloroplasts are synthesized in the

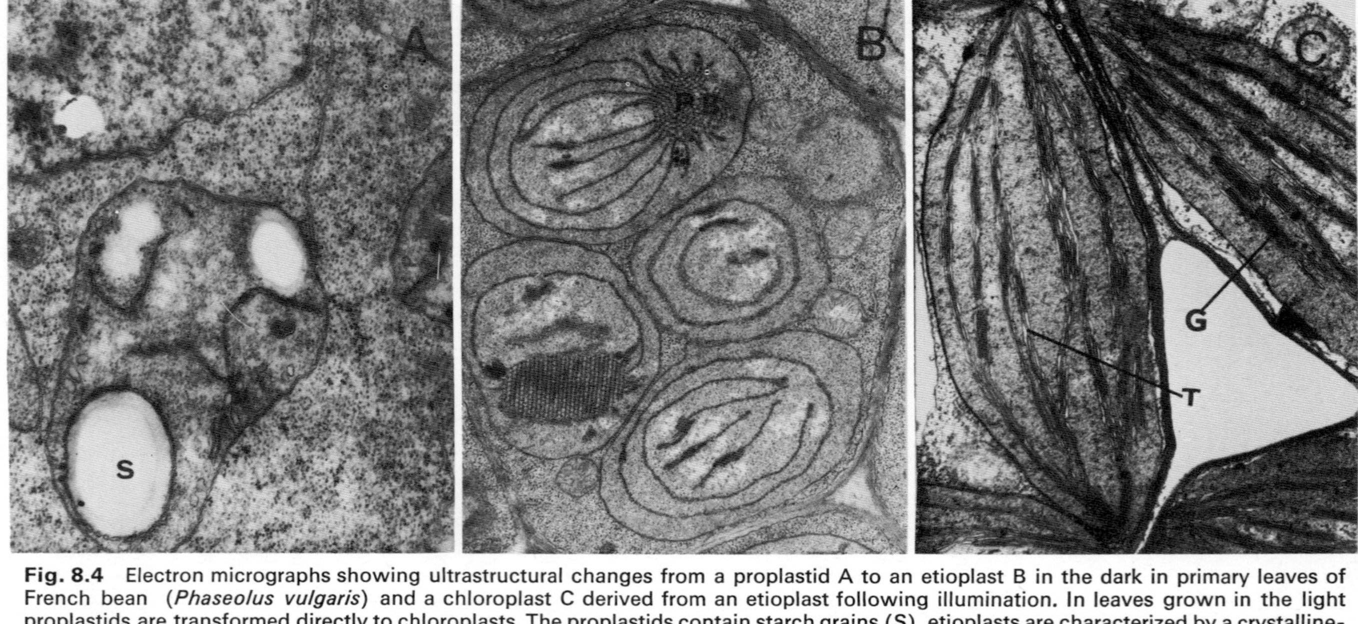

Fig. 8.4 Electron micrographs showing ultrastructural changes from a proplastid A to an etioplast B in the dark in primary leaves of French bean (*Phaseolus vulgaris*) and a chloroplast C derived from an etioplast following illumination. In leaves grown in the light proplastids are transformed directly to chloroplasts. The proplastids contain starch grains (S), etioplasts are characterized by a crystalline-like prolamellar body (PB) and chloroplasts contain membranes arranged as thylakoids (T) and grana (G). Magnifications: A, ×29 400; B, ×14 700 ; C, ×21 000. Electron micrographs by courtesy of J. W. Bradbeer.

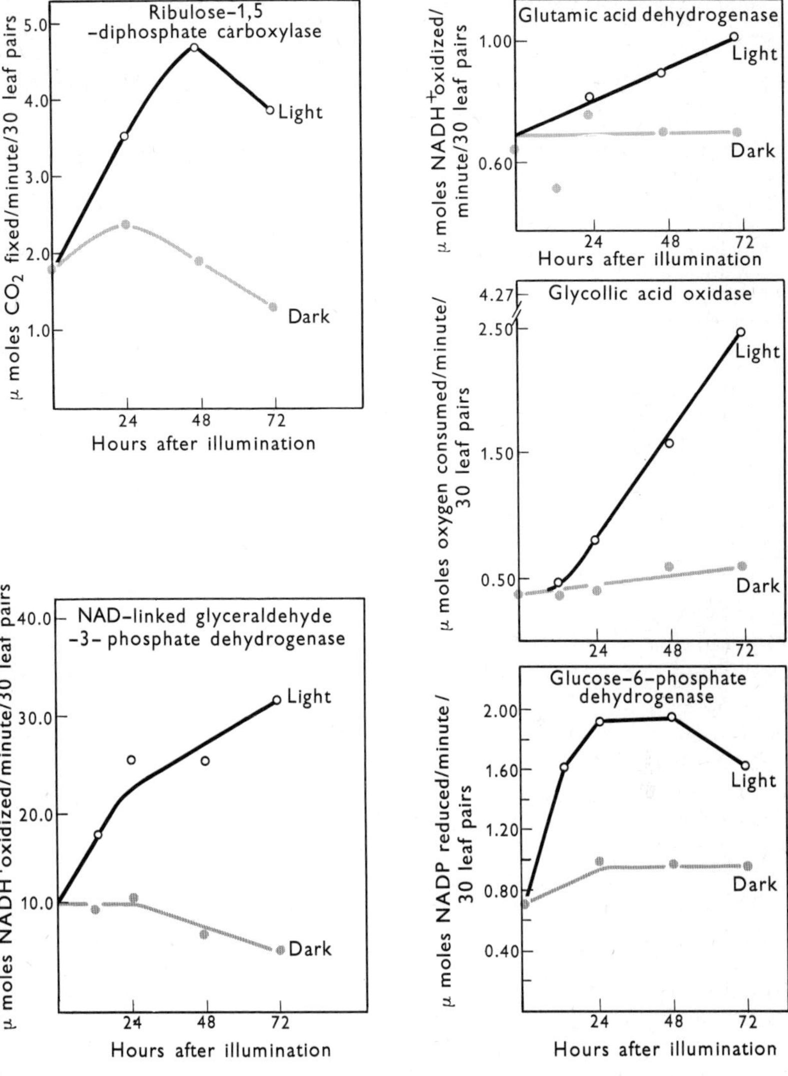

Fig. 8.5 Changes in activity of chloroplastic (ribulose-1,5-diphosphate carboxylase), mitochondrial (glutamic acid dehydrogenase), peroxisomal (glycollic acid oxidase) and a cytoplasmic enzyme (NAD-linked glyceraldehyde-3-phosphate dehydrogenase) following illumination of French bean (*Phaseolus vulgaris*) leaves. (From Filner and Klein.[184])

organelle and the synthetic activities of the 70s ribosomes to a large extent may be restricted to the production of membrane components.

Table 8.2 Sites of protein synthesis in leaves (data from Bradbeer[68] and Boulter, Ellis and Yarwood.[67])

Chloroplast 70s ribosomes	Cytoplasmic 80s ribosomes
Ribulose diphosphate carboxylase (large subunit)	Ribulose diphosphate carboxylase (small subunit)
Cytochrome *f*	Ribose phosphate isomerase
Cytochrome *b* 563	Phosphoribulo kinase
Cytochrome *b* 559	Phosphoglycerate kinase
	Triose phosphate dehydrogenase (NADP)
	Triose phosphate isomerase
	Fructose diphosphate aldolase
	Transketolase
	Ferredoxin
	Ferredoxin NADP reductase
	PEP carboxylase

Studies have been conducted to establish the mechanism by which illumination stimulates protein synthesis. The restricted synthetic capabilities of dark-grown plants can be related to a limited capacity of the ribosomal preparations from these plants to support amino acid incorporation in *in vitro* systems. In contrast ribosomal preparations from illuminated leaves are efficient in amino acid incorporation.[505, 694] This difference in capacity for directing peptide synthesis can at least in part be attributed to the fact that the ribosomal preparations from illuminated leaves contain a preponderance of polysomes whereas similar preparations from leaves of dark-grown plants contain principally monosomes.[505, 633, 695] These data are consistent with the speculation that illumination enhances mRNA production which allows for increased polysome formation. However, it is also possible that illumination in some manner increases the capacity of the ribosomes to associate with existing mRNA; in this case the light effect is exerted at the translational level.[634] That this possibility may occur is demonstrated by the finding that ribosomal preparations from illuminated leaves, in comparison to those from dark-grown leaves, have a superior capacity for polyphenylalanine formation when incubated with polyuridylic acid.[694]

In addition to the possibility of changes in ribosomal efficiency after illumination it appears that there may also be alterations in the distribution of various iso-accepting tRNA species[167] and these variations could also influence the translational capacity.

Illumination of leaves from dark-grown plants usually results in an increase in the level of RNA and an enhancement of the capacity to incorporate radioactive precursors into RNA. In view of the extensive chloroplast development which occurs during the greening, de-etiolation, phase it might be expected that the RNA synthesized after illumination would be principally chloroplastic. However, investigations[505] of this aspect indicate that leaves from dark-grown plants may possess relatively large amounts of chloroplast rRNA and ultrastructural studies demonstrate the occurrence of ribosomes in etioplasts. The increased nucleic acid synthesis which occurs after illumination, depending on leaf age, may be principally cytoplasmic or chloroplastic and cytoplasmic.[286, 572]

Clearly, illumination does not trigger the preferential synthesis of chloroplast ribosomal RNA; rather it appears that the production of this component is a function of physiological age. In so far as illumination accelerates leaf development it may also precipitate chloroplast rRNA synthesis. After illumination of older dark-grown leaves precursors are incorporated into cytoplasmic RNA.

The mechanism by which the capacity to incorporate precursors into RNA is enhanced by illumination has not been resolved. In barley (*Hordeum vulgare*) leaves illumination causes an increase in the level of soluble RNA polymerase whch might account for an enhanced capacity for RNA synthesis.[505] The observations that precursors are incorporated into ribosomal and soluble RNA together with the finding that polysome level is increased (suggesting mRNA synthesis) indicate that there is an apparent stimulation of the capacity to synthesize all RNA species.[505] It might be expected, in view of the enhanced production of various enzymes, that light would trigger the production of mRNA directing the synthesis of the enzymes. However, in many instances[68, 182, 184] low levels of enzyme activity are present in dark-grown tissues and thus illumination, although enhancing their production, does not initiate or induce the synthesis of these proteins.

In several instances the stimulating effect of illumination on leaf development can be replaced by supplying exogenous growth regulators. Kinetin increases the level of various photosynthetic enzymes in etiolated rye (*Secale cereale*) seedlings[182] and pretreatment in the dark with cytokinin benzyladenine increases the production of chlorophyll when the cotyledons of dark-grown cucumbers (*Cucumis sativus*) are illuminated.[4] Kinetin or gibberellins induce the unrolling of etiolated leaves of barley seedlings.[506] This hormone-induced unrolling resembles the leaf expansion stimulated by light in terms of increased protein and RNA synthesis and elevated polysome levels.

NITROGEN BUDGETS IN ANNUAL AND PERENNIAL PLANTS

The development of chloroplasts and the onset of photosynthesis which follows illumination of the seedling has marked repercussions on the nitrogen metabolism of the plant. Intermediates generated during photosynthesis can serve as additional sources of reductant to be utilized in the reduction of nitrate to ammonia. Photophosphorylation produces an alternate supply of ATP to provide an energy source for the various synthetic processes. Also intermediates of photosynthetic carbon cycles serve as an alternate source of carbon skeletons for the biosynthesis of various amino acids. Thus in leaf tissue the carbon entering the amino acids may be derived either directly from photosynthesis or respiratory metabolism. Likewise the nitrogen involved in the synthesis of nitrogenous constituents in the leaf may be derived from amides or amino acids or ureides transported into the leaf or it may arise from the reduction of nitrate within the leaf.

The relative contributions of inorganic nitrate or nitrogen in organic combinations to the synthesis of nitrogenous compounds in the leaf seem to vary between species and are in addition influenced by availability of nitrate in the soil solution. Detailed analyses of nitrogen transport and utilization in annual plants have been made by Pate and co-workers[477, 665] who have compared nitrogen assimilation in *Xanthium* and peas (*Pisum arvense*). In the former case nitrogen arriving in the aerial portions is principally in the form of nitrate; in contrast in peas the main nitrogenous compounds transported to the shoot may be amides or nitrate depending on the availability of nitrate in the soil solution.

The proposed scheme for nitrogen mobilization in the two types is depicted in Figs. 8.6 and 8.8.

Nitrogen metabolism in *Xanthium*

In *Xanthium* nitrate is absorbed (1) by the roots (Fig. 8.6); little of this nitrate is reduced or used for the assimilation into organic nitrogenous compounds; some nitrate may be stored (2), but most of it is transported to the shoot (3) in the xylem transpiration stream.

Older parts of the shoot axis can abstract nitrate from the xylem (4). This nitrate is stored largely in the parenchyma of the internodes and petioles(5). A small portion of this nitrate may be reduced by the nitrate reductase of these organs (6). Generally nitrate reductase activity is low in stem tissue. All of the nitrate entering laminae of mature leaves is reduced at once by an active reductase (7). Since free amino acids and proteins cease to accumulate once the leaf is fully expanded the products of nitrate assimilation in these mature leaves must be continually exported (8).

Nitrate cannot be detected in mature leaves suggesting its rapid reduction; however, nitrate reductase activity is low in mature leaves (Fig. 8.7) so that if nitrate entered the lamina in concentrations similar

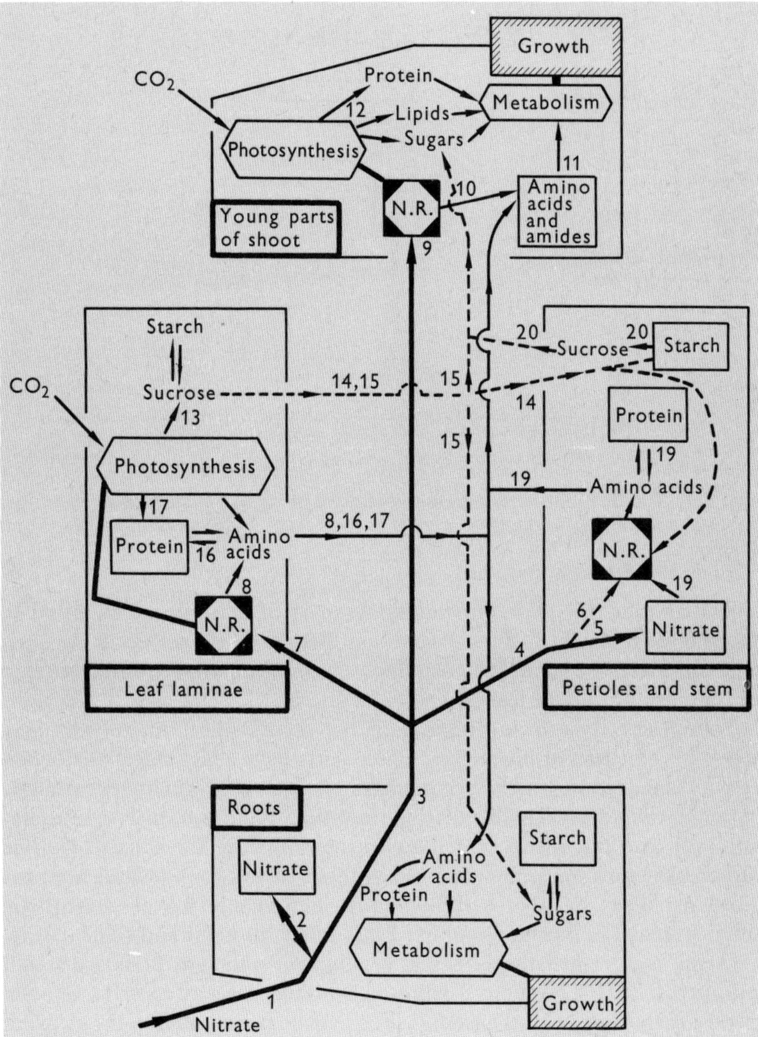

Fig. 8.6 The assimilation of nitrate in the cocklebur (*Xanthium pennsylvanicum*), a species where nitrate reduction (nitrate reductase, NR) is apparently restricted to the shoot system and where free nitrate is frequently stored in stem and root. (From Wallace and Pate,[665] Fig. 4, p. 223.)

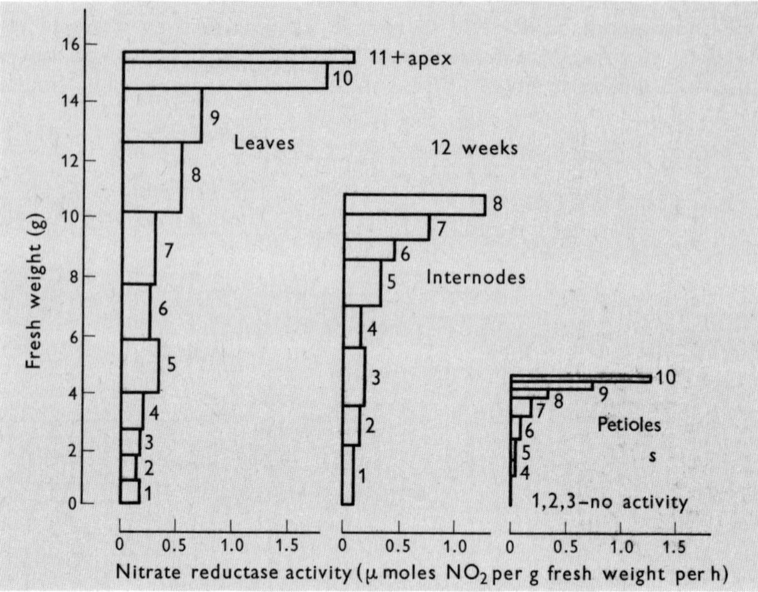

Fig. 8.7 The distribution of nitrate reductase activity in the cocklebur, *Xanthium pennsylvanicum.* No reductase activity was present in extracts of roots. (From Wallace and Pate,[665] Fig. 2, p. 219.)

to that in which it exists in the petiole or stem it would be expected to accumulate in the leaf. The observation that this does not occur suggests that there may be some stricture imposed onto the import of nitrate into the laminae of mature leaves.

Young actively growing regions of the shoot appear to be extremely active in reduction of nitrate (9). These parts have a high nitrate reductase activity and nitrate rarely accumulates. Amino constituents accumulate in young tissues presumably arising from the products of nitrate assimilation (10), and these may contribute to the growth of the shoot (11). The carbon skeletons for amino acid biosynthesis in the young leaves are provided primarily by photosynthesis (12); additionally the photosynthetic activity results in carbon being fixed into lipids, proteins and some sugars.

In the mature leaf the principal products of photosynthesis are starch and sucrose (13). The sucrose exported from the leaf may contribute to the reserve carbohydrate of the stem (14) or may be transported to the growing region of the root and shoot (15). Significantly this transported sugar does not contribute carbon skeletons for amino acid biosynthesis but serves principally for the synthesis of insoluble carbohydrates.

During leaf ageing there is a rapid decrease in soluble nitrogen and protein content. The amino acids derived from these sources may be transported in the phloem to the shoot and root meristem (16). Additionally, amino acids derived from the normal turnover of leaf proteins may be transported to growing areas of the plant (17). Starch, nitrate and protein accumulate in the root and stem axis and foreseeably these reserves could contribute amino acids (19) and sugars (20) to the younger parts of the plant during conditions of nutrient stress.

Nitrogen metabolism in field pea (*Pisum arvense*)

In nodulating peas and other legumes growing in the presence of free nitrate, three sites are available for the assimilation of inorganic nitrogen. These are the nitrogen fixing nodules, and the nitrate reductase in the root and in the shoot (Fig. 8.8).

The nitrate absorbed by the roots may be metabolized principally in the root by nitrate reductase NR_1 using reductant generated during the respiration of sugars recently arrived from the shoot (2B). The products of this nitrate assimilation are predominantly compounds closely related to aspartic and glutamic acids and the majority of them are transported to the shoot in the translocation stream (1). A small proportion of the products of nitrate assimilation are stored in the root and are transported to the growing region of the root (4). Additional nitrogenous compounds in the xylem stream are produced as a result of the nitrogen fixation in the nodules; again the products of this assimilation appear to be the amides glutamine and asparagine and these are transported to the shoot system (1). The xylem exudate also contains ureides (allantoic acid) and some amino acids. The nitrogen in these compounds could originate either from fixation or by secondary metabolism of the products of nitrate reduction. The carbon skeletons in these originate from the carbohydrates translocated from the aerial portions. Some of the nitrogenous products arising from the activities of nitrogen fixation in the nodule may be utilized to support the growth of the root distal to the nodule (4). However, this is probably of minor consequence in comparison with the nitrogenous compounds transported in the phloem to the growth centres (2B).

The older parts of the shoot can absorb metabolites from the xylem stream and the nitrogenous compounds can undergo metabolic rearrangement to produce other amides, amino acids, nucleotides, nucleic acids, proteins, etc. These products may turnover and be degraded and later in the life of the stem nitrogenous components may be released to be translocated to the shoot (2A) and roots (2B).

In mature leaf tissues the components imported by way of the xylem stream can serve as nitrogen source for the photosynthetic production of amino acids. The amides appear to be extremely important in this regard. Some of the nitrate (1) arriving in the mature leaf may be reduced by the

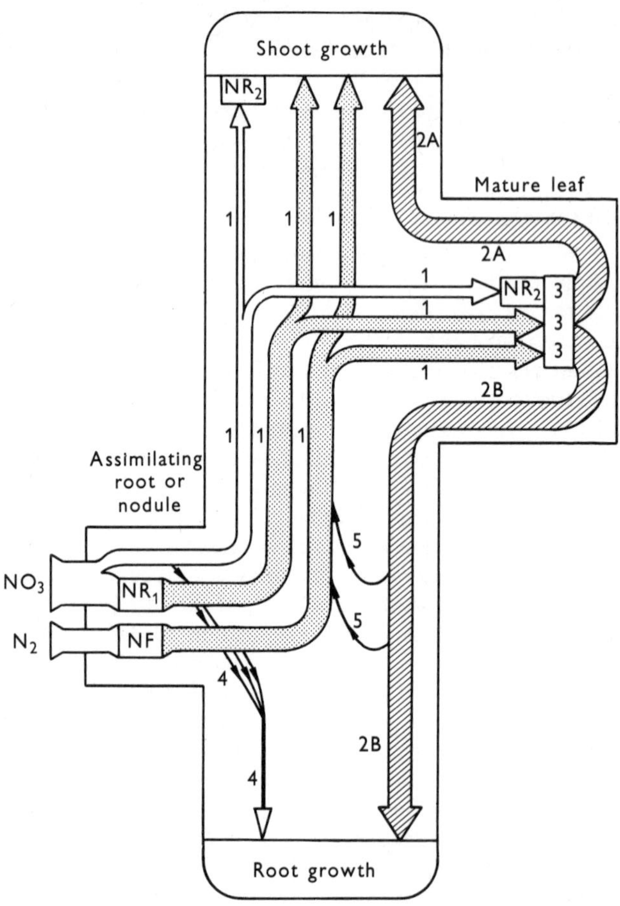

Fig. 8.8 Scheme illustrating the main features of the assimilatory and transport systems for nitrogen in the field pea (*Pisum arvense*). The situation represented is a young vegetative plant, effectively nodulated and growing in the presence of a moderately high level of nitrate in the medium. (**a**) Sites of assimilation: NF, nitrogen fixing nodules; NR_1 root-located nitrate reductase; NR_2 shoot-located nitrate reductase. (**b**) Pathways of transport; 1, xylem transport; 2A, 2B, upward and downward phloem transport respectively; 3, transfer from leaf to phloem of minor veins; 4, direct transfer to distal, non-assimilating parts of root; 5, transfer from downward translocation stream to upward stream of nitrogen in transpiration stream. (**c**) Compounds transported: ☐ free nitrate; ▩ amides, ureide + some amino acids; ▨ variety of amino acids + some amide. Note: The breadth of each pathway gives an approximate indication of its relative significance in transporting nitrogen. (From Oghoghorie and Pate,[477] Fig. 5, p. 45.)

nitrate reductase NR_2 and this furnishes another nitrogen source to be utilized in the synthesis of amino acids. In older mature leaves protein synthesis is minimal; in fact during ageing there is a decline in soluble nitrogen and protein. Thus the products of nitrogen metabolism in the mature leaf (3) are predominantly translocated as amino acids and some amides to the growing centres (2A, 2B).

The growing shoot thus has access to three principal sources of nitrogen: the nitrogenous components present in the xylem stream (1) originating from nitrogen fixation and nitrate reduction in the root; additional components, principally amino acids, are supplied by the phloem stream (2A); and some nitrate, translocated in the xylem stream (1), may also be metabolized by nitrate reductase (NR_2) in the growing shoot and furnish a further nitrogen source.

In marked contrast to the system described above for the supply of nitrogenous metabolites to the shoot, it appears that the nitrogen demands for root growth are supported primarily by materials translocated to the root in the phloem stream from the aerial portions.

These two systems, of *Xanthium* and pea (*Pisum arvense*), vary considerably in their nitrogen utilization and other intermediate forms are to be expected.

It might be speculated that in nitrogen-fixing legumes successive applications of nitrate will influence the cycling of nitrogenous compounds. This influence of nitrate will be due to two causes: on the one hand nitrate inhibits nodular nitrogen fixation so that there will be reduced input from this source and with more available nitrate increasing amounts of this nutrient will be translocated to the shoot and thus the nitrogen assimilation will be dominated by the shoot reductase system in the manner described for *Xanthium*.

Nitrogen metabolism in woody perennials

The growth of many woody perennials is characterized by a flush of growth which occurs in the spring of the year. In many genera, e.g. *Pinus*, *Picea*, and *Fagus*, this growth is limited to the expansion of those leaf primordia and internodes found in the winter bud. In others, *Betula* and *Populus*, shoot elongation may be prolonged as a result of the continuous production and expansion of new leaf primordia and nodes during the growing season. In other types, such as *Quercus*, there are recurrent flushes of growth which involve formation and expansion of the current year's buds after an initial growth flush from the winter bud has been completed.

In those species with one episode of spring growth the nitrogen demands for the expansion of leaves and stem elongation are apparently met by utilization of reserves stored predominantly in the barks of stems (Fig. 8.9). The principal nitrogen reserve component of the bark is protein. This protein nitrogen (particularly under conditions of nitrogen fertilization) is enriched in the amino acid arginine.[642] During the spring of the year and prior to bud break the protein content begins to decline and there

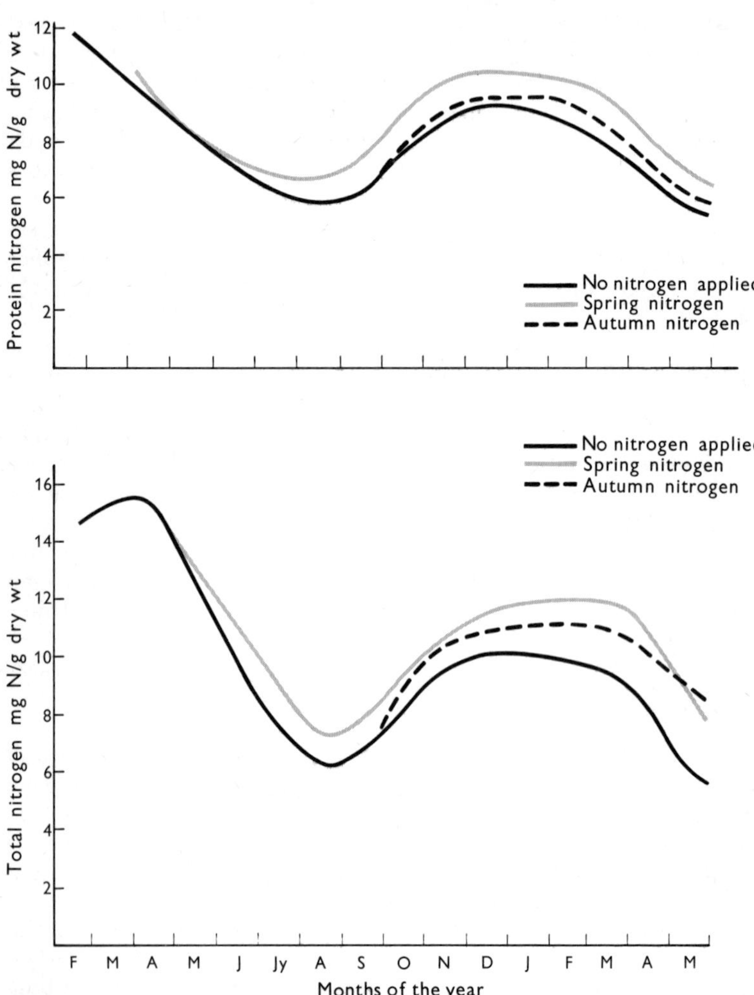

Fig. 8.9 Changes in total and protein nitrogen in the bark of apple trees during a year's growth cycle. The influence of exogenously applied nitrogen is also illustrated. (From Tromp,[642] Fig. 2a and Fig. 1, p. 145.)

is an increase in soluble nitrogen. This soluble nitrogen is composed principally of amino acids and amides while the amino acid composition differs from that of the proteins; it nevertheless seems that the increase in soluble nitrogen in the spring of the year to a great extent is due to hydrolysis of the reserve proteins in the bark.[642] The mechanism of protein hydrolysis has not been elucidated but it occurs after winter chilling and thus is somewhat analogous to the proteolysis induced following the cold treatment of certain seeds.

After bud break, the soluble nitrogen content of the bark begins to decline and the decrease in protein content continues. These decreases occur during the period of stem elongation and the exported nitrogen supports the growth of the shoot. It is not clear in what form the nitrogen is transported from the bark to the growing shoot; the asparagine content of the soluble amino acid pools increases during the spring suggesting that this may be an important, translocatory component.[642] Exogenously supplied labelled arginine is decomposed to carbon dioxide and ornithine[266] indicating that it might be subject to the same type of breakdown and metabolism which occur during seed germination.

In species with one flush of growth, it is found that application of nitrogenous fertilizer in the spring of the year has little influence on that year's shoot growth; as can be seen (Fig. 8.9), this fertilization has only a slight effect on the rate of depletion of the protein nitrogen from the bark.

At the termination of shoot growth in midsummer the nitrogen content of the bark begins to increase. This increase is due to an accumulation of protein and soluble nitrogenous components. In late summer and early autumn the amides glutamine and asparagine accumulate in the storage tissue but during late autumn and winter arginine becomes the main soluble nitrogenous component.[642, 643] It is speculated that the amide nitrogen is utilized in arginine biosynthesis. On the basis of observations[60] that nitrate nitrogen is seldom encountered in the xylem exudate, it appears that the bulk of the reduction of nitrate nitrogen occurs in the root system, and the reduction products, principally glutamine and asparagine, are translocated to the aerial parts. In view of the fact that growth is not stimulated by exogenous application of fertilizers during a specific growth period it seems that the products arising from nitrogen metabolism in the root are utilized principally for the formation of storage products and support the growth of the stem in the spring of the year.

This storage nitrogen may be formed directly *in situ* or it may be derived as a result of breakdown of the nitrogenous components in the leaf tissue which are transported back into the stem tissue during the autumn leaf senescence.

Under conditions of high fertilizer application nitrate does appear in the xylem exudate of some trees and the demonstrated[338] presence of nitrate reductase in leaf tissues of these trees indicates that some nitrate assimilation can occur in the leaves. However the components produced as a result of the metabolism of this nitrate do not contribute directly to the growth of the shoot in those species with one flush of spring elongation. Rather the products appear to be utilized for the production of reserve nitrogenous compounds to support the growth of the shoot in the following year.

ENVIRONMENTAL FACTORS INFLUENCING NITROGEN METABOLISM

Light

As indicated on p. 252, illumination has a major influence on the morphogenesis of plants which is associated with changes in nitrogen metabolism. In addition, during the growth of the illuminated plant light intensity can alter the synthesis or utilization of many nitrogenous compounds. Characteristically at low light intensity nitrate accumulates in leaf tissue and soluble amino acid levels are elevated while the rate of protein synthesis declines. Within the soluble amino acid pool there may be a change in the distribution of components; in darkness the amide asparagine tends to accumulate whereas in the light glutamine is the dominant amide. These changes in nitrogen metabolism which are influenced by light are complicated and they may occur on a diurnal basis during the flux from darkness to light during the night/day transition.[591]

The increase in nitrate level in darkened or shaded leaves was originally attributed to a lack of photosynthate to furnish the reductant required in the conversion of nitrate to ammonia by nitrate reductase. While this is a possibility another important consideration is that the tissue level of nitrate reductase is reduced during shading or darkness[42,44,243] (Fig. 8.10).

Thus irrespective of the availability of reductant, nitrate reduction might be expected to be minimal, owing to the low level of enzymes. Various speculations[42] have been made concerning the role of light in increasing the levels of nitrate reductase and publications have appeared describing the induction of the enzyme by light. However illuminated leaves grown in the absence of nitrate do not contain nitrate reductase. In contrast if leaves from plants grown in the dark or shade in the presence of nitrate are illuminated there is an increase in nitrate reductase. Nitrate reductase can be induced in the dark-grown tissue by nitrate; however, the enzyme levels produced are considerably lower than those encountered in illuminated tissues.

The most acceptable explanation of these observations is that illumination, as described earlier, enhances the capacity for protein synthesis. An

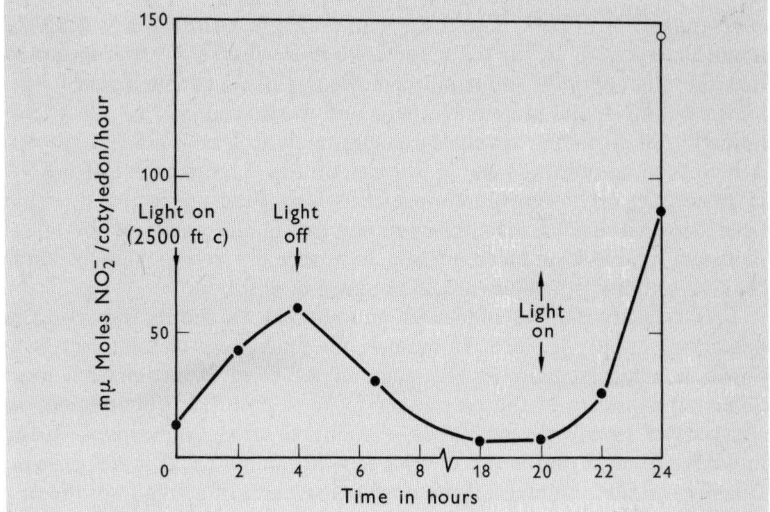

Fig. 8.10 The effect of dark and light treatments on the level of nitrate reductase activity extractable from radish (*Raphanus sativus*) cotyledons. (From Beevers *et al.*,[44] Fig. 1, p. 693.)

active protein synthesis is obviously a necessary prerequisite for the production of nitrate reductase following induction by nitrate. Thus, illumination does not induce nitrate reductase; rather by increasing protein synthesis it enhances the capacity for the expression of the inductive effects of substrate nitrate.

Light intensity will obviously influence the rate of photosynthesis, and thus an altered rate of carbon dioxide fixation would change the rate of provision of carbon skeletons for amino acid biosynthesis. The rate of photophosphorylation will regulate the production of ATP required as an energy source in the many synthetic processes.

Major alterations in growth habit of the plant may be brought about by alterations in the length of illumination relative to the dark period. The classic example of this influence of photoperiod is in the induction of flowering. Some plants are induced to flower by a reduction of the period of illumination below a critical daylength; these are the short day plants. In other types, the long day plants, flowering is induced when the photoperiod is extended beyond a certain critical minimum period. There are many other intermediate types and the characterization into short day and long day plants is an over-simplification. Nevertheless it is well established that in many species changes in the photoperiod can bring about a transition from vegetative growth to reproductive growth. The transition occurs in various aerial meristems which change from producing leaves to the development of flowers. The manner in which this alteration is achieved is not understood. In the early 1960's various workers[63,108,203] indicated that the photoperiodic induction of flowering could be prevented by inhibitors of RNA and protein synthesis and it was implied that the photo-induction of flowering involved the production of new mRNA species. Attempts to demonstrate the production of a new complement of mRNA in stem apices following photo-induction have been unsuccessful. It has been demonstrated[214] that there are changes in the nuclei of the apical meristem in photo-induced plants; however, the relationship of these changes to floral development has not been established.

The capacity for an inductive photoperiod to induce flowering is dependent upon plant age. In annuals the attainment of the ripeness to flower is achieved early and in some types floral induction can occur following exposure of the cotyledons to an appropriate photoperiod. In contrast, in woody perennials the capacity to flower in response to an inductive photoperiod is not developed until after several years' growth. It is not clear, at this stage, if the unresponsiveness of young trees lies in a deficiency in the capacity to produce a flower inducing principle in response to an inductive photoperiod or whether young apices are incapable of responding to the inductive component. Certainly apices may change during the ontogeny of the plant as demonstrated by the transition from a juvenile to a mature growth habit in many species which is accompanied by an altered leaf form. The basis for these divergent growth habits is not understood but they foreseeably will be associated with an altered gene expression.

Photoperiod also plays a key role in regulating other aspects of growth and development. The termination of shoot elongation in temperate woody perennials is triggered by the shortening of the photoperiod. There are changes in the levels of various endogenous growth regulators in the leaf tissue on the shoots during short days.[503,672] It is believed that these changes control the elongation of the shoot apex. The cessation of shoot growth is achieved by two principal mechanisms; in some instances a resting bud is formed at the shoot apex and in other cases there is an abortion of the shoot tip which is abscinded and a resting overwintering bud develops from the subtending lateral bud.[342] The mechanism by which the termination of bud elongation or abscission is achieved is not fully understood. One of the growth regulating components present in leaves is abscisic acid; when this compound is applied to actively growing apical buds it will induce dormancy,[174] and it also induces abscission in some tissues.[3] Abscisic acid has been shown to reduce the rate of incorporation of amino acids and nucleosides into protein and nucleic acids.[114,506] It is thus possible that the termination of stem growth is associated with a reduced capacity for synthesis of macromolecules.

However, although stem elongation is stopped, cellular proliferation and differentiation continue and the overwintering bud contains the leaf primordia and internodes responsible for growth in the following spring.

Photoperiod also influences cambial activity in temperate woody perennials. In some instances cambial growth ceases shortly after the termination of shoot elongation and it has been suggested that there is a relationship between the stimulation of cell division in the cambium and hormones produced by the actively growing shoot. However, it can be shown that following the termination of shoot growth by short day treatments the cambial division can be restored by long day treatments. It appears that the stimulus for cambial development originates in the leaf but moves only downward in the stem. This cambial stimulating factor is suspected to be auxin.[343] The relative proportion of xylem and phloem produced as the result of cambial divisions is further influenced by the levels of gibberellic acid and cytokinins.[671] Although the mechanisms by which these hormones control cambial division and differentiation of the products is not understood, it is clear that synthesis of various enzymes involved in cell division and production of cell wall constituents must be regulated by these compounds.

Changes in photoperiod frequently control the formation of perennating structures in many herbaceous perennials. These perennating organs, tubers, corms, bulbs, etc. are frequently the sites of deposition of storage reserves. Again, the onset of synthesis of these reserves must involve the production of enzymes required for their anabolism in the storage tissue. Frequently, the development of the storage structures occurs after the termination of shoot growth and thus there appears to be a diversion of metabolites to the underground reserve organs. Protein is frequently a prominent reserve component in tubers and bulbs. Presumably the amino acids utilized in the synthesis of this reserve are derived from nitrogenous constituents translocated from the aerial portions; however, the nature of the imported components has not been followed in detail.

Moisture

An inadequate water supply is often a limiting factor in plant growth. Under conditions of low soil moisture plants rapidly lose turgidity and wilt; in addition there are characteristic changes in the protein, nucleic acid (Fig. 8.11), and amino acid metabolism of the plant. If the moisture stress is extreme or prolonged the plant senesces and dies.

The decline in protein content of stressed plants is associated with an increase in soluble amino acids and it was originally speculated that water deficit increased proteolytic activity. No studies have been made of the levels of proteolytic enzymes in stressed leaves; however, the observations[30] that the protein amino acid composition changes during wilting

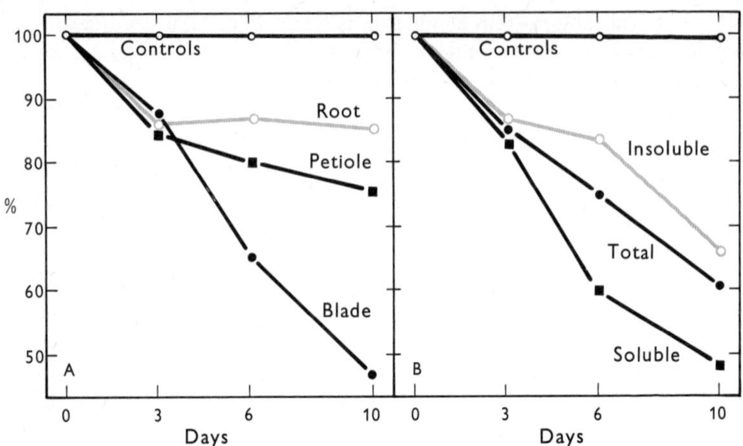

Fig. 8.11 Relative concentrations of RNA (left) in various sugar beet (*Beta vulgaris*) tissues and of protein (right) in young leaf blades during a single wilting cycle. Data for dry treatments are expressed as a percentage of the corresponding wet treatment values. (From Shah and Loomis,[556] Fig. 2, p. 243.)

might indicate that certain proteins are preferentially hydrolysed. As well as an accumulation of soluble amino nitrogen during moisture stress, it is found that there is a change in the composition of this fraction. Particularly striking in this regard are accumulations of proline and asparagine. Labelling and feeding studies[596] indicate that the carbon skeletons of the accumulating proline are derived from carbohydrate whereas the nitrogenous component is generated from other amino acids presumably by a transaminase acting on α-ketoglutarate (derived from the carbohydrate) to form glutamate which is converted to proline. The accumulated proline can thus be regarded as a carbon and nitrogen reserve and it appears to function in this role during rehydration of the leaf. Thus, if leaves are infiltrated with carbohydrate during rehydration the soluble proline remains high but some of the amino acid becomes incorporated into protein. In contrast in carbohydrate deficient leaves, proline is oxidized and its carbon skeletons are utilized in generation of organic acids and other amino acids.[594]

The decline in protein content and accumulation of amino acids which occur during wilting can also in part be accounted for by a decreased rate of protein synthesis, coupled with a sustained protein degradation. Wilted leaves have a reduced capacity to incorporate amino acids into protein suggesting a lowered synthetic capability.[47] This concept is supported by the observation that ribosomal preparations from wilted leaves have a lower polysome content than similar extracts from unstressed leaves.[274, 436]

The decline in polysome content is initially associated with an increase in monoribosome level and occurs within about 30 minutes of the imposition of water stress. The effect is reversible and upon rehydration there is an increase in polysome content.[274] The mechanism by which these changes in polysome level occur is not known; such changes may indicate a fluctuation in mRNA level. However, from the discussion on protein synthesis (Chapter 6) it should be apparent that polyribosome level, in addition to being regulated by mRNA level, is also controlled by levels of initiation factors, initiator tRNA, etc.

The observed[556] decline in tissue RNA during wilting which would obviously lead to a reduction in ribosome level is also indicative of a decreased capacity for protein synthesis. Again there are two possible mechanisms to account for the decline in RNA; on the one hand degradation of RNA may increase following moisture stress or there may be a reduced synthesis of the polynucleotide. No consideration seems to have been given to this latter possibility. In contrast it has been shown that there is an increase in ribonuclease level in stressed leaves which could partly account for the accelerated degradation of RNA during wilting.[160]

Many of the changes in nitrogen metabolism during moisture stress are similar to those occurring during senescence (discussed in Chapter 9). Since senescence can be prevented by application of growth regulators, it is suggested[296] that the onset of the wilting response and altered nitrogen metabolism may be due to a deficiency of growth regulators in the leaves. These growth regulators may be synthesized in the root and transported to the shoot in the xylem translocation stream. A reduction in moisture supply would restrict the xylem flow and effectively limit the availability of growth regulators in the aerial portions.

Temperature

Plants vary considerably in their response to temperature. At high temperatures, growth is usually arrested and if the exposure to elevated temperatures is prolonged then the plant dies. Similarly at low temperatures growth is also curtailed but there are variations in the plant's response to prolonged exposure to low temperatures. Many plants, particularly annuals, are killed by temperatures approaching 0°C. Other winter-hardy types can withstand temperatures down to −40°C or lower. However the capacity to withstand these extremely low temperatures is induced gradually. This induction of cold resistance or hardening seems to be mediated by two mechanisms, short day treatments and progressively lowering temperatures. The hardening response is associated with changes in tissue nitrogenous components.[677] Characteristically[376, 568] there is an increase in soluble protein content which is accompanied by a rise in RNA. In leaf tissue there is also an increase in soluble amino acids during the hardening process.[720] It has been speculated that the proteins synthesized during the hardening process would have more hydrophobic amino acid residues or more disulphide bridges which could ostensibly increase their resistance to inactivation by freezing. However, electrophoretic analyses,[210] although showing minor changes in soluble proteins during hardening, do not reveal major differences in the proteins from hardened or non-hardened plants.

Exposure to cold treatment is a necessary prerequisite for certain developmental processes in some plants. The requirement for a cold treatment to break the dormancy of certain seeds has been already discussed. In some species a cold treatment is necessary before the plant can be stimulated to flower by inductive photoperiod. This cold temperature treatment which makes the plant responsive to an inductive photoperiod is termed vernalization. The vernalization can occur in young seedlings and the response seems to be perceived in the embryo. Various studies have been conducted to determine the metabolic changes which occur during the required cold treatment which might be related to the onset of the capacity to respond to photoperiodic induction. The majority of the studies indicate that during the cold treatment there is an increase in protein and soluble amino acid content.[640, 641] However these changes are also induced in varieties which have no vernalization requirement and are thus probably a consequence of the hardening process and are unrelated to changes in flowering capability. It may be significant, however, that cold temperature treatment induces a greater accumulation of proline, glutamine, asparagine and soluble proteins in varieties requiring vernalization than in similar types which do not need a vernalizing treatment.[641]

In most woody perennials, a cold treatment is necessary to break the bud dormancy induced by short days. It may be recalled that the protein breakdown in the bark and accumulation of soluble amino acids preceded bud growth and it is possible that a cold treatment is necessary for the induction of proteolytic activity. Durzan's[168] studies indicate that soluble nitrogenous components can undergo extensive interconversion during cold treatment. In the buds of the spruce *Picea glauca* the arginine which accumulates in the fall of the year is progressively converted to proline. This accumulation of proline is similar to that which occurs in the other stress conditions (drought or vernalization) described earlier. The mechanism of arginine to proline conversion in the spruce has not been established but it presumably involves the enzyme arginase which would give rise to ornithine and urea. The interconversion of this ornithine to glutamic acid and proline can occur by reactions discussed in Chapter 2. The fate of the urea produced by arginase activity has not been established; however, the accumulation of asparagine which occurs at the same time as the proline build-up suggests that the urea nitrogen may be converted to the amido nitrogen of asparagine. In the spring of the year a decline in proline content is accompanied by an increase in glutamine[170] and labelling studies suggest that the carbon of proline is converted to glutamic acid and glutamine via Δ'-pyrroline-5-carboxylic acid and glutamic γ-semialdehyde.

Clearly, although winter buds are considered to be dormant, extensive metabolism occurs during the period of cold exposure.

9

Nitrogen Metabolism During Fruit Ripening and Leaf Senescence

FRUIT RIPENING

In many fruits the onset of ripening is characterized by a transient increase in respiratory activity. This has been termed the 'respiration climacteric' and usually precedes the softening and visible ripening of the fruit. This period appears to be associated with major metabolic changes and fruits have been extensively studied with a view to establishing both the cause of the respiratory burst and the nature of the changes in metabolism which occur during the ripening process.

There is a tendency to equate fruit ripening with senescence. However, it should be stressed[164] that ripening is the 'beginning of the end'; it precedes senescence and the metabolic changes occurring during ripening differ considerably from those which occur, for example, in senescing leaves. In fact the senescence of fruit, i.e. the deteriorative events which occur after ripening, have not been intensively studied.

RNA

During ripening it is generally found, in contrast to leaf senescence, that the RNA content is sustained or decreases only slowly. During the climacteric there may be a transient burst of RNA synthesis, and there is an increase in number of ribosomes during this time.[346] As ripening proceeds after the climacteric the relative proportion of polysomes to ribosomes has been reported to decrease.[401] It appears that the increased RNA synthesis during the climacteric may be an essential prerequisite for the ripening process; thus application of inhibitors of RNA synthesis prior to the climacteric prevents the onset of ripening.[201] Likewise treatment of unripe fruit with ethylene, which accelerates the ripening process, enhances the incorporation of precursors into RNA.[279]

Although the capacity to incorporate precursors into RNA changes during the climacteric it does not appear that the post climacteric decline in synthetic capability is associated with any significant loss of RNA. A relatively constant RNA level is maintained during ripening even though ribonuclease levels increase during the climacteric.[521] The physiological significance of the increase in ribonuclease is not clear but in some respects resembles the situation of intact leaves. Young expanding leaves with a high RNA content, in some instances, have more ribonuclease than older leaves.

Protein

Fruits are characteristically low in protein;[247] nevertheless, in tomatoes, apples, peas, and cantaloupes for example, there is a net increase in protein during ripening while in bananas and avocados there is no change in this component.[543] It has been shown by several authors that during the ripening period there is a shift in the pattern of proteins within the tissue and even in those fruits showing no net increase in protein content there may be an enhanced incorporation of amino acids.[121,237] These amino acids appear to be incorporated into specific enzyme proteins which are considered to play an important role in the ripening process. Application of the protein synthesis inhibitor cycloheximide early during the climacteric effectively inhibited ripening. However, if the administration of the inhibitors was delayed until late in the climacteric ripening proceeded; thus it appears that some enzymes essential for ripening are synthesized early in the climacteric.[70]

Many studies have been made of changes in enzyme level during the ripening process. Characteristically such hydrolytic enzymes as lipase, ribonuclease, acid phosphatase, cellulase, polygalacturonase, amylase, and invertase increase during ripening whereas many respiratory enzymes decline.[543]

Amino acids

During the ripening process there is usually an increase in the soluble amino acid content; however, there is considerable variation between fruit with regard to the dominant soluble amino acids. In bananas the amides asparagine and glutamine decline during the ripening period but histidine accumulates. Proline is the most prominent amino acid in many citrus fruits and its level increases during ripening. In contrast proline is virtually absent from ripe tomatoes.[88]

CHARACTERISTICS OF SENESCENCE

Senescence may be defined as the deteriorative events which precede the death of a mature cell.

In some instances the process may occur shortly after the maturation of certain individual cells; such is the situation in the production of xylem elements or the cells of the root cap. Senescence in these cells occurs while the remaining adjoining tissues remain healthy and viable. In other cases the cells of many tissues senesce more or less simultaneously, leading to the death of a particular organ; this determinate organ senescence is characteristic of leaves, fruits, and petals for example. In annual plants senescence is organismic and all the cells eventually senesce and the plants die.

A great deal of attention has focused on the senescence of leaves. Most of us are aware of the characteristic colour changes which are associated with the senescence of leaves of deciduous trees in autumn. In this situation the majority of the leaves senesce simultaneously; this has been described as synchronous senescence and seems to be triggered by photoperiodic conditions. This hypothesis can be simply confirmed by observing the leaves on limbs of trees of deciduous trees growing close to street lights. These leaves are retained for considerably longer periods of time than those on limbs removed from the immediate vicinity of the light.

In some annuals and herbaceous perennials the leaves senesce in a sequential manner. The leaves first formed by the young apex (i.e. those at the base of the stem) senesce early in the life of the plant and senescence proceeds from the base to the apex. In herbaceous perennials the senescence and death are confined to the aerial portions; the underground organs do not die but serve as perennating organs which function as a source of new growth. In many annuals the onset of senescence and the death of the whole plant occur after, and are triggered by, flowering. This senescence has been termed correlative senescence and in these instances the death of the plant can be prevented by removing the reproductive parts.

In considering these types of senescence it is apparent that the process is not a random event; clearly it is regulated internally in some manner. These orderly events of organ senescence provide a mechanism by which the internal cellular constituents may be depleted and transferred to other parts of the plant. The materials arising from the senescence of leaves and stems of annual plants are transferred to the developing fruit and seed. In woody perennials products from the senescing leaves serve as precursors for the synthesis of food reserves in the bark. The metabolites from the senescing aerial portions of herbaceous perennials can be used for the production of reserves in perennating structures. The senescence of leaves is so regulated that after the death of the leaf it is abscinded. This abscission of organs involves a specialized senescence of specific cells in which cell death is associated with the breakdown of adjoining cell walls and the formation of an abscission zone.[673]

BIOCHEMICAL CHANGES DURING LEAF SENESCENCE

Most detailed studies of senescence have been conducted on leaves. These organs form a useful experimental system in which to study senescence in that the characteristic symptoms of the process, a loss of chlorophyll, a decrease in protein and a decline in RNA content (Fig. 9.1), occur at an accelerated rate in detached leaves.[110] Moreover the senescence of the detached leaves can be regulated by application of various growth regulators (Fig. 9.2).

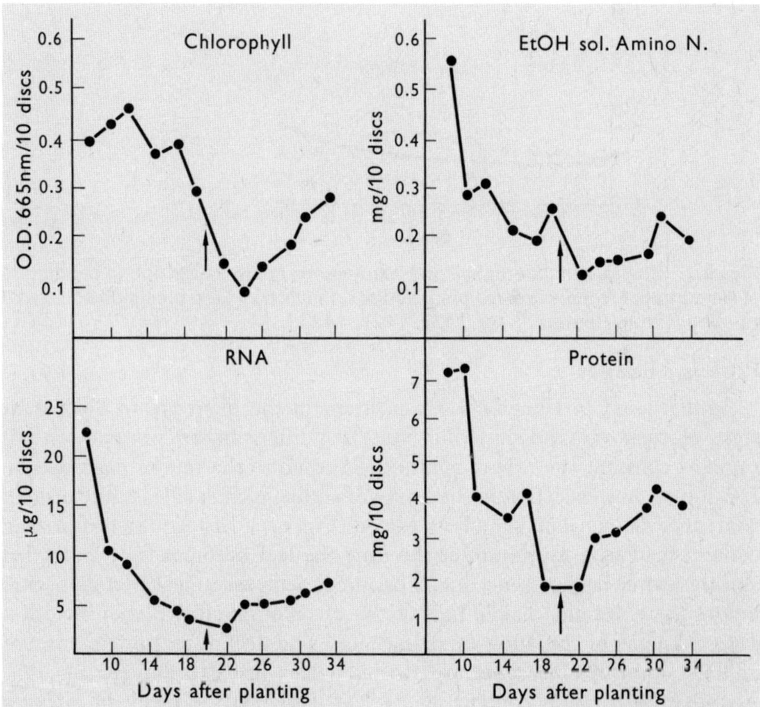

Fig. 9.1 The change of level of metabolites during senescence and regreening of primary leaves of *Phaseolus vulgaris*. ↑ Arrow indicates time of decapitation of stem portions above the senescing leaf. (Data from Chang.[103])

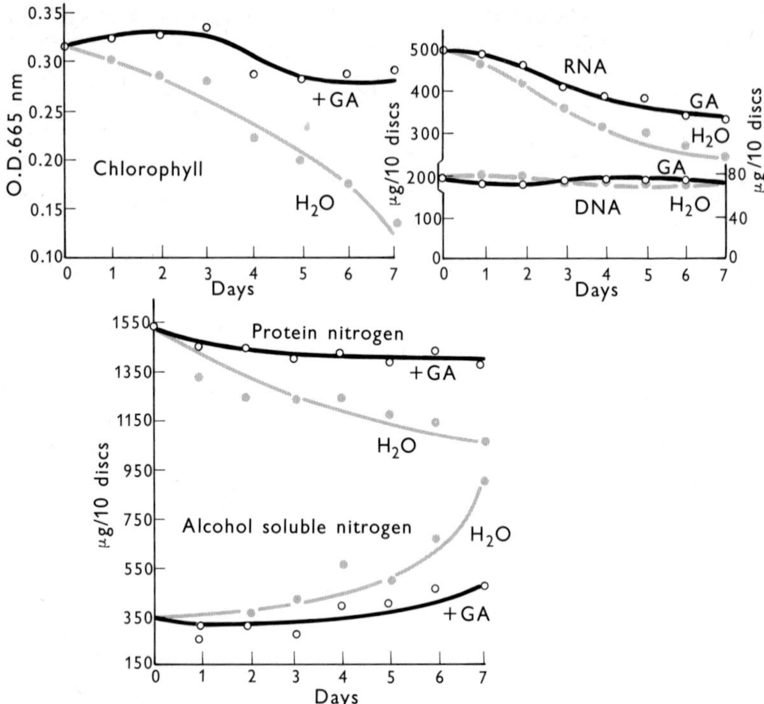

Fig. 9.2 Changes in chlorophyll and nitrogenous components during incubation of Nasturtium (*Tropaeolum majus*) leaf discs on water or 20 p.p.m. gibberellic acid solution. (From Beevers,[39] pp. 1420, 1422, 1423.)

Protein changes

As discussed in Chapter 6 the proteins in the plant are in a constant state of turnover and at steady state conditions when protein content remains constant the rate of synthesis is equal to the rate of degradation. It is found, however, that in the leaves of annual plants, at least, the protein content is not constant for a long period (Fig. 9.1). Instead the leaf protein content reaches a maximum at the time the leaf becomes fully expanded and thereafter begins to decline. The initial decrease in leaf protein occurs before there are any visible indications of chlorophyll depletion or other outward signs of the onset of senescence. The decline in protein content and the onset of senescence are accelerated by leaf detachment.

A reduced protein content of the leaf could be the result of either a reduced synthetic capacity or an increased protein degradation and experimental evidence has been provided which supports both these possibilities.

It is frequently observed[488,703] that the capacity to incorporate exogenously supplied labelled amino acids declines as detached leaves are incubated in water. In contrast, in detached leaves treated with various growth regulators which retard senescence the ability to incorporate amino acids into proteins is retained (Table 9.1).

Table 9.1 Changes in the capacity of senescing leaf discs of Nasturtium (*Tropaeolum majus*) to incorporate exogenously supplied ^{14}C-leucine into proteins. In the presence of the growth regulators kinetin and gibberellic acid which retard senescence the ability to incorporate amino acids is maintained. (Calculated from Beevers.[39])

Day	Treatment	Total EtOH sol. counts	Total protein counts	% incorporated
0	H_2O	11 930	4 875	29
1	H_2O	30 730	9 600	24
	GA 20 p.p.m.	19 550	13 550	41
	Kin 20 p.p.m.	18 990	13 240	42

On the basis of this type of data it has been concluded that senescence is caused by a failure or a decreased capacity for protein synthesis. However, the interpretation of the results of experiments of this type is difficult owing to the fact that in detached senescing leaves amino acids accumulate (Fig. 9.2). These may effectively dilute the exogenously supplied precursor and thus result in an apparent lower incorporation rate. It is also possible that the products synthesized from the labelled precursors are hydrolysed more rapidly in the senescing tissue in comparison with leaf discs treated with growth regulators. An accelerated degradation would give the impression of a reduced capacity for protein synthesis.

Several workers[349,432,562,620] in fact contend that senescence is attributable to an increased degradation. If the primary event of senescence is an increased proteolysis then it is important to understand the mechanism by which the proteins are degraded and the manner in which degradation is accelerated following leaf detachment. The observed accumulation of soluble amino acids in detached leaves is consistent with an extensive proteolysis; however there is very little information available concerning enzymes which might be responsible for the protein hydrolysis. In Chapter 6 it was indicated that proteolytic activity had been demonstrated in extracts from leaves[682] and the enzymes have been partially characterized; unfortunately very few studies have attempted to relate the measured proteolytic activity to leaf senescence. Anderson and Rowan[14] indicated that in tobacco leaves there was no increase in proteolytic activity on a fresh weight basis as the leaf aged. More recently it has been reported that senescent leaves of *Perilla*,[319] tobacco (*Nicotiana*),[144] and bean (*Phaseolus vulgaris*)[103] have lower protease levels than young leaves (Fig. 9.3). Also protease level does not increase during the senescence of the corolla of ephemeral flowers of *Ipomoea* (morning glory) even though proteins are depleted rapidly in this organ.[406]

In detached leaves there is an increase in proteolytic activity during senescence. Some authors[349,432,562,620] attribute the onset of senescence to a build-up of proteolytic activity. However, others[14,15,39] indicate that there is little relationship between protease level and senescence rate; furthermore accumulation of amino acids and decline in protein content may precede the build-up of proteolytic activity.

The differences between the detached and attached leaf with regard to changes in protease level during senescence may be related to changes in amino acid content. Martin and Thimann[404] have indicated that proteolytic activity may be induced in oat (*Avena*) leaf segments by various amino acids; serine was the most effective. In leaf segments or detached leaves amino acids accumulate during senescence and these foreseeably could induce the protease and account for the demonstrated higher proteolytic levels. In contrast, in attached senescing leaves amino acids do not accumulate and protease level is low; perhaps significantly protease activity is initially high in young leaves which have a high soluble amino acid content (Fig. 9.3).

Of course it is possible that the onset of senescence could be achieved without any increase in protease activity. Since proteins turnover the plant cells are equipped with the necessary machinery for protein degradation. An increased proteolysis could occur if there was a changed accessibility of the protease(s) to their substrate(s). Matile[405] has proposed that the vacuole of plant cells functions as a lysosome and is a repository of various acid hydrolases including proteases. Release of these proteases from the vacuole would obviously facilitate the breakdown of cytoplasmic proteins.

The question, however, arises whether or not the release of the hydrolases from the vacuole would allow for the selectivity of protein breakdown which occurs during senescence. This is a necessary requirement for other proteases also. The breakdown of proteins during senescence is a regulated event and not an uncontrolled autolysis. Axelrod and Jagendorf[20] initially demonstrated that the specific activity of various hydrolytic enzymes increased during the senescence of detached tobacco leaves at a time when the total protein content was declining. Recent studies with *Perilla*[319] and bean (*Phaseolus vulgaris*)[103] (Fig. 9.3) have demonstrated that during senescence the activity of various enzymes declines at different rates suggesting a differential rate of breakdown of the enzymatic proteins. Characteristically, photosynthesis declines earlier than respiratory activity. Ultrastructural studies[91] indicate that cellular breakdown follows a fairly

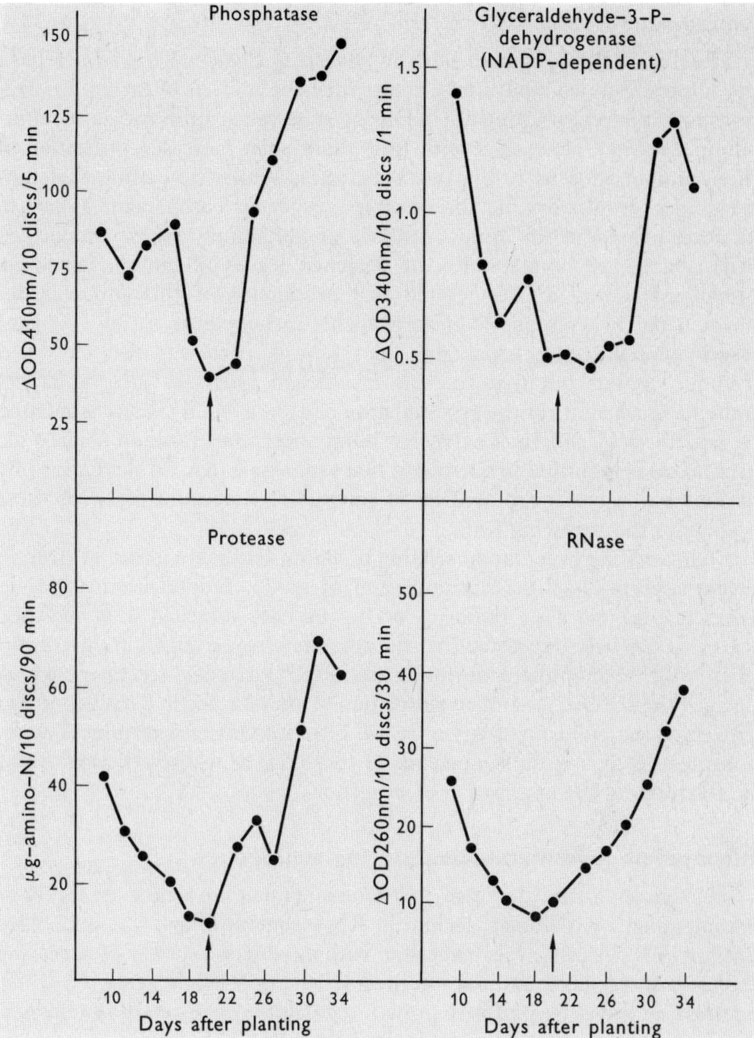

Fig. 9.3 The change of enzyme activities during senescence and regreening of primary leaves of *Phaseolus vulgaris*. The arrow indicates time of decapitation of aerial portions above the senescing leaf. (From Chang.[103])

well defined sequence in which chloroplasts and free ribosomes are the first organelles to degenerate followed by the mitochondria and finally a disruption of the plasmalemma and nucleus.

Amino acid metabolism

The metabolism of amino acids in senescing detached leaves has been the subject of much study and a comprehensive analysis of early investigations is provided by Chibnall.[110] During senescence amino acids accumulate in detached leaves and with time there is an increased utilization of these amino acids as respiratory substrates. While the carbon skeletons are used in respiratory metabolism the nitrogenous components appear to be accumulated in the amido residues of glutamine and asparagine. In early senescence or starvation of detached leaves glutamine formation predominates and it is suggested that production of this amide is dependent on the presence[715] of an available carbohydrate supply. At later stages of senescence asparagine biosynthesis predominates; however, Yemm[715] points out that there is insufficient aspartate present in leaf proteins to account for the accumulation of asparagine by direct amidation of aspartic acid and thus extensive interconversion of amino acids must occur. It was indicated in Chapter 2 that asparagine may be synthesized by various pathways and it will be of interest to determine if all of these operate in the senescing leaf.

An interesting observation relating to amino acid metabolism is given in recent reports[686] of the accumulation of auxin, indolyl acetic acid, in senescing leaves. This build-up occurs in both attached and detached leaves. It has been suggested that the auxin is synthesized from the tryptophan which accumulates during senescence.[561] Kinetin treatment which delayed the senescence and accumulation of amino acids in detached leaves prevented the build-up of auxin. It will be of interest to determine if auxin accumulates during the senescence of those leaves whose senescence can be retarded by the application of exogenous auxin.

Ribonucleic acid metabolism during senescence

The characteristic decrease in protein content in senescing leaves is accompanied by a similar decline in RNA content (Figs. 9.1, 9.2). This decline in RNA content is associated with a reduced capacity of senescing leaves to incorporate radioactive precursors into nucleic acids.[488, 703] In contrast, in leaves treated with growth regulators which retard senescence, the capacity to utilize precursors is sustained. This evidence suggests that the decline in RNA content during senescence is due to a decreased capacity for RNA synthesis and it has been postulated[191] that senescence is caused by a failure of the DNA to provide an effective template for RNA synthesis. Hormonal treatments which delay senescence would, on the basis of this hypothesis, appear to maintain the template in a functional state. In this regard Trewavas[636] has indicated that benzyladenine, a growth regulator which retards senescence, increased the rate of RNA synthesis in *Lemna* whereas abscisic acid, which accelerates the senescence of detached leaves, markedly reduced the production of RNA.

To date no attempts have been reported on tests to determine the template availability of the DNA in senescing leaves; a decline in RNA synthesis could equally be attributable to a decreased polymerase activity. By using chromatin from senescing barley leaves, Srivastava[586] concluded that the capacity for RNA synthesis was actually enhanced during senescence. However as senescence proceeded there was an increase in chromatin associated nucleases which degraded the newly synthesized RNA which could account for a decline in RNA and loss of cellular function.

If senescence is associated with a decrease in DNA template availability it might be expected that there would be a reduction in the synthesis of mRNA components. Initial attempts to demonstrate this utilized sucrose density gradient centrifugation, MAK column chromatography and, more recently, gel electrophoresis. Unfortunately these techniques have insufficient resolving power to identify mRNA specifically and thus it was not possible to associate the onset of senescence with a decreased mRNA synthesis.

An alternate approach to detect changes in mRNA level has been to determine the relative levels of polysomes to monosomes in ribosomal preparations. This approach has been used with limited success. Leaf tissue has a high ribonuclease content which makes the preparation of ribosomes and polysomes even more difficult than usual. However ribosomal preparations have been obtained from senescing leaves.[94, 172] It has been found that although polysome level declines during the growth of the leaf the mature leaves still contain cytoplasmic polysomes; however, chloroplast polysomes are present only in younger leaves.[94] The presence of cytoplasmic polysomes at later stages of senescence may indicate that mRNA is long lived or alternatively mRNA synthesis may be sustained even when the leaves are yellowing and senescing. The mRNA synthesized at this time could be coding for the production of hydrolytic enzymes during senescence.

Although polysomes may still be present, the total ribosome content of leaves declines during senescence. There seems to be a selectivity in the decrease; in the initial stages the proportion of chloroplast ribosomes declines, at later phases of senescence the free cytoplasmic ribosomes are depleted whereas the membrane-bound ribosomes are retained.[172] If, as suggested in Chapter 6, these bound ribosomes are synthesizing secretory proteins they may be involved in the production of those enzymes which Matile[405] indicates as being located in the vacuole.

Although unable to demonstrate changes in mRNA the studies in which RNA has been fractionated have been useful in characterizing the alterations in metabolism of component nucleic acids which occur during senescence. Senescent leaves have a reduced capacity to incorporate precursors into all RNA components detectable by gel electrophoresis.[495] Additionally it seems that during senescence there is an increase in the ratio of soluble RNA to ribosomal RNA, suggesting a greater lability or reduced synthesis of ribosomal RNA. Analyses of the changes in the levels of chloroplast and cytoplasmic ribosomal RNA indicate that the two classes may be degraded at different rates during senescence.[171] RNA extracted from senescent leaves of *Xanthium* contained no detectable chloroplast rRNA; in contrast the RNA prepared from yellow senescent leaves of *Vicia faba* and *Nicotiana tabacum* contained appreciable chloroplast rRNA.

It is interesting that the soluble RNA content is maintained during senescence. It has been proposed that declining protein content in senescing tissue might be the result of a reduced protein synthesis caused by a decreased translational ability associated with a changed tRNA complement.[610] There are changes in the iso-accepting leucyl-tRNA species during cotyledonary senescence of soybeans (*Glycine max*);[49] in addition the capacity of the leucyl-tRNA synthetase to acylate certain iso-accepting species is also altered during this period.[50, 316] Such information demonstrates the possibility that senescence, in part, may be associated with a loss of the specific anticodon in a particular iso-accepting tRNA species, required for the translation of mRNA. This event would lead to the cessation in the synthesis of a specific protein.

Instead of invoking a reduced synthesis, other workers attribute the decline in RNA content during senescence to an increased hydrolysis associated with an enhanced ribonuclease level. However, the evidence supporting this concept is extremely conflicting. In some detached leaves the ribonuclease does increase during senescence and this build-up is prevented by treatment of leaves with growth regulators which retard senescence.[576] In other instances the ribonuclease level does not increase during senescence of detached leaves.[587] In attached leaves ribonuclease level declines[322, 502] during senescence (Fig. 9.3). The overall rate of decline in this enzyme is slower than the decrease in protein content so that the specific activity of the enzyme (activity per unit protein) increases as senescence proceeds.[103] In the corolla of *Ipomoea*, on the other hand, ribonuclease level increased rapidly at the onset of senescence.[406]

The possibility exists, and the findings with attached leaves support the suggestion, that the decline in RNA content in senescing leaves can be achieved without any major increase in the tissue content of ribonuclease. The plant cells at maturity contain sufficient ribonuclease to degrade all of the cellular RNA in only a few hours.[373] The decline in RNA content could be achieved by increasing the accessibility of the existing enzyme to its substrate.

Few studies have attempted to characterize the ribonucleases that may be involved in the senescence process. At least two ribonucleases are present in leaf tissue, one soluble and the other particulate. It appears that in detached leaves there may be rapid increase in the soluble enzyme owing to an injury response unrelated to senescence.[711] It is claimed[648] that the particulate enzyme, which fits into the nuclease category of Wilson (see Chapter 6), may be more closely associated with senescence.

There is no massive accumulation of nucleotides in senescing leaves and the mechanism of metabolism of the products arising from nucleic acid hydrolysis has not been investigated. The mechanism of RNA cleavage during senescence has not been studied; if the hydrolysis is catalysed by an endonuclease it might be expected that intermediate size polynucleotides would be produced. However, gel electrophoresis of nucleic acids extracted from senescing leaves has failed to demonstrate the accumulation of intermediate molecular weight degradation products.

THE REGULATION OF SENESCENCE AND ITS REVERSAL

Original hypotheses explaining the acceleration of senescence, after leaf detachment, suggested that the onset of senescence occurred after the depletion of the carbohydrates. It was reasoned that in attached leaves the carbohydrate content of the leaf was sustained by imports from other leaves. However, Chibnall[110] in 1939 pointed out that the experimental data showed that there was no direct relationship between soluble carbohydrate status of the leaf and the onset of senescence.

Other workers[652] have suggested that the onset of senescence may be associated with a reduced respiratory capability; however it is found that many of the characteristic changes of senescence, such as chlorophyll loss, and decrease in protein and RNA precede the respiratory decline which occurs during the terminal phase.[305]

Senescence is delayed if roots form on the petioles of detached leaves. With considerable foresight, Chibnall[110] suggested that the roots supplied some hormone which was essential for maintaining the normal metabolism of leaves. Experimental support for this concept has been provided by the many observations that exogenously supplied plant growth regulators influence the senescence rate of detached leaves. Cytokinins, auxin and gibberellic acid retard the senescence of detached leaves or leaf discs of certain species whereas abscisic acid accelerates the process. Cytokinins and gibberellins are considered to be synthesized in the meristematic regions of the plant and thus leaf detachment effectively interrupts the normal flow of these hormones into the leaf.

These observations suggest that leaf senescence may be controlled by the balance of endogenous growth regulators and it has been shown that in many woody perennials there is a decrease in endogenous auxins and gibberellins and a higher level of growth inhibitors under the short days preceding leaf senescence.[503, 672] Under drought conditions a depleted flow of materials from the roots would reduce the availability of gibberellins[310] and cytokinins[296] normally supplied by the root exudate. A lowered level of these compounds in the leaf may account for the decline in RNA and protein characteristic of water stressed leaves.[296] Additionally, the senescence may be accelerated by the accumulation of abscisic acid in the water stressed leaves.[710] Senescing detached leaves show a decrease in gibberellin like components and an increase in abscisic acid like compounds.[111] Further support for the concept that endogenous growth regulators may control senescence is provided by the observation that there is a close correlation between longevity of cut roses and endogenous cytokinin activity.[410]

Many experiments have been conducted in an attempt to establish the mechanism by which growth regulators may control senescence. Essentially, two schemes are advanced; on the one hand it is suggested that the growth regulators which retard senescence sustain the synthesis of RNA and protein whereas abscisic acid curtails these processes. Others suggest that auxins, gibberellins and cytokinins retard senescence by preventing the build-up of degradative enzymes; conversely abscisic acid enhances hydrolytic activity.

In advancing any explanation for senescence and its control it must be recognized that the process is a regulated event and not an uncontrolled autolysis. Perhaps the clearest demonstration of this aspect comes from experiments showing suspended senescence and senescence reversal. The usual decline in protein, chlorophyll and RNA which occurs when detached leaves are floated on water can be suspended by additions of growth regulators after the onset of senescence. In some instances it has been reported that the leaves re-green. The process of senescence can be reversed if the aerial portions of the plant above a senescing leaf are removed.

The re-greening of the leaf is preceded by an increased level of protein and RNA (Fig. 9.1). Remarkably there is an increase in hydrolytic enzymes (Fig. 9.3), and this precedes the increase in level of the photosynthetic enzyme glyceraldehyde 3-phosphate dehydrogenase. Analytical studies by Callow and Woolhouse[95] indicate that during senescence reversal of leaves of *Perilla* synthesis of cytoplasmic ribosomal RNA precedes the commencement of rapid chloroplast ribosomal RNA production. Ultrastructural studies indicate that there is a rapid restoration of cytoplasmic integrity following decapitation and the re-greening which follows appears to involve the repair of existing chloroplasts rather than *de novo* production of new organelles.[103]

Clearly this reversal of senescence indicates that the machinery for regulating protein synthesis is retained intact until late stages of senescence. In both detached and attached leaves there is a point beyond which senescence cannot be arrested or reversed. This time may coincide with the breakdown of the tonoplast and the exposure of the cytoplasm to acid hydrolases associated with the vacuole. Disruption of the cytoplasm would obviously have a deleterious effect on the capacity for protein synthesis and the capability to carry out cellular repair. However, in spite of the apparent terminal nature of tonoplast rupture, ultrastructural studies indicate that many organelles are still recognizable following this event. Ultimately they break down, however; the nucleus and plasmalemma are retained until later stages and death of the cell is apparently coincident upon disruption of the plasmalemma and nuclear membrane.[91]

Bibliography

1a. AARNES, H. and ROGNES, S. E. (1974). Threonine sensitive aspartokinase and homoserine dehydrogenase from *Pisum sativum* (L.). *Phytochem.*, **13**, 2717–2724.

1. ABELES, F. B. (1972). Biosynthesis and mechanism of action of ethylene. *A. Rev. Pl. Physiol.*, **23**, 259–292.

2. ABELES, F. B. (1973). *Ethylene in Plant Biology*. Academic Press, New York.

3. ADDICOTT, F. T. (1970). Plant hormones in the control of abscission. *Biol. Rev.*, **45**, 385–424.

4. ADEPIPE, N. O. and FLETCHER, R. A. (1971). Benzyladenine as a regulator of chlorophyll synthesis in cucumber cotyledons. *Can. J. Bot.*, **49**, 59–61.

5. AGURELL, S., LUNDSTRÖM, J. and SANDBERG, F. (1967). Biosynthesis of mescaline in peyote. *Tetrahedron Lett.*, Pt. **2**, No. **26**, 2433–2435.

6. AKAZAWA, T. and BEEVERS, H. (1957). Mitochondria in the endosperm of the germinating castor bean: a developmental study. *Biochem. J.*, **67**, 115–118.

7. ALEEM, M. I. H. (1970). Oxidation of inorganic nitrogen compounds. *A. Rev. Pl. Physiol.*, **21**, 67–90.

8. ALEEM, M. I. H., HOCH, G. E. and VARNER, J. E. (1965). Water as the source of oxidant and reductant in bacterial chemosynthesis. *Proc. natn. Acad. Sci. U.S.A.*, **54**, 869–873.

9. ALEEM, M. I. H., LEES, H., LYRIC, R. and WEISS, D. (1962). Nitrohydroxylamine: the unknown intermediate in nitrification? *Biochem. biophys. Res. Commun.*, **7**, 126–127.

10. ALEEM, M. I. H. and LEES, H. (1963). Autotrophic enzyme systems. 1. Electron transport systems concerned with hydroxylamine oxidation in *Nitrosomonas*. *Can. J. Biochem. Physiol.*, **41**, 763–778.

11. ALEEM, M. I. H. and NASON, A. (1959). Nitrite oxidase, a particulate cytochrome electron transport system from *Nitrobacter*. *Biochem. biophys. Res. Commun.*, **1**, 323–327.

12. AMEN, R. D. (1968). A model of seed dormancy. *Bot. Rev.*, **34**, 1–31.

13. ANDERSON, J. N. and MARTIN, R. O. (1973). Identification of cadaverine in *Pisum sativum*. *Phytochem.*, **12**, 443–446.

14. ANDERSON, J. W. and ROWAN, K. S. (1965). Activity of peptidase in tobacco leaf tissue in relation to senescence. *Biochem. J.*, **97**, 741–746.

15. ANDERSON, J. W. and ROWAN, K. S. (1966). The effect of furfuryl amino-purine on senescence in tobacco leaf tissue after harvest. *Biochem. J.*, **98**, 401–404.

16. ANDERSON, L. A. and SMILLIE, R. M. (1966). Binding of chloramphenicol by ribosomes from chloroplasts. *Biochem. biophys. Res. Commun.*, **23**, 535–539.
17. ANDERSON, M. B. and CHERRY, J. H. (1969). Differences in leucyl-transfer RNA's and synthetase in soybean seedlings. *Proc. natn. Acad. Sci. U.S.A.*, **62**, 202–209.
18. APARICIO, P. J., CÁRDENAS, J., ZUMFT, W. G., MA VEGA, J., HERRERA, J., PANEQUE, A. and LOSADA, M. (1971). Molybdenum and iron as constituents of the enzymes of the nitrate reducing system from *Chlorella*. *Phytochem.*, **10**, 1487–1495.
19. ARNON, D. I. and STOUT, P. R. (1939). Molybdenum as an essential element for higher plants. *Pl. Physiol., Lancaster*, **14**, 599–602.
20. AXELROD, B. and JAGENDORF, A. T. (1951). The fate of phosphatase, invertase, and peroxidase in autolyzing leaves. *Pl. Physiol., Lancaster*, **26**, 406–410.
21. BAILEY, C. J. and BOULTER, D. (1972). The structure of vicilin of *Vicia faba*. *Phytochem.*, **11**, 59–64.
22. BAIN, J. M. and MERCER, F. V. (1966). Subcellular organization of the developing cotyledons of *Pisum sativum* L. *Aust. J. biol. Sci.*, **19**, 49–67.
23. BAIN, J. M. and MERCER, F. V. (1966). Subcellular organization of the cotyledons in germinating seeds and seedlings of *Pisum sativum* L. *Aust. J. biol. Sci.*, **19**, 69–84.
24. BAIN, J. M. and MERCER, F. V. (1966). The relationship of the axis and the cotyledons in germinating seeds and seedlings of *Pisum sativum* L. *Aust. J. biol. Sci.*, **19**, 85–96.
25. BALINSKY, D. and DAVIES, D. D. (1961). Aromatic biosynthesis in higher plants. I. Preparation and properties of dehydroshikimic reductase. *Biochem. J.*, **80**, 292–296.
26. BALINSKY, D. and DAVIES, D. D. (1962). Aromatic biosynthesis in higher plants. IV. The distribution of dehydroshikimic reductase and dehydroquinase. *J. exp. Bot.*, **13**, 414–421.
27. BALINSKY, D., DENNIS, A. W. and CLELAND, W. W. (1971). Kinetic and isotope-exchange studies on shikimate dehydrogenase from *Pisum sativum*. *Biochemistry*, **10**, 1947–1952.
28. BARKER, G. R. and DOUGLAS, T. (1960). The function of ribonuclease in germinating peas. *Nature, Lond.*, **188**, 943–944.
29. BARNES, R. L. (1959). Formation of allantoin and allantoic acid from adenine in leaves of *Acer saccharinum* L. *Nature, Lond.*, **184**, 1944.
30. BARNETT, N. M. and NAYLOR, A. W. (1966). Amino acid and protein metabolism in bermuda grass during water stress. *Pl. Physiol., Lancaster*, **41**, 1222–1230.
31. BATTERSBY, A. R. and HARPER, B. J. T. (1962). Alkaloid biosynthesis. Part I. The biosynthesis of papaverine. *J. chem. Soc.*, 3526–3533.
32. BATTERSBY, A. R. (1967). Phenol oxidations in the alkaloid field. In *Oxidative Coupling of Phenols*, ed. TAYLOR, W. I. and BATTERSBY, A. R., 142–175. Marcel Dekker Inc., New York.
33. BATTERSBY, A. R., BINKS, R. and HUXTABLE, R. (1967). Biosynthesis of cactus alkaloids. *Tetrahedron Lett.*, Pt. **1**, No. **6**, 563–565.
34. BAUR, A. H. and YANG, S. F. (1972). Methionine metabolism in apple tissue in relation to ethylene biosynthesis. *Phytochem.*, **11**, 3207–3214.
35. BEALE, S. I. and CASTELFRANCO, P. A. (1974). The biosynthesis of δ amino-levulinic acid in higher plants. II. Formation of ^{14}C-δ aminolevulinic acid from labeled precursors in greening plant tissues. *Pl. Physiol., Lancaster*, **53**, 297–303.
36. BEEVERS, H. (1951). An L-glutamic acid decarboxylase from barley. *Biochem. J.*, **48**, 132–137.
37. BEEVERS, H. (1961). Metabolic production of sucrose from fat. *Nature, Lond.*, **191**, 433–436.
38. BEEVERS, L. (1968). Protein degradation and proteolytic activity in the cotyledons of germinating pea seeds (*Pisum sativum*). *Phytochem.*, **7**, 1837–1844.
39. BEEVERS, L. (1968). Growth regulator control of senescence in leaf discs of

nasturtium (*Tropaeolum majus*). In *Biochemistry and Physiology of Plant Growth Substances*, ed. WIGHTMAN, F. and SETTERFIELD, G., 1417–1435. Runge Press, Ottawa.

40. BEEVERS, L. and GUERNSEY, F. S. (1966). Changes in some nitrogenous components during the germination of pea seeds. *Pl. Physiol., Lancaster*, **41**, 1455–1458.
41. BEEVERS, L. and HAGEMAN, R. H. (1969). Nitrate reduction in higher plants. *A. Rev. Pl. Physiol.*, **20**, 495–522.
42. BEEVERS, L. and HAGEMAN, R. H. (1972). The role of light in nitrate metabolism in higher plants. In *Photophysiology*, Vol. VII, ed. GIESE, A. C., 85–113. Academic Press, New York.
43. BEEVERS, L. and POULSON, R. (1972). Protein synthesis in cotyledons of *Pisum sativum* L. I. Changes in cell-free amino acid incorporation capacity during seed development and maturation. *Pl. Physiol., Lancaster*, **49**, 476–481.
44. BEEVERS, L., SCHRADER, L. E., FLESHER, D. and HAGEMAN, R. H. (1965). The role of light and nitrate in the induction of nitrate reductase in radish cotyledons and maize seedlings. *Pl. Physiol., Lancaster*, **40**, 691–698.
45. BEEVERS, L. and SPLITTSTOESSER, W. E. (1968). Protein and nucleic acid metabolism in germinating peas. *J. exp. Bot.*, **19**, 698–711.
46. BELSER, W. L., MURPHY, J. B., DELMER, D. P. and MILLS, S. E. (1971). End product control of tryptophan biosynthesis in extracts and intact cells of the higher plant *Nicotiana tabacum* var. Wisconsin 38. *Biochim. biophys. Acta*, **237**, 1–10.
47. BEN-ZIONI, A., ITAI, C. and VAADIA, Y. (1967). Water and salt stresses, kinetin and protein synthesis in tobacco leaves. *Pl. Physiol., Lancaster*, **42**, 361–365.
48. BERGERSEN, F. J. (1971). Biochemistry of symbiotic nitrogen fixation in legumes. *A. Rev. Pl. Physiol.*, **22**, 121–140.
49. BICK, M. D., LIEBKE, H., CHERRY, J. H. and STREHLER, B. L. (1970). Changes in leucyl- and tyrosyl-tRNA of soybean cotyledons during plant growth. *Biochim. biophys. Acta*, **204**, 175–182.
50. BICK, M. D. and STREHLER, B. L. (1971). Leucyl transfer RNA synthetase changes during soybean cotyledon senescence. *Proc. natn. Acad. Sci. U.S.A.*, **68**, 224–228.
51. BIDWELL, R. G. S., BARR, R. A. and STEWARD, F. C. (1964). Protein synthesis and turn-over in cultured plant tissue: sources of carbon for synthesis and the fate of the protein breakdown products. *Nature, Lond.*, **203**, 367–373.
52. BIRD, I. F., CORNELIUS, M. J., KEYS, A. J. and WHITTINGHAM, C. P. (1972). Oxidation and phosphorylation associated with the conversion of glycine to serine. *Phytochem.*, **11**, 1587–1594.
53. BIRMINGHAM, B. C. and MACLACHLAN, G. A. (1972). Generation and suppression of microsomal ribonuclease activity after treatments with auxin and cytokinin. *Pl. Physiol., Lancaster*, **49**, 371–375.
54. BLUMENTHAL-GOLDSCHMIDT, S., BUTLER, G. W. and CONN, E. E. (1963). Incorporation of hydrocyanic acid labelled with carbon-14 into asparagine in seedlings. *Nature, Lond.*, **197**, 718–719.
55. BLUMENTHAL, S. G., HENDRICKSON, H. R., ABROL, Y. P. and CONN, E. E. (1968). Cyanide metabolism in higher plants. III. The biosynthesis of β-cyanoalanine. *J. biol. Chem.*, **243**, 5302–5307.
56. BOGORAD, L. (1958). The enzymatic synthesis of porphyrins from porphobilinogen. *J. biol. Chem.*, **233**, 516–519.
57. BOGORAD, L. (1965). Chlorophyll biosynthesis. In *Chemistry and Biochemistry of Plant Pigments*, ed. GOODWIN, T. W., 29–74. Academic Press, New York.
58. BOGORAD, L. and JACOBSEN, A. B. (1964). Inhibition of greening of etiolated leaves by actinomycin D. *Biochem. biophys. Res. Commun.*, **14**, 113–117.
59. BOGORAD, L. and WOODCOCK, C. L. F. (1971). Rifamycins: the inhibition of plastid RNA synthesis *in vivo* and *in vitro* and variable effects on chlorophyll formation in maize leaves. In *Autonomy and Biogenesis of Mitochondria and*

Chloroplasts, ed. BOARDMAN, N. K., LINNANE, A. W. and SMILLIE, R. M., 92–97. North Holland Publishing Company, Amsterdam.

60. BOLLARD, E. C. (1956). Nitrogenous compounds in plant xylem sap. *Nature, Lond.*, **178**, 1189–1190.

61. BONE, D. H. (1959). Metabolism of citrulline and ornithine in mung bean mitochondria. *Pl. Physiol., Lancaster*, **34**, 171–175.

62. BONNER, J., HUANG, R. C. and GILDEN, R. V. (1963). Chromosomally directed protein synthesis. *Proc. natn. Acad. Sci. U.S.A.*, **50**, 893–900.

63. BONNER, J. and ZEEVAART, J. A. D. (1962). Ribonucleic acid synthesis in the bud an essential component of floral induction in *Xanthium. Pl. Physiol., Lancaster*, **37**, 43–49.

64. BOTTOMLEY, W., SMITH, J. H. and BOGORAD, L. (1971). RNA polymerases of maize: partial purification and properties of the chloroplast enzyme. *Proc. natn. Acad. Sci. U.S.A.*, **68**, 2412–2416.

65. BOTTOMLEY, W., SPENCER, D., WHEELER, A. and WHITFELD, P. R. (1971). The effect of a range of RNA polymerase inhibitors on RNA synthesis in higher plant chloroplasts and nuclei. *Archs. Biochem. Biophys.*, **143**, 269–275.

66. BOULTER, D. (1970). Protein synthesis in plants. *A. Rev. Pl. Physiol.*, **21**, 91–114.

67. BOULTER, D., ELLIS, R. J. and YARWOOD, A. (1972). Biochemistry of protein synthesis in plants. *Biol. Rev.*, **47**, 113–175.

68. BRADBEER, J. W. (1973). The synthesis of chloroplast enzymes. In *Biosynthesis and its Control in Plants*, ed. MILBORROW, B. V., 279–302. Academic Press, London.

69. BRADBEER, J. W. and FLOYD, V. M. (1964). Nucleotide synthesis in hazel seeds during after-ripening. *Nature, Lond.*, **201**, 99–100.

70. BRADY, C. J., PALMER, J. K., O'CONNELL, P. B. A. and SMILLIE, R. M. (1970). An increase in protein synthesis during ripening of the banana fruit. *Phytochem.*, **9**, 1037–1047.

71. BRAWERMAN, G., MENDECKI, T. and LEE, S. Y. (1972). A procedure for the isolation of mammalian messenger ribonucleic acid. *Biochemistry*, **11**, 637–641.

72. BREIDENBACH, R. W., CASTELFRANCO, P. and CRIDDLE, R. S. (1967). Biogenesis of mitochondria in germinating peanut cotyledons. II. Changes in cytochromes and mitochondrial DNA. *Pl. Physiol., Lancaster*, **42**, 1035–1041.

73. BRIARTY, L. G., COULT, D. A. and BOULTER, D. (1969). Protein bodies of developing seeds of *Vicia faba. J. exp. Bot.*, **20**, 358–372.

74. BRIARTY, L. G., COULT, D. A. and BOULTER, D. (1970). Protein bodies of germinating seeds of *Vicia faba. J. exp. Bot.*, **21**, 513–524.

75. BRITTEN, R. J. and KOHNE, D. E. (1968). Repeated sequences in DNA. *Science, N.Y.*, **161**, 529–540.

76. BROWN, E. G. (1963). Purine and pyrimidine derivatives in mature pea seeds. *Biochem. J.*, **88**, 498–504.

77. BROWN, E. G. (1965). Changes in the free nucleotide and nucleoside pattern of pea seeds in relation to germination. *Biochem. J.*, **95**, 509–514.

78. BROWN, E. G. and MANGAT, B. S. (1970). Studies on the free nucleotide pool and RNA components of detached leaves of *Phaseolus vulgaris* during root development. *Phytochem.*, **9**, 1859–1868.

79. BROWN, E. G. and SHORT, K. C. (1969). The changing nucleotide pattern of sycamore cells during culture in suspension. *Phytochem.*, **8**, 1365–1372.

80. BROWN, E. G. and SILVER, A. V. (1966). The natural occurrence of a uracil 5-peptide and its metabolic relationship to guanosine 5′-monophosphate. *Biochim. biophys. Acta*, **119**, 1–10.

81. BROWN, R. and BROADBENT, D. (1950). The development of cells in the growing zones of the root. *J. exp. Bot.*, **1**, 249–263.

82. BRYAN, J. K. (1969). Studies on the catalytic and regulatory properties of homoserine dehydrogenase of *Zea mays* roots. *Biochim. biophys. Acta*, **171**, 205–216.

83. BRYAN, P. A., CAWLEY, R. D., BRUNNER, C. E. and BRYAN, J. K. (1970). Isolation

and characterization of a lysine-sensitive aspartokinase from a multicellular plant. *Biochem. biophys. Res. Commun.*, **41**, 1211–1217.

84. BUCHOWITZ, J. and REIFER, I. (1961). The conversion of orotic acid to pyrimidine derivatives in plant material. *Acta biochim. pol.*, **8**, 24–34.

85. BULEN, W. A., BURNS, R. C. and LECOMTE, J. R. (1965). Nitrogen fixation: hydrosulfite as electron donor with cell-free preparations of *Azotobacter vinelandii* and *Rhodospirillum rubrum*. *Proc. natn. Acad. Sci. U.S.A.*, **53**, 532–539.

86. BURKARD, G., ECLANCHER, B. and WEIL, J. H. (1969). Presence of *n*-formylmethionyl-transfer RNA in bean chloroplasts. *FEBS Letters*, **4**, 285–287.

86a. BURKARD, G. and KELLER, E. B. (1974). Poly (A) polymerase and Poly (G) polymerase in wheat chloroplasts. *Proc. natn. Acad. Sci. U.S.A.*, **71**, 389–393.

87. BURRIS, R. H. and MILLER, C. E. (1941). Application of ^{15}N to the study of biological nitrogen fixation. *Science, N.Y.*, **93**, 114–115.

88. BURROUGHS, L. F. (1970). Amino acids. In *The Biochemistry of Fruits and Their Products*, Vol. 1, ed. HULME, A. C., 119–146. Academic Press, London.

89. BURTON, E. G. and SAKAMI, W. (1969). The formation of methionine from the monoglutamate form of methyltetra-hydrofolate by higher plants. *Biochem. biophys. Res. Commun.*, **36**, 228–233.

90. BUTLER, G. W., BAILEY, R. W. and KENNEDY, L. D. (1965). Studies on the glucosidase 'linamarase'. *Phytochem.*, **4**, 369–381.

91. BUTLER, R. D. and SIMON, E. W. (1970). Ultrastructural aspects of senescence in plants. In *Advances in Gerontological Research*, Vol. III, ed. STREHLER, B. L., 73–129. Academic Press, New York.

92. BUTTROSE, M. S. (1963). Ultrastructure of the developing wheat endosperm. *Aust. J. biol. Sci.*, **16**, 305–317.

93. BYERS, M. (1971). The amino acid composition of some leaf protein preparations. In *Leaf Protein IBP Handbook*, No. 20, ed. PIRIE, N. W., 95–114. Blackwell Scientific Publications, London.

94. CALLOW, J. A., CALLOW, M. E. and WOOLHOUSE, H. W. (1972). *In vitro* protein synthesis, ribosomal RNA synthesis and polyribosomes in senescing leaves of *Perilla*. *Cell Differentiation*, **1**, 79–92.

95. CALLOW, M. E. and WOOLHOUSE, H. W. (1973). Changes in nucleic-acid metabolism in regreening leaves of *Perilla*. *J. exp. Bot.*, **24**, 285–294.

96. CAMERON, E. C. and MAZELIS, M. (1971). A nonproteolytic 'trypsin-like' enzyme. *Pl. Physiol., Lancaster*, **48**, 278–281.

97. CARNAHAN, J. E., MORTENSON, L. E., MOWER, H. F. and CASTLE, J. E. (1960). Nitrogen fixation in cell-free extracts of *Clostridium pasteurianum*. *Biochim. biophys. Acta*, **44**, 520–535.

98. CASTRIC, P. A., FARNDEN, K. J. F. and CONN, E. E. (1972). Cyanide metabolism in higher plants V. The formation of asparagine from β-cyanoalanine. *Archs. Biochem. Biophys.*, **152**, 62–69.

99. CATSIMPOOLAS, N., CAMPBELL, T. G. and MEYER, E. W. (1968). Immunochemical study of changes in reserve proteins of germinating soybean seeds. *Pl. Physiol., Lancaster*, **43**, 799–805.

100. CATSIMPOOLAS, N., ROGERS, D. A., CIRCLE, S. J. and MEYER, E. W. (1967). Purification and structural studies of the 11S component of soybean proteins. *Cereal Chem.*, **44**, 631–637.

101. CHANDLER, J. L. R. and GHOLSON, R. K. (1972). Nicotinic acid decarboxylation in tobacco roots. *Phytochem.*, **11**, 239–242.

102. CHANDRA, G. R. and DUYNSTEE, E. E. (1971). Methylation of ribonucleic acids and hormone-induced α-amylase synthesis in the aleurone cells. *Biochim. biophys. Acta*, **232**, 514–523.

103. CHANG, T. C. (1973). *Senescence reversal in leaves of* Phaseolus vulgaris *L.* M.S. thesis, University of Oklahoma.

104. CHEN, D., SARID, S. and KATCHALSKI, E. (1968). Studies of the nature of messenger RNA in germinating wheat embryos. *Proc. natn. Acad. Sci. U.S.A.*, **60**, 902–909.

105. CHEN, S. C. and CHANG, J. L. L. (1972). Does gibberellic acid stimulate seed germination via amylase synthesis? *Pl. Physiol., Lancaster*, **49**, 441–442.
106. CHERRY, J. H. (1962). Nucleic acid determination in storage tissues of higher plants. *Pl. Physiol., Lancaster*, **37**, 670–678.
107. CHERRY, J. H. (1963). Nucleic acid, mitochondria, and enzyme changes in cotyledons of peanut seeds during germination. *Pl. Physiol., Lancaster*, **38**, 440–446.
108. CHERRY, J. H. and VAN HUYSTEE, R. (1965). Effects of 5-fluorouracil on photoperiodic induction and nucleic acid metabolism of *Xanthium*. *Pl. Physiol., Lancaster*, **40**, 987–993.
109. CHESHIRE, R. M. and MIFLIN, B. J. (1973). The regulation of aspartokinase. *Pl. Physiol., Lancaster*, **51** (Suppl.), 54.
110. CHIBNALL, A. C. (1939). *Protein Metabolism in the Plant*. Yale University Press, New Haven, U.S.A.
111. CHIN, T. Y. and BEEVERS, L. (1970). Changes in endogenous growth regulators in nasturtium leaves during senescence. *Planta*, **92**, 178–188.
112. CHIN, T. Y., POULSON, R. and BEEVERS, L. (1972). The influence of axis removal on protein metabolism in cotyledons of *Pisum sativum* L. *Pl. Physiol., Lancaster*, **49**, 482–489.
113. CHOU, K. and SPLITTSTOESSER, W. E. (1972). Changes in amino acid content and the metabolism of γ-aminobutyrate in *Cucurbita moschata* seedlings. *Physiologia Pl.*, **26**, 110–114.
114. CHRISPEELS, M. J. and VARNER, J. E. (1967). Hormonal control of enzyme synthesis: on the mode of action of gibberellic acid and abscisin in aleurone layers of barley. *Pl. Physiol., Lancaster*, **42**, 1008–1016.
115. CHRISPEELS, M. J., VATTER, A. E., MADDEN, D. M. and HANSON, J. B. (1966). The structural organization of ribonucleic acid associated with soybean microsomes. *J. exp. Bot.*, **17**, 492–501.
116. CHRISTIANSON, D. D., NIELSEN, H. C., KHOO, U., WOLF, M. J. and WALL, J. S. (1969). Isolation and chemical composition of protein bodies and matrix proteins in corn endosperm. *Cereal Chem.*, **46**, 372–381.
117. CHROBOCZEK, H. and CHERRY, J. H. (1966). Characterization of nucleic acids in peanut cotyledons. *J. molec. Biol.*, **19**, 28–37.
118. CLANDININ, M. T. and COSSINS, E. A. (1972). Localization and interconversion of tetrahydropteroylglutamates in isolated pea mitochondria. *Biochem. J.*, **128**, 29–40.
119. CLARK, A. H. and MANN, P. J. G. (1957). The oxidation of tryptamine to 3-indoleacetaldehyde by plant amine oxidase. *Biochem. J.*, **65**, 763–764.
120. CLARK, M. C., PAGE, O. T. and FISHER, M. G. (1972). Purification and properties of *N*-ribosyladenine ribohydrolase from potato leaves. *Phytochem.*, **11**, 3413–3419.
121. CLEMENTS, R. L. (1970). Protein patterns in fruits. In *Biochemistry of Fruits and Their Products*, Vol. I, ed. HULME, A. C., 159–178. Academic Press, London.
122. CLOWES, F. A. L. and JUNIPER, B. E. (1968). *Plant Cells*. Oxford University Press.
123. CONN, E. E. and AKAZAWA, T. (1958). Biosynthesis of *p*-hydroxybenzaldehyde. *Fed. Proc.*, **17**, 205.
124. CONN, E. E. and BUTLER, G. W. (1969). The biosynthesis of cyanogenic glycosides and other simple nitrogen compounds. In *Perspectives in Phytochemistry*, ed. HARBORNE, J. B. and SWAIN, T., 47–74. Academic Press, London.
125. COSSINS, E. A. and CABALLERO, A. (1970). Further studies of intermediary metabolism in radish cotyledons. The utilization of glycine, glyoxylate, and acetate by tissue disks. *Can. J. Bot.*, **48**, 1767–1774.
126. COSSINS, E. A. and SHAH, S. P. J. (1972). Pteroylglutamates of higher plant tissues. *Phytochem.*, **11**, 587–593.
127. COTTON, R. G. H. and GIBSON, F. (1968). The biosynthesis of phenylalanine and tyrosine in the pea (*Pisum sativum*): Chorismate mutase. *Biochim. biophys. Acta*, **156**, 187–189.
128. COUDRAY, Y., QUETIER, F. and GUILLE, E. (1970). New compilation of satellite DNA's. *Biochim. biophys. Acta*, **217**, 259–267.

129. COWLES, J. R. and KEY, J. L. (1972). Demonstration of two tyrosyl-tRNA synthetases of pea roots. *Biochim. biophys. Acta*, **281**, 33–44.
130. CROCKER, W. and BARTON, L. V. (1953). *Physiology of Seeds*. Chronica Botanica, Waltham, Mass.
131. CUTTING, J. A. and SCHULMAN, H. M. (1968). Leghemoglobin—concerning the sites of biosynthesis of its components. *Fed. Proc.*, **27**, 768.
132. DALLING, M. J., TOLBERT, N. E. and HAGEMAN, R. H. (1972). Intracellular location of nitrate reductase and nitrite reductase. II. Wheat roots. *Biochim. biophys. Acta*, **283**, 513–519.
133. DALTON, H. and MORTENSON, L. E. (1972). Dinitrogen (N_2) fixation (with a biochemical emphasis). *Bact. Rev.*, **36**, 231–260.
134. DARNELL, J. E., PHILLIPSON, L., WALL, R. and ADESMIK, M. (1971). Polyadenylic acid sequences: role in conversion of nuclear RNA into messenger RNA. *Science, N.Y.*, **174**, 507–510.
135. DART, P. J. and MERCER, F. V. (1963). Development of the bacteroid in the root nodule of the barrel medic (*Medicago tribuloides* Desr.) and subterranean clover (*Trifolium subterraneum* L.). *Arch. Mikrobiol.*, **46**, 382–401.
136. DAS, N. K., PATUA, K. and SKOOG, F. (1956). Initiation of mitosis and cell division by kinetin and indoleacetic acid in excised tobacco pith tissue. *Physiologia Pl.*, **9**, 640–651.
137. DATKO, A. H., GIOVANELLI, J. and MUDD, S. H. (1974). Homocysteine biosynthesis in green plants. *J. biol. Chem.*, **249**, 1139–1155.
138. DAUSSANT, J., NEUCERE, N. J. and YATSU, L. Y. (1969). Immunochemical studies on *Arachis hypogaea* proteins with particular reference to the reserve proteins. I. Characterization, distribution and properties of α arachin and α conarachin. *Pl. Physiol., Lancaster*, **44**, 471–479.
139. DAUSSANT, J., NEUCERE, N. J. and CONKERTON, E. J. (1969). Immunochemical studies on *Arachis hypogaea* proteins with particular reference to the reserve proteins. II. Protein modification during germination. *Pl. Physiol., Lancaster*, **44**, 480–484.
140. DAVIES, M. E. (1971). Regulation of histidine biosynthesis in cultured plant cells: evidence from studies on amitrole toxicity. *Phytochem.*, **10**, 783–788.
141. DAVIS, E. and MCLACHLAN, G. A. (1969). Generation of cellulase activity during protein synthesis by pea microsomes *in vitro*. *Archs. Biochem. Biophys.*, **129**, 581–587.
142. DAYHOFF, M. O. (1972). *Atlas of Protein Sequence and Structure*. National Biomedical Research Foundation, Silver Spring, Maryland.
143. DECHARY, J. M. (1970). Seed proteases and protease inhibitors. *Econ. Bot.*, **24**, 113–122.
144. DE JONG, D. W. (1972). Detergent extraction of enzymes from tobacco leaves varying in maturity. *Pl. Physiol., Lancaster*, **50**, 733–737.
145. DELMER, D. P. and MILLS, S. E. (1968). Tryptophan synthase from *Nicotiana tabacum*. *Biochim. biophys. Acta*, **167**, 431–443.
146. DELMER, D. P. and MILLS, S. E. (1968). Tryptophan biosynthesis in cell cultures of *Nicotiana tabacum*. *Pl. Physiol., Lancaster*, **43**, 81–87.
147. DENG, Q. and IVES, D. H. (1972). Modes of nucleoside phosphorylation in plants: studies on the apparent thymidine kinase and true uridine kinase of seedlings. *Biochim. biophys. Acta*, **277**, 235–244.
148. DEVER, J. E., BANDURSKI, R. S. and KIVILAAN, A. (1968). Partial chemical characterization of corn root cell walls. *Pl. Physiol., Lancaster*, **43**, 50–56.
149. DHARMAWARDENE, M. W. N., STEWART, W. D. P. and STANLEY, S. O. (1972). Nitrogenase activity, amino acid pool patterns and amination in blue-green algae. *Planta*, **108**, 133–145.
149a. DIECKERT, J. W. and DIECKERT, M. C. (1972). The deposition of vacuolar proteins in oil seeds. In *Symposium in Seed Proteins*, ed. INGLETT, G. E., 52–85. AVI Publishing Company, Inc., Conn., U.S.A.
150. DILWORTH, M. J. (1966). Acetylene reduction by nitrogen-fixing preparations from *Clostridium pasteurianum*. *Biochim. biophys. Acta*, **127**, 285–294.

151. DIMLER, R. J. (1966). Alcohol-insoluble proteins of cereal grains. *Fed. Proc.*, **25**, 1670–1675.

152. DIXON, R. O. D. (1967). The origin of the membrane envelope surrounding the bacteria and bacteroids and the presence of glycogen in clover root nodules. *Arch. Mikrobiol.*, **56**, 156–166.

153. DIXON, R. O. D. and FOWDEN, L. (1961). γ-Aminobutyric acid metabolism in plants. Part 2. Metabolism in higher plants. *Ann. Bot.*, *N.S.* **25**, 513–530.

154. DODD, W. A. and COSSINS, E. A. (1970). Homocysteine-dependent transmethylases catalyzing the synthesis of methionine in germinating pea seeds. *Biochim. biophys. Acta*, **201**, 461–470.

155. DONEY, R. C. and THOMPSON, J. F. (1971). The recovery of radioactivity in methionine from labeled S-methyl-L-cysteine in leaves of *Phaseolus vulgaris*. *Phytochem.*, **10**, 1745–1750.

156. DOUGALL, D. K. (1970). Threonine deaminase from Paul's Scarlet rose tissue cultures. *Phytochem.*, **9**, 959–964.

156a. DOUGALL, D. K. (1974). Evidence for the presence of glutamate synthase in extracts of carrot cell cultures. *Biochem. biophys. Res. Commun.*, **58**, 639–646.

157. DOUGALL, D. K. and FULTON, M. M. (1967). Biosynthesis of protein amino acids in plant tissue culture. III. Studies on the biosynthesis of arginine. *Pl. Physiol., Lancaster*, **42**, 387–390.

158. DOUGALL, D. K. and FULTON, M. M. (1967). Biosynthesis of protein amino acids in plant tissue culture. IV. Isotope competition experiments using glucose-U-^{14}C and potential intermediates. *Pl. Physiol., Lancaster*, **42**, 941–945.

159. DOUGALL, D. K. and SHIMBAYASHI, K. (1960). Factors affecting growth of tobacco callus tissue and its incorporation of tyrosine. *Pl. Physiol., Lancaster*, **35**, 396–404.

160. DOVE, L. D. (1967). Ribonuclease activity in stressed tomato leaflets. *Pl. Physiol., Lancaster*, **42**, 1176–1178.

161. DOVE, L. D. (1973). Ribonucleases in vascular plants: cellular distribution and changes during development. *Phytochem.*, **12**, 2561–2570.

162. DRENTH, J., JANSONIUS, J. N., KOEKOEK, R., SWEN, H. M. and WOLTHERS, B. G. (1968). Structure of Papain. *Nature, Lond.*, **218**, 929–932.

163. DUDOCK, B. S., KATZ, G., TAYLOR, E. K. and HOLLEY, R. W. (1969). Primary structure of wheat germ phenylalanine transfer RNA. *Proc. natn. Acad. Sci. U.S.A.*, **62**, 941–945.

164. DULL, G. G. (1971). The pineapple: General. In *Biochemistry of Fruits and Their Products*, Vol. 2, ed. HULME, A. C., 303–304. Academic Press, New York.

165. DUNHAM, V. L. and CHERRY, J. H. (1973). Multiple DNA polymerase activity solubilized from higher plant chromatin. *Biochem. biophys. Res. Commun.*, **54**, 403–410.

166. DUNHAM, V. L. and CHERRY, J. H. (1973). Solubilization and characterization of RNA polymerase from a higher plant. *Phytochem.*, **12**, 1897–1902.

167. DURE, L. S. and MERRICK, W. C. (1971). Synthesis of chloroplast tRNA species during plant seed embryogenesis and germination. In *Autonomy and Biogenesis of Mitochondria and Chloroplasts*, ed. BOARDMAN, N. K., LINNANE, A. W. and SMILLIE, R. M., 413–421. North Holland Publishing Company, Amsterdam.

168. DURZAN, D. J. (1968). Nitrogen metabolism of *Picea glauca*. I. Seasonal changes of free amino acids in buds, shoot apices, and leaves, and the metabolism of uniformly labelled ^{14}C-L-Arginine by buds during the onset of dormancy. *Can. J. Bot.*, **46**, 909–919.

169. DURZAN, D. J. (1969). Nitrogen metabolism of *Picea glauca*. IV. Metabolism of uniformly labelled ^{14}C-L-arginine, [carbamyl-^{14}C]-L-citrulline and [1,2,3,4-^{14}C]-γ-guanidinobutyric acid during diurnal changes in the soluble and protein nitrogen associated with the onset of expansion of spruce buds. *Can. J. Biochem.*, **47**, 771–783.

170. DURZAN, D. J. (1973). Nitrogen metabolism of *Picea glauca*. V. Metabolism of

uniformly labelled ^{14}C-L-proline and ^{14}C-L-glutamine by dormant buds in late fall. *Can. J. Bot.*, **51**, 359–369.

171. DYER, T. A. and OSBORNE, D. J. (1971). Leaf nucleic acids. II. Metabolism during senescence and the effect of kinetin. *J. exp. Bot.*, **22**, 552–560.

172. EILAM, Y., BUTLER, R. D. and SIMON, E. W. (1971). Ribosomes and polysomes in cucumber leaves during growth and senescence. *Pl. Physiol., Lancaster*, **47**, 317–323.

173. EILAM, Y. and KLEIN, S. (1962). The effect of light intensity and sucrose feeding on the fine structure in chloroplasts and on the chlorophyll content of etiolated leaves. *J. Cell Biol.*, **14**, 169–180.

174. EL-ANTABLY, H. M. M., WAREING, P. F. and HILLMAN, J. (1967). Some physiological responses to D,L-abscisin (Dormin). *Planta*, **73**, 74–90.

175. ELLIS, R. J. (1969). Chloroplast ribosomes: stereospecificity of inhibition by chloramphenicol. *Science, N.Y.*, **163**, 477–478.

176. ELLIS, R. J. (1970). Further similarities between chloroplast and bacterial ribosomes. *Planta*, **91**, 329–335.

177. ELLIS, R. J. and DAVIES, D. D. (1961). Glutamic-oxaloacetic transaminase of cauliflower. I. Purification and specificity. *Biochem. J.*, **78**, 615–623.

178. ERICSON, M. C. and CHRISPEELS, M. J. (1973). Isolation and characterization of glucosamine-containing storage glyco-proteins from the cotyledons of *Phaseolus aureus*. *Pl. Physiol., Lancaster*, **52**, 98–104.

179. EVANS, W. R. and AXELROD, B. (1961). Pyrimidine metabolism in germinating seedlings. *Pl. Physiol., Lancaster*, **36**, 9–13.

180. EVINS, W. H. and VARNER, J. E. (1972). Hormonal control of polyribosome formation in barley aleurone layers. *Pl. Physiol., Lancaster*, **49**, 348–352.

181. FAMBROUGH, D. M., FUJIMURA, F. and BONNER, J. (1968). Quantitative distribution of histone components in the pea plant. *Biochemistry*, **7**, 575–584.

182. FEIERABEND, J. (1969). Der Einfluss von Cytokininen auf die Bildung von Photosyntheseenzymen in Roggenkeimlingen. *Planta*, **84**, 11–29.

183. FERGUSON, A. R. and BOLLARD, E. G. (1969). Nitrogen metabolism of *Spirodela oligorrhiza*. I. Utilization of ammonium, nitrate and nitrite. *Planta*, **88**, 344–352.

184. FILNER, B. and KLEIN, A. O. (1968). Changes in enzymatic activities in etiolated bean seedling leaves after a brief illumination. *Pl. Physiol., Lancaster*, **43**, 1587–1596.

185. FILNER, P. (1965). Semi-conservative replication of DNA in a higher plant cell. *Expl. Cell Res.*, **39**, 33–39.

186. FILNER, P. and VARNER, J. E. (1967). A test for de novo synthesis of enzymes: density labelling with H_2O^{18} of barley α-amylase induced by gibberellic acid. *Proc. natn. Acad. Sci. U.S.A.*, **58**, 1520–1526.

187. FILNER, P., WRAY, J. L. and VARNER, J. E. (1969). Enzyme induction in higher plants. *Science, N.Y.*, **165**, 358–367.

188. FINLAYSON, A. J. and MCCONNELL, W. B. (1960). Studies on wheat plants using ^{14}C-compounds. XIV. Conversion of [1,7-$^{14}C_2$]-αα′-diaminopimelic acid to [1-^{14}C] lysine. *Biochim. biophys. Acta*, **45**, 622–623.

189. FLETCHER, J. S. (1972). Heterogeneous populations of mitochondria in higher plant cells. *Nature, Lond.*, **238**, 466–467.

190. FLETCHER, J. S. and BEEVERS, H. (1971). Influence of cycloheximide on the synthesis and utilization of amino acids in suspension cultures. *Pl. Physiol., Lancaster*, **48**, 261–264.

191. FLETCHER, R. A. and OSBORNE, D. J. (1965). Regulation of protein and nucleic acid synthesis by gibberellin during leaf senescence. *Nature, Lond.*, **207**, 1176–1177.

192. FLINN, A. M. and PATE, J. S. (1968). Biochemical and physiological changes during maturation of fruit of the field pea (*Pisum arvense* L.). *Ann. Bot., N.S.*, **32**, 479–495.

193. FOLKES, B. F. and YEMM, E. W. (1965). The amino acid content of the proteins of barley grains. *Biochem. J.*, **62**, 4–11.

194. FOLKES, B. F. and YEMM, E. W. (1958). The respiration of barley plants. X.

Respiration and the metabolism of amino-acids and proteins in germinating grain. *New Phytol.*, **57**, 106–131.

195. FOREST, J. C. and WIGHTMAN, F. (1972). Amino acid metabolism in plants. II. Transamination reactions of free protein amino acids in cell-free extracts of cotyledons and growing tissues of bushbean seedlings (*Phaseolus vulgaris* L.). *Can. J. Biochem.*, **50**, 538–542.

196. FOREST, J. C. and WIGHTMAN, F. (1973). Amino acid metabolism in plants. IV. Kinetic studies with a multispecific aminotransferase purified from bush-bean seedlings. *Can. J. Biochem.*, **51**, 332–343.

197. FORTI, G., TOGNOLI, C. and PARISI, B. (1962). Purification from pea leaves of a phosphatase that attacks nucleotides. *Biochim. biophys. Acta*, **62**, 251–260.

198. FOWDEN, L. (1959). New amino acids of plants. *Biol. Rev.*, **33**, 393–441.

199. FOWDEN, L. (1965). Origins of the amino acids. In *Plant Biochemistry*, ed. BONNER, J. and VARNER, J. E., 361–390. Academic Press, New York.

200. FOWDEN, L. (1973). The non-protein amino acids of plants: concepts of biosynthetic control. In *Biosynthesis and its Control in Plants*, ed. MILBORROW, B. V., 323–339. Academic Press, London.

200a. FOWLER, M. W., JESSUP, W. and SARKISSIAN, G. S. (1974). Glutamate synthetase type activity in higher plants. *FEBS Letters*, **46**, 340–342.

201. FRENKEL, C., KLEIN, I. and DILLEY, D. R. (1968). Protein synthesis in relation to ripening of pome fruits. *Pl. Physiol., Lancaster*, **43**, 1146–1153.

202. FURUHASHI, K. and YATAZAWA, M. (1970). Methionine–lysine–threonine–isoleucine interrelationships in the amino acid nutrition of rice callus tissue. *Pl. Cell Physiol., Tokyo*, **11**, 569–578.

203. GALUN, E., GRESSEL, J. and KEYNAN, A. (1964). Suppression of floral induction by actinomycin D—An inhibitor of messenger RNA synthesis. *Life Sciences*, **3**, 911–915.

204. GAMBORG, O. L. (1966). Aromatic metabolism in plants. II. Enzymes of the shikimate pathway in suspension cultures of plant cells. *Can. J. Biochem.*, **44**, 791–799.

205. GAMBORG, O. L. and KEELEY, F. W. (1966). Aromatic metabolism in plants. I. A study of the prephenate dehydrogenase from bean plants. *Biochim. biophys. Acta*, **115**, 65–72.

206. GANDER, J. E. (1958). *In vivo* biosynthesis of glycosidic cyanide in sorghum. *Fed. Proc.*, **17**, 226.

207. GASIOR, E. and MOLDAVE, K. (1972). Evidence for a soluble protein factor specific for the interaction between amino acylated transfer RNA's and the 40S subunit of mammalian ribosomes. *J. molec. Biol.*, **66**, 391–402.

208. GATICA, M. and ALLENDE, J. E. (1971). The presence of peptidyl transferase in wheat embryo ribosomes. *Biochim. biophys. Acta*, **228**, 732–735.

209. GERHARDT, B. P. and BEEVERS, H. (1970). Developmental studies on glyoxysomes in *Ricinus* endosperm. *J. Cell Biol.*, **44**, 94–102.

210. GERLOFF, E. D., STAHMAN, M. A. and SMITH, D. (1967). Soluble proteins in alfalfa roots as related to cold hardiness. *Pl. Physiol., Lancaster*, **42**, 895–899.

211. GIANNATTASIO, M., MANDATO, E. and MACCHIA, V. (1974). Content of 3′,5′-cyclic AMP and cyclic AMP phosphodiesterase in dormant and activated tissues of Jerusalem artichoke tubers. *Biochem. biophys. Res. Commun.*, **57**, 365–371.

212. GIBBS, M. and SCHIFF, J. A. (1960). Chemosynthesis: the energy relations of chemo autotrophic organisms. In *Plant Physiology*, Vol. 1B, ed. STEWARD, F. C., 299–319. Academic Press, New York.

213. GIBSON, R. A., SCHNEIDER, A. E. and WIGHTMAN, F. (1972). Biosynthesis and metabolism of indol-3yl-acetic acid. II. *In vivo* experiments with ^{14}C-labelled precursors of IAA in tomato and barley shoots. *J. exp. Bot.*, **23**, 381–399.

214. GIFFORD, E. M. and STEWART, K. D. (1965). Ultrastructure of vegetative and reproductive apices of *Chenopodium album*. *Science, N.Y.*, **149**, 75–77.

215. GILBERT, M. L. and GALSKY, A. G. (1972). The action of cyclic-AMP on GA

controlled responses. III. Characteristics of barley endosperm acid phosphatase induction by gibberellic acid and cyclic 3′,5′-adenosine monophosphate. *Pl. Cell Physiol., Tokyo*, **13**, 867–873.

216. GIOVANELLI, J. and MUDD, S. H. (1966). Enzymatic synthesis of cystathionine by extracts of spinach, requiring *o*-acetylhomoserine or *o*-succinylhomoserine. *Biochem. biophys. Res. Commun.*, **25**, 366–371.

217. GIOVANELLI, J. and MUDD, S. H. (1967). Synthesis of homocysteine and cysteine by enzyme extracts of spinach. *Biochem. biophys. Res. Commun.*, **27**, 150–156.

218. GIRI, K. V., KRISHNASWAMY, P. R. and RAO, N. A. (1958). Studies on plant flavokinase. *Biochem. J.*, **70**, 66–71 .

219. GIRI, K. V., RAO, N. A., CAMA, H. R. and KUMAR, S. A. (1960). Studies on flavin-adenine dinucleotide-synthesizing enzyme in plants. *Biochem. J.*, **75**, 381–386.

220. GIVAN, C. V., GIVAN, A. L. and LEECH, R. M. (1970). Photoreduction of α-ketoglutarate to glutamate by *Vicia faba* chloroplasts. *Pl. Physiol., Lancaster*, **45**, 624–630.

221. GLAZER, A. N. and SMITH, E. L. (1971). Papain and other plant sulfhydryl proteolytic enzymes. In *The Enzymes*, Vol. 3, *Hydrolysis: Peptide Bonds*, ed. BOYER, P. D., 501–554. Academic Press, New York.

222. GODOVARI, H. R. and WAYGOOD, E. R. (1970). NAD metabolism in plants. I. Intermediates of biosynthesis in wheat leaves and the effect of benzimidazole. *Can. J. Bot.*, **48**, 2267–2275.

223. GOLINSKA, B. and LEGOCKI, A. B. (1973). Purification and some properties of elongation factor 1 from wheat germ. *Biochim. biophys. Acta*, **324**, 156–170.

224. GOODCHILD, D. J. and BERGERSEN, F. J. (1966). Electron microscopy of the infection and subsequent development of soybean nodule cells. *J. Bact.*, **92**, 204–213.

225. GRAHAM, J. S. D. (1963). Starch-gel electrophoresis of wheat flour proteins. *Aust. J. biol. Sci.*, **16**, 342–349.

226. GRAHAM, J. S. D., MORTON, R. K. and RAISON, J. K. (1963). Isolation and characterization of protein bodies from developing wheat endosperms. *Aust. J. biol. Sci.*, **16**, 375–383.

227. GRAHAM, T. A. and GUNNING, B. E. S. (1970). Localization of legumin and vicilin in bean cotyledon cells using fluorescent antibodies. *Nature, Lond.*, **228**, 81–82.

228. GRANICK, S. (1959). Magnesium porphyrins formed by barley seedlings treated with δ-amino levulinic acid. *Pl. Physiol., Lancaster*, **34** (Suppl.), xviii.

229. GRANT, D. R. and LAWRENCE, J. M. (1964). Effects of sodium dodecyl sulfate and other dissociating reagents on the globulins of peas. *Archs. Biochem. Biophys.*, **108**, 552–561.

230. GRANT, D. R. and VOELKERT, E. (1971). The formation of homoserine from methionine in germinating peas. *Can. J. Biochem.*, **49**, 795–798.

231. GREENBERG, J. R. and PERRY, R. P. (1972). Relative occurrence of polyadenylic acid sequences in messenger and heterogeneous nuclear RNA of L cells as determined by poly(u)-hydroxylapatite chromatography. *J. molec. Biol.*, **72**, 91–98.

232. GREGORY, F. G. and SEN, P. K. (1937). Physiological studies in plant nutrition. VI. The relation of respiration rate to the carbohydrate and nitrogen metabolism of the barley leaf, as determined by nitrogen and potassium deficiency. *Ann. Bot., N.S.*, **1**, 521–561.

233. GUARDIOLA, J. L. and SUTCLIFFE, J. F. (1971). Control of protein hydrolysis in the cotyledons of germinating pea (*Pisum sativum* L.) seeds. *Ann. Bot., N.S.*, **35**, 791–807.

234. GUDERIAN, R. H., PULLIAM, R. L. and GORDON, M. P. (1972). Characterization and fractionation of tobacco leaf transfer RNA. *Biochim. biophys. Acta*, **262**, 50–65.

235. GUILFOYLE, T. J. and HANSON, J. B. (1973). Increased activity of chromatin-bound ribonucleic acid polymerase from soybean hypocotyl with spermidine and high ionic strength. *Pl. Physiol., Lancaster*, **51**, 1022–1025.

236. GYLDENHOLM, A. O. (1968). Macromolecular physiology of plastids. V. On the nucleic acid metabolism during chloroplast development. *Hereditas*, **59**, 142–168.

237. HAARD, N. F. (1973). Upsurge of particulate peroxidase in ripening banana fruit. *Phytochem.*, **12**, 555–560.

238. HABER, A. H., CARRIER, W. L. and FOARD, D. E. (1961). Metabolic studies of gamma-irradiated wheat plants growing without cell division. *Am. J. Bot.*, **48**, 431–438.

239. HABER, A. H. and FOARD, D. E. (1964). Further studies of gamma-irradiated wheat and their relevance to use of mitotic inhibition for developmental studies. *Am. J. Bot.*, **51**, 151–159.

240. HABERLANDT, G. (1890). Die Kleberschicht des Gras-Endosperms als Diastase ausscheidendes Drüsengewebe. *Ber. dt. bot. Ges.*, **8**, 40–48.

241. HADIYEV, D., MEHTA, S. L. and ZALIK, S. (1969). Nucleic acids and ribonucleases of wheat leaves and chloroplasts. *Can. J. Biochem.*, **47**, 273–283.

242. HAGEMAN, R. H., CRESSWELL, C. F. and HEWITT, E. J. (1962). Reduction of nitrate, nitrite and hydroxylamine to ammonia by enzymes extracted from higher plants. *Nature, Lond.*, **193**, 247–250.

243. HAGEMAN, R. H. and FLESHER, D. (1960). Nitrate reductase activity in corn seedlings as affected by light and nitrate content of nutrient media. *Pl. Physiol., Lancaster*, **35**, 700–708.

244. HAGUE, D. R. and KOFOID, E. C. (1971). The coding properties of lysine-accepting transfer ribonucleic acids from black-eyed peas. *Pl. Physiol., Lancaster*, **48**, 305–311.

245. HALL, R. H. (1970). N^6-(Δ^2-isopentenyl) adenosine: chemical reactions, biosynthesis, metabolism and significance to the structure and function of tRNA. In *Progress in Nucleic Acid Research and Molecular Biology*, Vol. 10, ed. DAVIDSON, J. N. and COHN, W. E., 57–86. Academic Press, New York.

246. HALLAM, N. D., ROBERTS, B. E. and OSBORNE, D. J. (1973). Embryogenesis and germination in rye (*Secale cereale* L.) III. Fine structure and biochemistry of the non-viable embryo. *Planta*, **110**, 279–290.

247. HANSEN, E. (1970). Proteins. In *Biochemistry of Fruits and Their Products*, Vol. 1, ed. HULME, A. C., 147–158. Academic Press, London and New York.

248. HANSON, K. R. and HAVIR, E. A. (1972). Mechanism and properties of phenylalanine ammonia lyase from higher plants. In *Recent Advances in Phytochemistry*, Vol. 4, ed. RUNECKLES, V. C. and WATKIN, J. E., 45–86. Appleton-Century-Crofts, New York.

249. HAREL, E. and KLEIN, S. (1972). Light dependent formation of δ-amino levulinic acid in etiolated leaves of higher plants. *Biochem. biophys. Res. Commun.*, **49**, 364–370.

250. HARLAND, J., JACKSON, J. F. and YEOMAN, M. M. (1973). Changes in some enzymes involved in DNA biosynthesis following induction of division in cultured plant cells. *J. Cell Sci.*, **13**, 121–138.

251. HARTLEY, B. S. (1960). Proteolytic enzymes. *A. Rev. Biochem.*, **29**, 45–74.

252. HARTLEY, M. R. and ELLIS, R. J. (1973). Ribonucleic acid synthesis in chloroplasts. *Biochem. J.*, **134**, 249–262.

253. HARVEY, B. M. R. and OAKS, A. (1974). The hydrolysis of endosperm protein in *Zea mays*. *Pl. Physiol., Lancaster*, **53**, 453–457.

254. HASSE, K., RATYCH, O. T. and SALNIKOW, J. (1967). Transaminierung und Decarboxylierung von Ornithin und Lysin in höheren Pflanzen. Hoppe-Seyler's *Z. physiol. Chem.*, **348**, 843–851.

255. HASSID, W. Z. (1967). Transformation of sugars in plants. *A. Rev. Pl. Physiol.*, **18**, 253–280.

256. HAYSTEAD, A. and STEWART, W. D. P. (1972). Characteristics of the nitrogenase system of the blue-green alga *Anabaena cylindrica*. *Arch. Mikrobiol.*, **82**, 325–336.

257. HEBER, U. W. and SANTARIUS, K. A. (1965). Compartmentation and reduction of pyridine nucleotides in relation to photosynthesis. *Biochim. biophys. Acta*, **109**, 390–408.

258. HEDLEY, C. L. and STODDART, J. L. (1971). Light stimulation of alanine aminotransferase activity in dark grown leaves of *Lolium temulentum* as related to chlorophyll formation. *Planta*, **100**, 309–324.

259. HELLEBURST, J. A. and BIDWELL, R. G. S. (1963). Protein turnover in wheat and snapdragon leaves. *Can. J. Bot.*, **41**, 969–983.

260. HEVESY, G., LINDERSTROM-LANG, K., KESTON, A. S. and OLSEN, C. (1940). Exchange of nitrogen atoms in the leaves of sunflower. *C. r. Trav. Lab. Carlsberg* (Sér. Chim.), **23**, 213–218.

261. HEWISH, D. R., WHELDRAKE, J. F. and WELLS, J. R. E. (1971). Incorporation of P^{32} into ribosomal RNA, transfer RNA and inositol hexaphosphate in germinating pea cotyledons. *Biochim. biophys. Acta*, **228**, 509–516.

262. HEWITT, E. J. and BETTS, G. F. (1963). The reduction of nitrite and hydroxylamine by ferredoxin and chloroplast grana from *Cucurbita pepo*. *Biochem. J.*, **89**, 20p.

263. HEWITT, E. J. and JONES, E. W. (1947). The production of molybdenum deficiency in plants in sand culture with special reference to tomato and *Brassica* crops. *J. Pomol. Hort. Sci.*, **23**, 254–262.

264. HEYES, J. K. (1960). Nucleic acid changes during cell expansion in the root. *Proc. R. Soc., B*, **152**, 218–230.

265. HILL, J. M. and MANN, P. J. G. (1968). Some properties of plant diamine oxidase, a copper-containing enzyme. In *Recent Aspects of Nitrogen Metabolism in Plants*, ed. HEWITT, E. J. and CUTTING, C. V., 149–161. Academic Press, London.

266. HILL-COTTINGHAM, D. G. and LLOYD-JONES, C. P. (1973). Metabolism of carbon-14 labelled arginine, citrulline and ornithine in intact apple stems. *Physiologia Pl.*, **29**, 125–128.

267. HILL-COTTINGHAM, D. G. and LLOYD-JONES, C. P. (1973). A technique for studying the adsorption, absorption and metabolism of amino acids in intact apple stem tissue. *Physiologia Pl.*, **28**, 443–446.

268. HOBDAY, S. M., THURMAN, D. A. and BARBER, D. J. (1973). Proteolytic and trypsin inhibitory activities in extracts of germinating *Pisum sativum* seeds. *Phytochem.*, **12**, 1041–1046.

269. HOFMAN, T. and LEES, H. (1953). The biochemistry of the nitrifying organisms. 4. The respiration and intermediary metabolism of *Nitrosomonas*. *Biochem. J.*, **54**, 579–583.

270. HOLDGATE, D. P. and GOODWIN, T. W. (1965). Quantitative extraction and estimation of plant nucleic acids. *Phytochem.*, **4**, 831–843.

271. HORGEN, P. A. and KEY, J. L. (1973). The DNA-directed RNA polymerases of soybean. *Biochim. biophys. Acta*, **294**, 227–235.

272. HOSOI, K., YOSHIDA, S. and HASEGAWA, M. (1970). L-tyrosine carboxyl lyase of barley roots. *Pl. Cell Physiol., Tokyo*, **11**, 899–906.

273. HOTTA, Y. and STERN, H. (1961). Deamination of deoxycytidine and 5 methyldeoxycytidine in developing anthers of *Lilium longiflorum*. *J. biophys. biochem. Cytol.*, **9**, 279–284.

274. HSIAO, T. C. (1970). Rapid changes in levels of polyribosomes in *Zea mays* in response to water stress. *Pl. Physiol., Lancaster*, **46**, 281–285.

275. HUANG, R. C. and BONNER, J. (1962). Histone, a suppressor of chromosomal RNA synthesis. *Proc. natn. Acad. Sci. U.S.A.*, **48**, 1216–1222.

276. HUCKLESBY, D. P., DALLING, M. J. and HAGEMAN, R. H. (1972). Some properties of two forms of nitrite reductase from corn (*Zea mays*) scutellum. *Planta*, **104**, 220–233.

277. HUCKLESBY, D. P. and HEWITT, E. J. (1970). Nitrite and hydroxylamine reduction in higher plants. *Biochem. J.*, **119**, 615–627.

278. HUGHES, D. W. and GENEST, K. (1973). Alkaloids. In *Phytochemistry*, Vol. II, ed. MILLER, L. P., 118–170. Van Nostrand–Reinhold Company, New York.

279. HULME, A. C., RHODES, M. J. C. and WOOLTORTON, L. S. C. (1971). The relationship between ethylene and the synthesis of RNA and protein in ripening seeds. *Phytochem.*, **10**, 749–756.

280. IHLE, J. N. and DURE, L. S. III (1972). The developmental biochemistry of

cottonseed embryogenesis and germination. I. Purification and properties of a carboxypeptidase from germinating cotyledons. *J. biol. Chem.*, **247**, 5034–5040.

281. IHLE, J. N. and DURE, L. S. III (1972). The developmental biochemistry of cottonseed embryogenesis and germination. II. Catalytic properties of the cotton carboxypeptidase. *J. biol. Chem.*, **247**, 5041–5047.

282. IHLE, J. N. and DURE, L. S. III (1972). The developmental biochemistry of cottonseed embryogenesis and germination. III. Regulation of the biosynthesis of enzymes utilized in germination. *J. biol. Chem.*, **247**, 5048–5055.

283. INAMOTOMI, K. and SLAUGHTER, J. C. (1971). The role of glutamate decarboxylase and γ-amino butyric acid in germinating barley. *J. exp. Bot.*, **22**, 561–571.

284. INGLE, J. (1963). The extraction and estimation of nucleotides and nucleic acids from plant material. *Phytochem.*, **2**, 353–370.

285. INGLE, J. (1968). Synthesis and stability of chloroplast-RNAs. *Pl. Physiol., Lancaster*, **43**, 1448–1453.

286. INGLE, J. (1968). The effect of light and inhibitors on chloroplast and cytoplasmic RNA synthesis. *Pl. Physiol., Lancaster*, **43**, 1850–1854.

287. INGLE, J. (1973). The regulation of ribosomal RNA synthesis. In *Biosynthesis and Its Control in Plants*, ed. MILBORROW, B. V., 69–91. Academic Press, London.

288. INGLE, J., BEEVERS, L. and HAGEMAN, R. H. (1964). Metabolic changes associated with the germination of corn. I. Changes in weight and metabolites and their redistribution in the embryo axis, scutellum and endosperm. *Pl. Physiol., Lancaster*, **39**, 735–740.

289. INGLE, J., BEITZ, D. and HAGEMAN, R. H. (1965). Changes in composition during development and maturation of maize seeds. *Pl. Physiol., Lancaster*, **40**, 835–839.

290. INGLE, J. and HAGEMAN, R. H. (1965). Metabolic changes associated with the germination of corn. II. Nucleic acid metabolism. *Pl. Physiol., Lancaster*, **40**, 48–53.

291. INGLE, J. and HAGEMAN, R. H. (1965). Metabolic changes associated with the germination of corn. III. Effects of gibberellic acid on endosperm metabolism. *Pl. Physiol., Lancaster*, **40**, 672–675.

292. INGLE, J., JOY, K. W. and HAGEMAN, R. H. (1966). The regulation of the enzymes involved in the assimilation of nitrate by higher plants. *Biochem. J.*, **100**, 577–588.

293. INGLE, J., KEY, J. L. and HOLM, R. E. (1965). Demonstration and characterization of a DNA-like RNA in excised plant tissue. *J. molec. Biol.*, **11**, 730–746.

294. INGLE, J., POSSINGHAM, J. V., WELLS, R., LEAVER, C. J. and LOENING, U. E. (1970). The properties of chloroplast ribosomal-RNA. *Symp. Soc. exp. Biol.*, **24**, 303–325.

295. INGLE, J. and SINCLAIR, J. (1972). Ribosomal RNA genes and plant development. *Nature, Lond.*, **235**, 30–32.

296. ITAI, C. and VAADIA, Y. (1965). Kinetin-like activity in root exudate of water-stressed sunflower plants. *Physiologia Pl.*, **18**, 941–944.

297. IWAI, K., NAKAGAWA, S. and OKINAKA, O. (1963). Isolation and identification of glycinamide ribonucleotide accumulated in pea seedlings in a 'folate-deficient' state. *Biochim. biophys. Acta*, **68**, 152–154.

298. IWAI, K., SUZUKI, N. and MIZOGUCHI, S. (1967). Purification and properties of formyltetrahydrofolate synthetase from spinach. *Agric. Biol. Chem.*, **31**, 267–274.

299. IWASAKI, H. and MATSUBARA, T. (1972). A nitrite reductase from *Achromobacter cycloclastes*. *J. Biochem., Tokyo*, **71**, 645–652.

300. IWASAKI, H., SHIDARA, S., SUZUKI, H. and MORI, T. (1963). Studies on denitrification. VII. Further purification and properties of denitrifying enzyme. *J. Biochem., Tokyo*, **53**, 299–303.

301. JACKSON, P., BOULTER, D. and THURMAN, D. A. (1969). A comparison of some

properties of vicilin and legumin from seeds of *Pisum sativum*, *Vicia faba* and *Cicer arietinum*. *New Phytol.*, **68**, 25–33.

302. JACKSON, W. A., FLESHER, D. and HAGEMAN, R. H. (1973). Nitrate uptake by dark grown corn seedlings: some characteristics of apparent induction. *Pl. Physiol., Lancaster*, **51**, 120–127.

303. JACOB, F. and MONOD, J. (1961). Genetic regulatory mechanisms in the synthesis of proteins. *J. molec. Biol.*, **3**, 318–356.

304. JACOBSEN, J. V. and VARNER, J. E. (1967). Gibberellic acid-induced synthesis of protease by isolated aleurone layers of barley. *Pl. Physiol., Lancaster*, **42**, 1596–1600.

304a. JACOBSEN, J. V. and ZWAR, J. A. (1974). Gibberellic acid causes increased synthesis of RNA which contains poly (A) in barley aleurone tissue. *Proc. natn. Acad. Sci. U.S.A.*, **71**, 3290–3293.

305. JAMES, W. O. (1953). *Plant Respiration*. Clarendon Press, Oxford.

306. JARVIS, B. C., FRANKLAND, B. and CHERRY, J. H. (1968). Increased DNA template and RNA polymerase associated with the breaking of seed dormancy. *Pl. Physiol., Lancaster*, **43**, 1734–1736.

307. JENDRISAK, J. and BECKER, W. M. (1973). Isolation, purification and characterization of RNA polymerases from wheat germ. *Biochim. biophys. Acta*, **319**, 48–54.

308. JENNINGS, A. C. and MORTON, R. K. (1963). Amino acids and protein synthesis in developing wheat endosperm. *Aust. J. biol. Sci.*, **16**, 384–394.

309. JONES, B. L., NAGABHUSHAN, N., GULYAS, A. and ZALIK, S. (1972). Two dimensional acrylamide gel electrophoresis of wheat leaf cytoplasmic and chloroplast ribosomal proteins. *FEBS Letters*, **23**, 167–170.

310. JONES, R. L. and PHILLIPS, I. D. J. (1966). Organs of gibberellin synthesis in light-grown sunflower plants. *Pl. Physiol., Lancaster*, **41**, 1381–1386.

311. JONES, V. M. and BOULTER, D. (1968). Arginine metabolism in germinating seeds of some members of the Leguminosae. *New Phytol.*, **67**, 925–934.

312. JORDAN, D. C., GRINYER, I. and COULTER, W. H. (1963). Electron microscopy of infection threads and bacteria in young root nodules of *Medicago sativa*. *J. Bact.*, **86**, 125–137.

313. JOY, K. W. and HAGEMAN, R. H. (1966). The purification and properties of nitrite reductase from higher plants, and its dependence on ferredoxin. *Biochem. J.*, **100**, 263–273.

314. JULIANO, B. O. and VARNER, J. E. (1969). Enzymic degradation of starch granules in the cotyledons of germinating peas. *Pl. Physiol., Lancaster*, **44**, 886–892.

315. KAHN, A. A., HEIT, C. E. and LIPPOLD, P. C. (1968). Increase in nucleic acid synthesizing capacity during cold treatment of dormant pear embryos. *Biochem. biophys. Res. Commun.*, **33**, 391–396.

316. KANABUS, J. and CHERRY, J. H. (1971). Isolation of an organ-specific leucyl-tRNA synthetase from soybean seedling. *Proc. natn. Acad. Sci. U.S.A.*, **68**, 873–876.

317. KANAMORI, M. and WIXOM, R. L. (1963). Studies in valine biosynthesis. V. Characteristics of the purified dihydroxyacid dehydratase from spinach leaves. *J. biol. Chem.*, **238**, 998–1005.

318. KANAZAWA, T., KANAZAWA, K., KIRK, M. R. and BASSHAM, J. A. (1972). Regulatory effects of ammonia on carbon metabolism in *Chlorella pyrenoidosa* during photosynthesis and respiration. *Biochim. biophys. Acta*, **256**, 656–669.

319. KANNANGARA, C. G. and WOOLHOUSE, H. W. (1968). Changes in the enzyme activity of soluble protein fractions in the course of foliar senescence in *Perilla frutescens* (L.) Britt. *New Phytol.*, **67**, 533–542.

320. KAPOOR, M. and WAYGOOD, E. R. (1962). Initial steps of purine biosynthesis in wheat germ. *Biochem. biophys. Res. Commun.*, **9**, 7–18.

321. KEATES, R. A. B. (1973). Cyclic nucleotide-independent protein kinase from pea shoots. *Biochem. biophys. Res. Commun.*, **54**, 655–661.

322. KESSLER, B. and ENGELBERG, N. (1962). Ribonucleic acid and ribonuclease activity in developing leaves. *Biochim. biophys. Acta*, **55**, 70–82.

323. KEY, J. L. (1964). Ribonucleic acid and protein synthesis as essential processes for cell elongation. *Pl. Physiol., Lancaster*, **39**, 365–370.

324. KEY, J. L. (1966). Effect of purine and pyrimidine analogues on growth and RNA metabolism in the soybean hypocotyl—the selective action of 5-fluorouracil. *Pl. Physiol., Lancaster*, **41**, 1257–1264.

325. KEY, J. L. (1969). Hormones and nucleic acid metabolism. *A. Rev. Pl. Physiol.*, **20**, 449–474.

326. KEY, J. L. and INGLE, J. (1964). Requirements for the synthesis of DNA-like RNA for growth of excised plant tissue. *Proc. natn. Acad. Sci., U.S.A.* **52**, 1382–1388.

327. KEY, J. L., LEAVER, C. J., COWLES, J. R. and ANDERSON, J. M. (1972). Characterization of short time labelled adenosine monophosphate-rich ribonucleic acids of soybean. *Pl. Physiol., Lancaster*, **49**, 783–788.

328. KHOO, U. and WOLF, M. G. (1970). Origin and development of protein granules in maize endosperm. *Am. J. Bot.*, **57**, 1042–1050.

329. KING, J., WANG, D. and WAYGOOD, E. R. (1965). Biosynthesis of nucleotides in wheat. II. Pyrimidines from ^{14}C-labelled compounds. *Can. J. Biochem.*, **43**, 237–244.

330. KING, J. and WAYGOOD, E. R. (1968). Glyoxylate aminotransferases from wheat leaves. *Can. J. Biochem.*, **46**, 771–779.

331. KIRK, J. T. O. and ALLEN, R. L. (1965). Dependence of chloroplast pigment synthesis on protein synthesis: effect of actidione. *Biochem. biophys. Res. Commun.*, **21**, 523–530.

332. KIRK, J. T. O. and TILNEY-BASSETT, R. A. E. (1967). *The Plastids.* W. H. Freeman and Co., London and San Francisco.

333. KIRK, P. R. and LEECH, R. M. (1972). Amino acid biosynthesis by isolated chloroplasts during photosynthesis. *Pl. Physiol., Lancaster*, **50**, 228–234.

334. KISAKI, T., YOSHIDA, N. and IMAI, A. (1971). Glycine decarboxylase and serine formation in spinach leaf mitochondrial preparation with reference to photorespiration. *Pl. Cell Physiol., Tokyo*, **12**, 275–288.

335. KLECZKOWSKI, K. and COHEN, P. P. (1964). Purification of ornithine transcarbamylase from pea seedlings. *Archs. Biochem. Biophys.*, **107**, 271–278.

336. KLEIN, S., BRYAN, G., and BOGORAD, L. (1964). Early stages in the development of plastid fine structure in red and far red light. *J. Cell Biol.*, **22**, 433–442.

337. KLEPPER, L., FLESHER, D. and HAGEMAN, R. H. (1971). Generation of reduced nicotinamide adenine dinucleotide for nitrate reduction in green leaves. *Pl. Physiol., Lancaster*, **48**, 580–590.

338. KLEPPER, L. and HAGEMAN, R. H. (1969). The occurrence of nitrate reductase in apple leaves. *Pl. Physiol., Lancaster*, **44**, 110–114.

339. KLUCAS, R. V. and EVANS, H. J. (1968). An electron donor system for nitrogenase-dependent acetylene reduction by extracts of soybean nodules. *Pl. Physiol., Lancaster*, **43**, 1458–1460.

340. KOCH, B., WONG, P., RUSSELL, S. A., HOWARD, R. and EVANS, H. J. (1970). Purification and some properties of a non-haem iron protein from the bacteroids of soya-bean (*Glycine max* Merr.) nodules. *Biochem. J.*, **118**, 773–781.

341. KOLLOFFEL, C. and STROBAND, H. W. J. (1973). Ornithine carbamyl transferase activity from the cotyledons of developing and germinating seeds of *Vicia faba. Phytochem.*, **12**, 2635–2638.

342. KOZLOWSKI, T. T. (1971). *Growth and Development of Trees*, Vol. I. *Seed germination, ontogeny and shoot growth.* Academic Press, New York.

343. KOZLOWSKI, T. T. (1971). *Growth and Development of Trees*, Vol. II. *Cambial growth, root growth and reproductive growth.* Academic Press, New York.

344. KREBS, H. A. and HENSELEIT, K. (1932). Untersuchungen über die Harnstoffbildungim Tierkörper. *Hoppe-Seyler's Z. physiol. Chem.* **210**, 33–66.

345. KRETOVICH, W. L. (1965). Some problems of amino acid and amide biosynthesis in plants. *A. Rev. Pl. Physiol.*, **16**, 141–154.

346. KU, L. L. and ROMANI, R. J. (1970). The ribosomes of pear fruit—their synthesis during the climacteric and the age-related compensatory response to ionizing radiation. *Pl. Physiol., Lancaster*, **45**, 401–407.

347. KULASOORIYA, S. A., LANG, N. J. and FAY, P. (1972). The heterocysts of blue-green algae. III. Differentiation and nitrogenase activity. *Proc. R. Soc., B*, **181**, 199–209.

348. KUNITZ, M. (1946). Crystalline soybean trypsin inhibitor. II. General properties. *J. gen. Physiol.*, **30**, 291–310.

349. KURAISHI, S. (1968). The effect of kinetin on protein level of *Brassica* leaf discs. *Physiologia Pl.*, **21**, 78–83.

350. KUTACEK, M. and KEFELI, V. I. (1968). The present knowledge of indole compounds in plants of the Brassicaceae family. In *Biochemistry and Physiology of Plant Growth Substances*, ed. WIGHTMAN, F. and SETTERFIELD, G., 127–152. Runge Press, Ottawa, Canada.

351. LAMPORT, D. T. A. (1969). The isolation and partial characterization of hydroxyproline rich glycopeptides obtained by enzymic degradation of primary cell walls. *Biochemistry*, **8**, 1155–1163.

352. LAMPORT, D. T. A. (1970). Cell wall metabolism. *A. Rev. Pl. Physiol.*, **21**, 235–270.

353. LARSON, L. A. and BEEVERS, H. (1965). Amino acid metabolism in young pea seedlings. *Pl. Physiol., Lancaster*, **40**, 424–432.

354. LEA, P. J. and NORRIS, R. D. (1972). tRNA and aminoacyl-tRNA synthetases from plants (review article). *Phytochem.*, **11**, 2897–2920.

354a. LEA, P. J. and MIFLIN, B. J. (1974). Alternative route for nitrogen assimilation in higher plants. *Nature, Lond.*, **251**, 614–616.

355. LEAVER, C. J. and HARMEY, M. A. (1972). Isolation and characterization of mitochondrial ribosomes from higher plants. *Biochem. J.*, **129**, 37–38.

356. LEAVER, C. J. and KEY, J. L. (1970). Ribosomal RNA synthesis in plants. *J. molec. Biol.*, **49**, 671–680.

357. LEES, H. (1955). *Biochemistry of Autotrophic Bacteria*. Butterworth, London.

358. LEES, H. and SIMPSON, J. R. (1957). The biochemistry of the nitrifying organisms. 5. Nitrite oxidation by *Nitrobacter*. *Biochem. J.*, **65**, 297–305.

359. LEETE, E. (1958). The biogenesis of the Nicotiana alkaloids. VI. The piperidine ring of anabasine. *J. Am. chem. Soc.*, **80**, 4393–4394.

360. LEETE, E. (1962). The stereospecific incorporation of ornithine into the tropine moiety of hyoscyamine. *J. Am. chem. Soc.*, **84**, 55–57.

361. LEETE, E. (1967). Biosynthesis of Nicotiana alkaloids. XI. Investigation of tautomerism in *N*-methyl-Δ'-pyrrolinium chloride and its incorporation into nicotine. *J. Am. chem. Soc.*, **89**, 7081–7084.

362. LEETE, E., BOWMAN, R. M. and MANUEL, M. F. (1971). The occurrence of the NIH shift during the formation of *N*-methyl tyramine from phenylalanine. *Phytochem.*, **10**, 3029–3033.

363. LEETE, E., GROS, E. G. and GILBERTSON, T. J. (1964). Biosynthesis of the pyrrolidine ring of nicotine: feeding experiments with N^{15} labelled ornithine-2-C^{14}. *Tetrahedron Lett.*, Pt. **1**, No. **11**, 587–592.

364. LEETE, E., GROS, E. G. and GILBERTSON, T. J. (1964). The biosynthesis of anabasine—origin of the nitrogen of the piperidine ring. *J. Am. chem. Soc.*, **86**, 3907–3908.

365. LEETE, E. and MARION, L. (1953). The biogenesis of alkaloids. VII. The formation of hordenine and *N*-methyltyramine from tyrosine in barley. *Can. J. Chem.*, **31**, 126–128.

366. LEETE, E. and SIEGFRIED, K. J. (1957). The biogenesis of nicotine. III. Further observations on the incorporation of ornithine into the pyrrolidine ring. *J. Am. chem. Soc.*, **79**, 4529–4531.

367. LEDOUX, L. and HUART, R. (1962). Nucleic acid and protein metabolism in barley seedlings. IV. Translocation of ribonucleic acids. *Biochim. biophys. Acta*, **61**, 185–196.

368. LEGOCKI, A. B. and MARCUS, A. (1970). Polypeptide synthesis in extracts of wheat germ. *J. biol. chem.*, **245**, 2814–2818.

369. LEIS, J. P. and KELLER, E. B. (1971). *N*-formylmethionyl-$tRNA_f$ of wheat chloroplasts. Its synthesis by wheat transformylase. *Biochemistry*, **10**, 889–894.

370. LENGYEL, P. and SOLL, D. (1969). Mechanism of protein synthesis. *Bact. Rev.*, **33**, 264–301.

371. LETHAM, D. S., SHANNON, J. S. and MCDONALD, I. R. (1964). The structure of zeatin, a factor inducing cell division. *Proc. chem. Soc.*, 230–231.

372. LEVIN, D. H., KYNER, D. and ACS, G. (1973). Protein initiation in eukaryotes: formation and function of a ternary complex composed of a partially purified ribosomal factor, methionyl transfer RNA_f, and guanosine triphosphate. *Proc. natn. Acad. Sci. U.S.A.*, **70**, 41–45.

373. LEWINGTON, R. J., TALBOT, M. and SIMON, E. W. (1967). The yellowing of attached and detached cucumber cotyledons. *J. exp. Bot.*, **18**, 526–534.

374. LEWIS, O. A. M., NIEMAN, E. and MUNZ, A. (1970). Origin of amino acids in *Datura stramonium* seeds. *Ann. Bot., N.S.*, **34**, 843–848.

375. LEWIS, O. A. M. and PATE, J. S. (1973). The significance of transpirationally derived nitrogen in protein synthesis in fruiting plants of pea (*Pisum sativum*). *J. exp. Bot.*, **24**, 596–606.

376. LI, P. H. and WEISER, C. J. (1967). Evaluation of extraction and assay methods for nucleic acids from Red-Osier dogwood and RNA, DNA, and protein changes during cold acclimation. *Proc. Am. Hort. Soc.*, **91**, 716–727.

377. LIBBENGA, K. R. and HARKES, P. A. A. (1973). Initial proliferation of cortical cells in the formation of root nodules of *Pisum sativum* L. *Planta*, **114**, 17–28.

378. LIBBERT, E., WICHNER, S., DUERST, E., KUNERT, R., KAISER, W., MANICKI, A., MATEUFFEL, R., RIECKE, E. and SCHRODER, R. (1968). Auxin content and auxin synthesis in sterile and non sterile plants with special regard to the influence of epiphytic bacteria. In *Biochemistry and Physiology of Plant Growth Substances*, ed. WIGHTMAN, F. and SETTERFIELD, G., 213–230. Runge Press, Ottawa, Canada.

379. LIEBERMAN, M. and MAPSON, L. W. (1964). Genesis and biogenesis of ethylene. *Nature, Lond.*, **204**, 343–345.

380. LIGNOWSKI, E. M., SPLITTSTOESSER, W. E. and CHOU, K. (1971). The change in arginine levels and the metabolism of urea and ornithine in *Cucurbita moschata* seedlings. *Physiologia Pl.*, **25**, 225–229.

381. LIN, C. Y. and KEY, J. L. (1971). Dissociation of N_2 gas-induced monomeric ribosomes and functioning of the derived subunits in protein synthesis in pea. *Pl. Physiol., Lancaster*, **48**, 547–552.

382. LINDELL, T. J., WEINBERG, F., MORRIS, P. W., ROEDER, R. G. and RUTTER, W. J. (1970). Specific inhibition of nuclear RNA polymerase II by α-amanitin. *Science, N.Y.*, **170**, 447–448.

383. LINKO, P. and MILNER, M. (1959). Enzyme activation in wheat grains in relation to water content. Glutamic acid–alanine transaminase, and glutamic acid decarboxylase. *Pl. Physiol., Lancaster*, **34**, 392–396.

384. LIS, H. and SHARON, N. (1973). The biochemistry of plant lectins (phytohaemagglutinins). *A. Rev. Biochem.*, **42**, 541–574.

385. LIS, H., SHARON, N. and KATCHALSKI, E. (1966). Soybean hemagglutinin, a plant glycoprotein. I. Isolation of a glycopeptide. *J. biol. Chem.*, **241**, 684–689.

386. LOENING, U. E. (1968). RNA structure and metabolism. *A. Rev. Pl. Physiol.*, **19**, 37–70.

387. LOENING, U. E. and INGLE, J. (1967). Diversity of RNA components in green plant tissues. *Nature, Lond.*, **215**, 363–367.

388. LOOMIS, W. D. and BATTAILE, J. (1966). Plant phenolic compounds and the isolation of plant enzymes. *Phytochem.*, **5**, 423–438.

389. LORD, J. M., KAGAWA, T. and BEEVERS, H. (1972). Intracellular distribution of enzymes of the cytidine diphosphate choline pathway in castor bean endosperm. *Proc. natn. Acad Sci. U.S.A.*, **69**, 2429–2432.

390. LOSADA, M., PANEQUE, A., RAMIREZ, J. M. and DEL CAMPO, F. F. (1963). Mechanism of nitrate reduction in chloroplasts. *Biochem. biophys. Res. Commun.*, **10**, 298–303.

391. LOWRY, O. H., ROSEBROUGH, N. J., FARR, A. L. and RANDALL, R. J. (1951). Protein measurement with the folin phenol reagent. *J. biol. Chem.*, **193**, 265–275.

392. LUCAS-LENARD, J. and LIPMAN, F. (1971). Protein biosynthesis. *A. Rev. Biochem.*, **40**, 409–448.

393. LUCKNER, M. (1972). *Secondary Metabolism in Plants and Animals.* Academic Press, New York.

394. LUND, H. A., VATTER, A. E. and HANSON, J. B. (1958). Biochemical and cytological changes accompanying growth and differentiation in the roots of *Zea mays. J. biophys. biochem. Cytol.*, **4**, 87–97.

394a. MAGALHAES, A. C., NEYRA, C. A. and HAGEMAN, R. H. (1974). Nitrite assimilation and amino nitrogen synthesis in isolated spinach chloroplasts. *Pl. Physiol., Lancaster*, **53**, 411–415.

395. MAHADEVAN, S. (1973). Role of oximes in nitrogen metabolism in plants. *A. Rev. Pl. Physiol.*, **24**, 69–88.

396. MAHL, M. C. and WILSON, P. W. (1968). Nitrogen fixation by cell-free extracts of *Klebsiella pneumoniae. Can. J. Microbiol.*, **14**, 33–38.

397. MANAHAN, C. O., APP, A. A. and STILL, C. C. (1973). The presence of poly-adenylate sequences in the ribonucleic acid of a higher plant. *Biochem. biophys. Res. Commun.*, **53**, 588–595.

397a. MANS, R. J. and WALTER, T. J. (1971). Transfer RNA primed oligoadenylate synthesis in maize seedlings, II. Primer, substrate and metal specificities and size of product. *Biochim. biophys. Acta*, **247**, 113–121.

398. MARCUS, A. and FEELEY, J. (1964). Activation of protein synthesis in the imbibition phase of seed germination. *Proc. natn. Acad. Sci. U.S.A.*, **51**, 1075–1079.

399. MARCUS, A. and FEELEY, J. (1965). Protein synthesis in imbibed seeds. II. Polysome formation during imbibition. *J. biol. Chem.*, **240**, 1675–1680.

400. MARCUS, A., WEEKS, D. P., LEIS, J. P. and KELLER, E. B. (1970). Protein chain initiation by methionyl-tRNA in wheat embryo. *Proc. natn. Acad. Sci. U.S.A.*, **67**, 1681–1687.

401. MAREI, N. and ROMANI, R. (1971). Ribosomes from fig fruits: physical properties and the occurrence of transient dimers. *Biochim. biophys. Acta*, **247**, 280–290.

402. MARGULIES, M. M. (1970). *In vitro* protein synthesis by plastids from *Phaseolus vulgaris.* V. Incorporation of ^{14}C-leucine into a protein fraction containing ribulose 5-diphosphate carboxylase. *Pl. Physiol., Lancaster*, **46**, 136–141.

403. MARRÉ, E. (1967). Ribosome and enzyme changes during maturation and germination of the castor bean seed. In *Current Topics in Developmental Biology*, Vol. 2, ed. MOSCONA, A. A. and MONROY, A. V., 75–105. Academic Press, New York.

404. MARTIN, C. and THIMANN, K. V. (1972). The role of protein synthesis in the senescence of leaves. I. The formation of protease. *Pl. Physiol., Lancaster*, **49**, 64–71.

405. MATILE, P. (1968). Lysosomes of root tip cells in corn seedlings. *Planta*, **79**, 181–196.

406. MATILE, P. and WINKENBACH, F. (1971). Function of lysosomes and lysosomal enzymes in the senescing corolla of the morning glory (*Ipomoea purpurea*). *J. exp. Bot.*, **22**, 759–771.

407. MATSUBARA, T. (1970). Studies on denitrification, XII. Gas production from amines and nitrite. *J. Biochem., Tokyo*, **67**, 229–235.

408. MATSUBARA, T. and IWASAKI, H. (1971). Enzymatic steps of dissimilatory nitrite reduction in *Alcaligenes faecalis. J. Biochem., Tokyo*, **69**, 859–868.

409. MATSUSHITA, S. (1959). On the ribonuclease during the germination of wheat. *Mem. Res. Inst. Food Sci. Kyoto Univ.*, **17**, 23–28.

410. MAYAK, S. and HALEVY, A. H. (1970). Cytokinin activity in rose petals and its relation to senescence. *Pl. Physiol., Lancaster*, **46**, 497–499.

411. MAYER, F. C., BIKEL, L. and HASSID, W. Z. (1968). Pathway of uridine diphosphate *N*-acetyl-D-glucosamine biosynthesis in *Phaseolus aureus. Pl. Physiol., Lancaster*, **43**, 1097–1107.

412. MAZLIAK, P. (1973). Lipid metabolism in plants. *A. Rev. Pl. Physiol.*, **24**, 287–310.

413. MAZUS, B. and BUCHOWICZ, J. (1972). Activity of enzymes involved in pyrimidine metabolism in the germinating wheat grains. *Phytochem.*, **11**, 77–82.

414. MAZUS, B. and BUCHOWICZ, J. (1972). RNA polymerase activity in resting and germinating wheat seeds. *Phytochem.*, **11**, 2443–2446.

415. MCKEE, H. S., NESTEL, L. and ROBERTSON, R. N. (1955). Physiology of pea fruits. II. Soluble nitrogenous constituents in the developing fruit. *Aust. J. biol. Sci.*, **8**, 467–475.

416. MCKEE, H. S., ROBERTSON, R. N. and LEE, J. B. (1955). Physiology of pea fruits. *Aust. J. biol. Sci.*, **8**, 137–163.

417. MCNARY, J. E. and BURRIS, R. H. (1962). Energy requirements for nitrogen fixation by cell-free preparations from *Clostridium pasteurianum. J. Bact.*, **84**, 598–599.

418. MERRICK, W. C. and DURE, L. S. (1971). Specific transformylation of one methionyl-tRNA from cotton seedling chloroplasts by endogenous and *Escherichia coli* transformylases. *Proc. natn. Acad. Sci. U.S.A.*, **68**, 641–644.

419. MERRICK, W. C., LUBSEN, N. H. and ANDERSON, W. F. (1973). A ribosome dissociation factor from rabbit reticulocytes distinct from initiation factor M_3. *Proc. natn. Acad. Sci. U.S.A.*, **70**, 2220–2223.

420. MESELSON, M. and STAHL, F. W. (1958). The replication of DNA in *Escherichia coli. Proc. natn. Acad. Sci. U.S.A.*, **44**, 671–677.

421. MEYER, V. and SCHULZE, E. (1884). Ueber die Einwirkung von Hydroxylaminsalzen auf Pflanzen. *Ber. dt. chem. Ges.*, **17**, 1554–1558.

422. MIFLIN, B. J. (1969). The inhibitory effects of various amino acids on the growth of barley seedlings. *J. exp. Bot.*, **20**, 810–819.

423. MIFLIN, B. J. (1969). Acetolactate synthetase from barley seedlings. *Phytochem.*, **8**, 2271–2276.

424. MIKOLA, J. and KOLEHMAINEN, L. (1972). Localization and activity of various peptidases in germinating barley. *Planta*, **104**, 167–177.

425. MILLBANK, J. W. (1972). Nitrogen metabolism in lichens. IV. The nitrogenase activity of the *Nostoc* phycobiont in *Peltigera canina. New Phytol.*, **71**, 1–10.

426. MILLERD, A., SIMON, M. and STERN, H. (1971). Legumin synthesis in developing cotyledons of *Vicia faba. Pl. Physiol., Lancaster*, **48**, 419–425.

427. MILLERD, A. and WHITFELD, P. R. (1973). Deoxyribonucleic acid and ribonucleic acid synthesis during the cell expansion phase of cotyledon development in *Vicia faba* L. *Pl. Physiol., Lancaster*, **51**, 1005–1010.

428. MINAMIKAWA, T. (1967). A study on 3-deoxy-D-arabino-heptulosonic acid 7-phosphate synthase in higher plants. *Pl. Cell Physiol., Tokyo*, **8**, 695–707.

429. MINAMIKAWA, T., OYAMA, I. and TOSHIDA, S. (1968). Alicyclic acid metabolism in plants. 2. Enzymes related to the shikimate pathway in developing mung bean seedlings. *Pl. Cell Physiol., Tokyo*, **9**, 451–460.

430. MINOTTI, P. L., WILLIAMS, D. C. and JACKSON, W. A. (1969). The influence of ammonium on nitrate reduction in wheat seedlings. *Planta*, **86**, 267–271.

431. MIYATA, M., MATSUBARA, T. and MORI, T. (1969). Studies on denitrification. XI. Some properties of nitric oxide reductase. *J. Biochem., Tokyo*, **66**, 759–765.

432. MIZRAHI, Y., AMIR, J. and RICHMOND, A. E. (1970). The mode of action of kinetin in maintaining the protein content of detached *Tropaeolum majus* leaves. *New Phytol.*, **69**, 355–361.

433. MIZUSAKI, S., TANABE, Y., NOGUCHI, M. and TAMAKI, E. (1973). Changes in the activities of ornithine decarboxylase, putrescine *N*-methyl transferase and *N*-methyl putrescine oxidase in tobacco roots in relation to nicotine biosynthesis. *Pl. Cell Physiol., Tokyo*, **14**, 103–110.

434. MOHR, H. (1969). Photomorphogenesis. In *Physiology of Plant Growth and Development*, ed. WILKINS, M. B., 509–558. McGraw-Hill, New York.

435. MOHR, H. (1972). *Lectures in Photomorphogenesis.* Springer Verlag, Berlin.

435a. MOLLER, B. L. (1974). Lysine biosynthesis in barley (*Hordeum vulgare* L.). *Pl. Physiol., Lancaster*, **54**, 638–643.

436. MORILLA, C. A., BOYER, J. S. and HAGEMAN, R. H. (1973). Nitrate reductase activity and polyribosomal content of corn (*Zea mays* L.) having low leaf water potentials. *Pl. Physiol., Lancaster*, **51**, 817–824.

437. MORRÉ, D. J., NYQUIST, S. and RIVERA, E. (1970). Lecithin biosynthetic enzymes of onion stem and the distribution of phosphorylcholine-cytidyl transferase among cell fractions, *Pl. Physiol., Lancaster*, **45**, 800–804.

438. MORRIS, C. J., THOMPSON, J. F. and JOHNSON, C. M. (1969). Metabolism of glutamic acid and *N*-acetylglutamic acid in leaf discs and cell-free extracts of higher plants. *Pl. Physiol., Lancaster*, **44**, 1023–1026.

439. MORTENSON, L. E. (1964). Ferredoxin requirement for nitrogen fixation by extracts of *Clostridium pasteurianum. Biochim. biophys. Acta*, **81**, 473–478.

440. MORTENSON, L. E. (1964). Ferredoxin and ATP requirements for nitrogen fixation by cell-free extracts of *Clostridium pasteurianum. Proc. natn. Acad. Sci. U.S.A.*, **52**, 272–279.

441. MORTON, R. K., PALK, B. A. and RAISON, J. K. (1964). Intracellular components associated with protein synthesis in developing wheat endosperm. *Biochem. J.*, **91**, 522–528.

442. MORTON, R. K. and RAISON, J. K. (1963). A complete intracellular unit for incorporation of amino-acid into storage protein utilizing adenosine triphosphate generated from phytate. *Nature, Lond.*, **200**, 429–433.

443. MORTON, R. K. and RAISON, J. K. (1964). The separate incorporation of amino acids into storage and soluble proteins by two independent systems isolated from developing wheat endosperm. *Biochem. J.*, **91**, 528–539.

444. MORY, Y. Y., CHEN, D. and SARID, S. (1974). Deoxyribonucleic acid polymerase from wheat embryos. *Pl. Physiol., Lancaster*, **53**, 377–381.

445. MOSSÉ, J. (1966). Alcohol-soluble proteins of cereal grains. *Fed. Proc.*, **25**, 1663–1669.

446. MULLINIX, K. P., STRAIN, G. C. and BOGORAD, L. (1973). RNA polymerase of maize purification and molecular structure of DNA dependent RNA polymerase. II. *Proc. natn. Acad. Sci. U.S.A.*, **70**, 2386–2390.

446a. MURPHY, M. J., SIEGEL, L. M., TOVE, S. R. and KAMIN, H. (1974). Siroheme; A new prosthetic group participating in six-electron reduction reactions catalyzed by both sulfite and nitrite reductases. *Proc. natn. Acad. Sci. U.S.A.*, **71**, 612–616.

447. NAGABHUSHAN, N., GULYAS, A. and ZALIK, S. (1974). Comparison of plant cytoplasmic ribosomal proteins by two-dimensional polyacrylamide gel electrophoresis. *Pl. Physiol., Lancaster*, **53**, 516–518.

448. NAGATANI, H., SHIMIZU, M. and VALENTINE, R. C. (1971). The mechanism of ammonia assimilation in nitrogen fixing bacteria. *Arch. Mikrobiol.*, **79**, 164–175.

449. NAIR, P. M. (1969). Asparagine synthetase from γ-irradiated potatoes. *Archs. Biochem. Biophys.*, **133**, 208–215.

450. NANDI, D. L. and WAYGOOD, E. R. (1967). Biosynthesis of porphyrins in wheat leaves. II. 5 amino laevulinate hydro-lyase. *Can. J. Biochem.*, **45**, 327–336.

451. NANDY, M. and GANGULI, N. C. (1961). Biological synthesis of 5-dehydroshikimic acid by a plant extract. *Biochim. biophys. Acta*, **48**, 608–610.

452. NASON, A., ANTOINE, A. D., KETCHUM, P. A., FRAZIER, W. A. and LEE, D. K. (1970). Formation of assimilatory nitrate reductase by *in vitro* intercistronic complementation in *Neurospora crassa. Proc. natn. Acad. Sci. U.S.A.*, **65**, 137–144.

453. NASON, A. and EVANS, H. J. (1953). Triphosphopyridine nucleotide-nitrate reductase in *Neurospora. J. biol. Chem.*, **202**, 655–673.

454. NAWA, Y. and ASHAI, T. (1973). Relationship between the water content of pea cotyledons and mitochondrial development during the early stages of germination. *Pl. Cell Physiol., Tokyo*, **14**, 607–610.

455. NAYLOR, J. M. (1966). Dormancy studies in seed of *Avena fatua*. 5. On the response of aleurone cells to gibberellic acid. *Can. J. Bot.*, **44**, 19–32.

456. NELSON, D. W. and BREMNER, J. M. (1972). Preservation of soil samples for inorganic nitrogen analyses. *Agron. J.*, **64**, 196–199.

457. NEWTON, J. W., WILSON, P. W. and BURRIS, R. H. (1953). Direct demonstration of ammonia as an intermediate in nitrogen fixation by *Azotobacter. J. biol. Chem.*, **204**, 445–451.

458. NEWTON, N. (1969). The two-haem nitrite reductase of *Micrococcus denitrificans*. *Biochim. biophys. Acta*, **185**, 316–331.
459. NICHOLAS, D. J. D. and JONES, O. T. G. (1960). Oxidation of hydroxylamine in cell-free extracts of *Nitrosomonas europaea*. *Nature, Lond.*, **185**, 512–514.
460. NICHOLLS, P. B. and MURRAY, A. W. (1968). Adenine phosphoribosyl transferase in plant tissues: some effects of kinetin on enzymic activity. *Pl. Physiol., Lancaster*, **43**, 645–648.
461. NIRENBERG, M. (1970). The flow of information from gene to protein. In *Aspects of Protein Biosynthesis*, ed. ANFINSEN, C. B., 215–246. Academic Press, New York and London.
462. NOGUCHI, M., KOIWAI, A. and TAMAKI, E. (1966). Studies on nitrogen metabolism in tobacco plants. VII. Δ-pyrroline-5-carboxylate reductase from tobacco leaves. *Agric. Biol. Chem.*, **30**, 452–456.
463. NOODÉN, L. D. and THIMANN, K. V. (1963). Evidence for a requirement for protein synthesis for auxin induced cell enlargement. *Proc. natn. Acad. Sci. U.S.A.*, **50**, 194–200.
464. NORRIS, R. D., LEA, P. J. and FOWDEN, L. (1973). Aminoacyl-tRNA synthetases in *Triticum aestivum* L. during seed development and germination. *J. exp. Bot.*, **24**, 615–625.
465. NOTTON, B. A. and HEWITT, E. J. (1971). The role of tungsten in the inhibition of nitrate reductase activity in spinach (*Spinacia oleracea* L.) leaves. *Biochem. biophys. Res. Commun.*, **44**, 702–710.
466. NOTTON, B. A. and HEWITT, E. J. (1971). Incorporation of radioactive molybdenum into protein during nitrate reductase formation and effect of molybdenum on nitrate reductase and diaphorase activities of spinach (*Spinacia oleracea* L.). *Pl. Cell Physiol., Tokyo*, **12**, 465–477.
467. NOZZOLILLO, C., PAUL, K. B. and GODIN, C. (1971). The fate of L-phenylalanine fed into germinating pea seeds (*Pisum sativum* L.) var. Alaska during imbibition. *Pl. Physiol., Lancaster*, **47**, 119–123.
468. NYGAARD, P. (1972). Deoxyribonucleotide pools in plant tissue cultures. *Physiologia Pl.*, **26**, 29–33.
469. NYGAARD, P. (1973). Nucleotide metabolism during pine pollen germination. *Physiologia Pl.*, **28**, 361–371.
470. OAKS, A. (1965). The synthesis of leucine in maize embryos. *Biochim. biophys. Acta*, **111**, 79–89.
471. OAKS, A. (1965). The soluble leucine pool in maize root tips. *Pl. Physiol., Lancaster*, **40**, 142–149.
472. OAKS, A. (1965). The effect of leucine on the biosynthesis of leucine in maize root tips. *Pl. Physiol., Lancaster*, **40**, 149–155.
473. OAKS, A. (1967). Asparagine synthesis in *Zea mays*. *Biochim. biophys. Acta*, **141**, 436–439.
474. OAKS, A. and JOHNSON, F. J. (1970). Effect of sugars on amino acid biosynthesis in maize root tips. *Can. J. Bot.*, **48**, 117–124.
475. O'BRIEN, T. J., JARVIS, B. C., CHERRY, J. H. and HANSON, J. B. (1968). The effect of 2,4-D on RNA synthesis by soybean hypocotyl chromatin. In *Biochemistry and Physiology of Plant Growth Substances*, ed. WIGHTMAN, F. and SETTERFIELD, G., 747–760. Runge Press, Ottawa.
476. ODMARK, G. (1969). Studies on deoxyribonucleotide synthesis in *Vicia faba*. *Physiologia Pl.*, **22**, 161–170.
477. OGHOGHORIE, C. G. O. and PATE, J. S. (1972). Exploration of the nitrogen transport system of a nodulated legume using ^{15}N. *Planta*, **104**, 35–49.
478. OGREN, W. L. and KROGMANN, D. W. (1965). Studies on pyridine nucleotide in photosynthetic tissue. *J. biol. Chem.*, **240**, 4603–4608.
479. OLNEY, H. O. and POLLOCK, B. M. (1960). Studies of rest period. II. Nitrogen and phosphorus changes in embryonic organs of after-ripening cherry seed. *Pl. Physiol., Lancaster*, **35**, 970–975.
480. O'NEAL, D. and NAYLOR, A. W. (1968). Purine and pyrimidine nucleotide inhibition of carbamyl phosphate synthetase from pea seedlings. *Biochem. biophys. Res. Commun.*, **31**, 322–327.

481. O'NEAL, T. C. and NAYLOR, A. W. (1969). Partial purification and properties of carbamyl phosphate synthetase of Alaska Pea (*Pisum sativum* L. cultivar Alaska). Purification and general properties. *Biochem. J.*, **113**, 271–279.

482. ONG, B. L. and JACKSON, J. F. (1972). Aspartate transcarbamylase from *Phaseolus aureus*. Partial purification and properties. *Biochem. J.*, **129**, 571–581.

483. ONG, B. L. and JACKSON, J. F. (1972). Pyrimidine nucleotide biosynthesis in *Phaseolus aureus*. Enzymic aspects of the control of carbamyl phosphate synthesis and utilization. *Biochem. J.*, **129**, 583–593.

484. OOTA, Y. and TAKATA, K. (1961). Cytoplasmic ribonucleic acids in bean germ tissues. In *Recent Advances in Botany*. 2: *The Plant Protoplast*, 1059–1063. University of Toronto Press, Toronto.

485. ÖPIK, H. (1968). Development of cotyledon cell structure in ripening *Phaseolus vulgaris* seeds. *J. exp. Bot.*, **19**, 64–76.

486. ÖPIK, H. and SIMON, E. W. (1963). Water content and respiration rate of bean cotyledons. *J. exp. Bot.*, **14**, 299–310.

487. ORY, R. L. and HENNINGSEN, K. W. (1969). Enzymes associated with protein bodies isolated from ungerminated barley seeds. *Pl. Physiol., Lancaster*, **44**, 1488–1498.

488. OSBORNE, D. J. (1962). Effect of kinetin on protein and nucleic acid metabolism in *Xanthium* leaves during senescence. *Pl. Physiol., Lancaster*, **37**, 595–602.

489. OSBORNE, T. B. (1924). *The Vegetable Proteins*. Longmans Green, New York.

490. OTA, A. (1965). Oxidative phosphorylation coupled with nitrate respiration. III. Coupling factors and mechanism of oxidative phosphorylation. *J. Biochem., Tokyo*, **58**, 137–144.

491. PAGE, O. T. and CLARK, M. C. (1968). Activity of *N*-ribosyl-adenosine ribohydrolase in potato leaves in response to *Phytophthora* infection. *Can. J. Bot.*, **46**, 979–985.

492. PALMIANO, E. P. and JULIANO, B. O. (1972). Biochemical changes in the rice grain during germination. *Pl. Physiol., Lancaster*, **49**, 751–756.

493. PANDITA, J. M. and DURZAN, D. J. (1970). Metabolism of guanine-8-^{14}C by germinating Jack Pine seedlings. *Proc. Can. Soc. Pl. Physiol.*, **10**, 62.

494. PANEQUE, A., RAMIREZ, J. M., DEL CAMPO, F. F. and LOSADA, M. (1964). Light and dark reduction of nitrite in a reconstituted enzymic system. *J. biol. Chem.*, **239**, 1737–1741.

495. PARANJOTHY, K. and WAREING, P. F. (1971). The effects of abscisic acid, kinetin, and 5-fluorouracil on ribonucleic acid and protein synthesis in senescing radish leaf discs. *Planta*, **99**, 112–119.

496. PATE, J. S. and GUNNING, B. E. S. (1972). Transfer cells. *A. Rev. Pl. Physiol.*, **23**, 173–196.

497. PAYNE, J. F. and BALA, K. (1972). RNA polymerase activity in germinating onion seeds. *Phytochem.*, **11**, 3105–3110.

498. PAYNE, P. I. and BOULTER, D. (1969). Free and membrane bound ribosomes of the cotyledons of *Vicia faba* L. I. Seed development. *Planta*, **84**, 263–271.

499. PAYNE, P. I. and DYER, T. A. (1972). Plant 5.8s RNA is a component of 80s but not 70s ribosomes. *Nature, Lond.*, **235**, 145–147.

500. PENNER, D. and ASHTON, F. M. (1967). Hormonal control of proteinase activity in squash cotyledons. *Pl. Physiol., Lancaster*, **42**, 791–796.

501. PHELPS, R. H. and SEQUEIRA, L. (1967). Synthesis of indoleacetic acid via tryptamine by a cell-free system from tobacco terminal buds. *Pl. Physiol., Lancaster*, **42**, 1161–1163.

502. PHILLIPS, D. R. and FLETCHER, R. A. (1969). Ribonuclease in leaves of *Phaseolus vulgaris* during maturation and senescence. *Physiologia Pl.*, **22**, 764–767.

503. PHILLIPS, I. D. J. and WAREING, P. F. (1959). Studies in dormancy of sycamore. II. The effect of day length on the natural growth-inhibitor content of the shoot. *J. exp. Bot.*, **10**, 504–514.

504. POSTGATE, J. R. (1969). A discussion on nitrogen fixation. *Proc. R. Soc.*, B, **172**, 355.

505. POULSON, R. and BEEVERS, L. (1970). Nucleic acid metabolism during greening and unrolling of barley leaf segments. *Pl. Physiol., Lancaster*, **46**, 315–319.

506. POULSON, R. and BEEVERS, L. (1970). Effects of growth regulators on ribonucleic acid metabolism of barley leaf segments. *Pl. Physiol., Lancaster*, **46**, 782–785.

507. POULSON, R. and BEEVERS, L. (1973). RNA metabolism during the development of cotyledons of *Pisum sativum* L. *Biochim. biophys. Acta*, **308**, 381–389.

508. PRICE, L. and KLEIN, W. H. (1961). Red, far red response and chlorophyll synthesis. *Pl. Physiol., Lancaster*, **36**, 733–735.

509. PRIDHAM, J. B. (1965). Low molecular weight phenols in higher plants. *A. Rev. Pl. Physiol.*, **16**, 13–36.

510. PRITCHARD, P. M., GILBERT, J. M., SHAFRITZ, D. A. and ANDERSON, W. F. (1970). Factors for the initiation of haemoglobin synthesis by reticulocyte ribosomes. *Nature, Lond.*, **226**, 511–514.

511. PUSZTAI, A. (1964). Hexosamines in the seeds of higher plants (Spermatophytes). *Nature, Lond.*, **201**, 1328–1329.

512. PUSZTAI, A. and DUNCAN, I. (1971). Changes in proteolytic enzyme activities and transformation of nitrogenous compounds in the germinating seeds of kidney bean (*Phaseolus vulgaris*). *Planta*, **96**, 317–325.

513. RABSON, R., MANS, R. J. and NOVELLI, G. D. (1961). Changes in cell-free amino acid incorporating activity during maturation of maize kernels. *Archs. Biochem. Biophys.*, **93**, 555–562.

514. RACUSEN, D. and FOOTE, M. (1960). Amino acid turnover and protein synthesis in leaves. *Archs. Biochem. Biophys.*, **51**, 68–78.

515. RAY, P. M., SHININGER, T. L. and RAY, M. M. (1969). Isolation of β-glucan synthetase particles from plant cells and identification with Golgi membranes. *Proc. natn. Acad. Sci. U.S.A.*, **64**, 605–612.

516. REBEIZ, C. A. and CASTELFRANCO, P. A. (1973). Protochlorophyll and chlorophyll biosynthesis in cell-free systems from higher plants. *A. Rev. Pl. Physiol.*, **26**, 129–172.

517. REBEIZ, C. A., YAGHI, M., ABOU-HAIDAR, M. and CASTELFRANCO, P. A. (1970). Protochlorophyll biosynthesis in cucumber (*Cucumis sativus*) cotyledons. *Pl. Physiol., Lancaster*, **46**, 57–63.

518. REHFELD, D. W. and TOLBERT, N. E. (1972). Aminotransferases in peroxisomes from spinach leaves. *J. biol. Chem.*, **247**, 4803–4811.

519. REINBOTHE, H. and MOTHES, K. (1962). Urea, ureides and guanidines in plants. *A. Rev. Pl. Physiol.*, **13**, 129–150.

520. RENNER, E. D. and BECKER, G. E. (1970). Production of nitric oxide and nitrous oxide during denitrification by *Corynebacterium nephridii*. *J. Bact.*, **101**, 821–826.

521. RHODES, M. J. C. and WOOLTORTON, L. S. C. (1967). The respiration climacteric in apple fruits. The action of hydrolytic enzymes in peel tissue during the climacteric period in fruit detached from the tree. *Phytochem.*, **6**, 1–12.

522. RICE, E. L. and PANCHOLY, S. K. (1972). Inhibition of nitrification by climax ecosystems. *Am. J. Bot.*, **59**, 1033–1040.

523. RIJVEN, A. H. G. C. and ZWAR, J. A. (1973). Methylation patterns of ribonucleic acids from chloroplasts and cytoplasm of fenugreek (*Trigonella foenum-graecum* L.) cotyledons. *Biochim. biophys. Acta*, **299**, 564–567.

524. RITENOUR, G. L., JOY, K. W., BUNNING, J. and HAGEMAN, R. H. (1967). Intracellular localization of nitrate reductase, nitrite reductase, and glutamic acid dehydrogenase in green leaf tissue. *Pl. Physiol., Lancaster*, **42**, 233–237.

525. ROBERN, H., WANG, D. and WAYGOOD, E. R. (1965). Biosynthesis of nucleotides in wheat. I. Purines from ^{14}C-labelled compounds. *Can. J. Biochem.*, **43**, 225–235.

526. ROBERTS, B. E., PAYNE, P. I. and OSBORNE, D. J. (1973). Protein synthesis and the viability of rye grains. Loss of activity of protein-synthesizing systems *in vitro* associated with a loss of viability. *Biochem. J.*, **131**, 275–286.

527. ROBERTS, E. H. (1972). *Viability of Seeds*. Syracuse University Press.

528. ROBERTS, G. R., KEYS, A. J. and WHITTINGHAM, C. P. (1970). The transport of

photosynthetic products from the chloroplasts of tobacco leaves. *J. exp. Bot.*, **21**, 683–692.

529. ROBERTS, R. M. (1970). The incorporation of D-glucosamine-^{14}C into root tissues of higher plants. *Pl. Physiol., Lancaster*, **45**, 263–267.

530. ROBERTS, R. M., CETORELLI, J. J., KIRBY, E. G. and ERICSON, M. (1972). Location of glycoproteins that contain glucosamine in plant tissues. *Pl. Physiol., Lancaster*, **50**, 521–535.

531. ROBINSON, E. and BROWN, R. (1952). The development of the enzyme complement in growing root cells. *J. exp. Bot.*, **3**, 356–374.

532. ROBINSON, F. A. (1966). *The Vitamin Cofactors of Enzyme Systems*. Pergamon Press, London.

533. ROBINSON, J. M. and STOCKING, C. R. (1968). Oxygen evolution and the permeability of the outer envelope of isolated whole chloroplasts. *Pl. Physiol., Lancaster*, **43**, 1597–1604.

534. ROBISON, G. W., BUTCHER, R. W. and SUTHERLAND, E. W. (1971). *Cyclic AMP*. Academic Press, London.

535. ROGERS, M. E., LOENING, U. E. and FRASER, R. S. S. (1970). Ribosomal RNA precursors in plants. *J. molec. Biol.*, **49**, 681–692.

536. ROGNES, S. E. (1970). A glutamine-dependent asparagine synthetase from yellow lupine seedlings. *FEBS Letters*, **10**, 62–66.

537. ROSENTHAL, G. A. (1970). Investigations of canavanine biochemistry in the Jack Bean plant *Canavalia ensiformis* (L.) DC. I. Canavanine utilization in the developing plant. *Pl. Physiol., Lancaster*, **46**, 273–276.

538. ROSENTHAL, G. A. and NAYLOR, A. W. (1969). Purification and general properties of argininosuccinate lyase from Jack Bean, *Canavalia ensiformis* (L.) DC. *Biochem. J.*, **112**, 415–419.

539. ROSS, C., CODDINGTON, R. L., MURRAY, M. G. and BLEDSOE, C. S. (1971). Pyrimidine metabolism in cotyledons of germinating Alaska peas. *Pl. Physiol., Lancaster*, **47**, 71–75.

540. ROSS, C. and COLE, C. Y. (1968). Metabolism of cytidine and uridine in bean leaves. *Pl. Physiol., Lancaster*, **43**, 1227–1231.

541. ROSS, C. and MURRAY, M. G. (1971). Development of pyrimidine-metabolizing enzymes in cotyledons of germinating peas. *Pl. Physiol., Lancaster*, **48**, 626–630.

542. RYAN, C. J. (1973). Proteolytic enzymes and their inhibitors in plants. *A. Rev. Pl. Physiol.*, **24**, 173–196.

543. SACHER, J. (1973). Senescence and post harvest physiology. *A. Rev. Pl. Physiol.*, **24**, 197–224.

544. SADAVA, D. and CHRISPEELS, M. J. (1971). Hydroxyproline biosynthesis in plant cells. Peptidyl proline hydroxylase from carrot discs. *Biochim. biophys. Acta*, **227**, 278–287.

544a. SAGHER, D., EDELMAN, M. and JAKOB, K. M. (1974). Poly (A) associated RNA in plants. *Biochim. biophys. Acta*, **349**, 32–38.

545. SANDERSON, G. W. (1966). 5-Dehydroshikimate reductase in the tea plant (*Camellia sinensis* L.). *Biochem. J.*, **98**, 248–252.

546. SASAOKA, K. (1961). Studies on homoserine dehydrogenase in pea seedlings. *Pl. Cell Physiol., Tokyo*, **2**, 231–242.

547. SASAOKA, K. and INAGAKI, H. (1960). Occurrence of aspartic β-semi-aldehyde dehydrogenase in pea seedlings. *Mem. Res. Inst. Food Sci. Kyoto Univ.*, **21**, 12–16.

548. SCHARPE, A. and VAN PARIJS, R. (1973). The formation of polyploid cells in ripening cotyledons of *Pisum sativum* L. in relation to ribosome and protein synthesis. *J. exp. Bot.*, **24**, 216–222.

549. SCHONHERR, O. T. and WANKA, F. (1971). An investigation of DNA polymerase in synchronously growing *Chlorella* cells. *Biochim. biophys. Acta*, **232**, 83–93.

550. SCHRADER, L. E., RITENOUR, G. L., EILRICH, G. L. and HAGEMAN, R. H. (1968). Some characteristics of nitrate reductase from higher plants. *Pl. Physiol., Lancaster*, **43**, 930–940.

551. SCHWARZ, O. J. and FITES, R. C. (1970). Thymidine kinase from peanut seedlings. *Phytochem.*, **9**, 1405–1414.
552. SCOTT, N. S. and INGLE, J. (1973). The genes for cytoplasmic ribosomal ribonucleic acid in higher plants. *Pl. Physiol., Lancaster*, **51**, 677–684.
553. SEAL, S. N., BEWLEY, J. D. and MARCUS, A. (1972). Protein chain initiation in wheat embryo. *J. biol. Chem.*, **247**, 2592–2597.
554. SEELY, M. K., CRIDDLE, R. S. and CONN, E. E. (1966). The metabolism of aromatic compounds in higher plants. VIII. On the requirement of hydroxynitrile lyase for flavin. *J. biol. Chem.*, **241**, 4457–4462.
555. SHAFRITZ, D. A., PRICHARD, P. M., GILBERT, J. M., MERRICK, W. C. and ANDERSON, W. F. (1972). Separation of reticulocyte initiation factor M_2 activity into two components. *Proc. natn. Acad. Sci. U.S.A.*, **69**, 983–987.
556. SHAH, C. B. and LOOMIS, R. S. (1965). Ribonucleic acid and protein metabolism in sugar beet during drought. *Physiologia Pl.*, **18**, 240–254.
557. SHAIN, Y. and MAYER, A. M. (1968). Activation of enzymes during germination: amylopectin-1,6-glucosidase in peas. *Physiologia Pl.*, **21**, 765–776.
558. SHAIN, Y. and MAYER, A. M. (1968). Activation of enzymes during germination—trypsin-like enzyme in lettuce. *Phytochem.*, **7**, 1491–1498.
559. SHANNON, L. M., KAY, E. and LEW, J. Y. (1966). Peroxidase isozymes from horseradish roots. *J. biol. Chem.*, **241**, 2166–2172.
560. SHARGOOL, P. D. (1971). Purification of argininosuccinate synthetase from cotyledons of germinating peas. *Phytochem.*, **10**, 2029–2032.
561. SHELDRAKE, A. R. (1973). The production of hormones in higher plants. *Biol. Rev.*, **48**, 509–561.
562. SHIBAOKA, H. and THIMANN, K. V. (1970). Antagonisms between kinetin and amino acids. Experiments on the mode of action of cytokinins. *Pl. Physiol., Lancaster*, **46**, 212–220.
563. SHIMURA, Y. and VOGEL, H. J. (1966). Diaminopimelate decarboxylase of *Lemna perpusilla:* partial purification and some properties. *Biochim. biophys. Acta*, **118**, 396–404.
564. SIEGEL, J. N. and GENTILE, A. C. (1966). Effect of 3-amino-1,2,4-triazole on histidine metabolism in algae. *Pl. Physiol., Lancaster*, **41**, 670–672.
565. SIEGELMAN, H. W., CHAPMAN, D. J. and COLE, W. J. (1968). The bile pigments of plants. In *Porphyrins and Related Compounds*, ed. GOODWIN, T. W., 107–120. Academic Press, New York.
566. SILVER, A. V. and GILMORE, V. (1969). The metabolism of purines and their derivatives in seedlings of *Pisum sativum*. *Phytochem.*, **8**, 2295–2299.
567. SILVER, W. S. (1969). Biology and ecology of nitrogen fixation by symbiotic associations of non-leguminous plants. *Proc. R. Soc., B*, **172**, 389–400.
568. SIMINOVITCH, D. (1963). Evidence from increase in ribonucleic acid and protein synthesis in autumn for increase in protoplasm during the frost hardening of black locust bark cells. *Can. J. Bot.*, **41**, 1301–1308.
569. SKOOG, F. and ARMSTRONG, D. J. (1970). Cytokinins. *A. Rev. Pl. Physiol.*, **21**, 359–384.
570. SMITH, D. L. (1973). Nucleic acid, protein, and starch synthesis in developing cotyledons of *Pisum arvense* L. *Ann. Bot., N.S.*, **37**, 795–804.
571. SMITH, H. (1973). Regulatory mechanisms in the photocontrol of flavonoid biosynthesis. In *Biosynthesis and its Control in Plants*, ed. MILBORROW, B. V., 303–321. Academic Press, London.
572. SMITH, H., STEWART, G. R. and BERRY, D. R. (1970). The effects of light on plastid ribosomal RNA and enzymes at different stages of barley etioplast development. *Phytochem.*, **9**, 977–984.
573. SMITH, R. V., NOY, R. J. and EVANS, M. C. W. (1971). Physiological electron donor systems to the nitrogenase of the blue-green alga *Anabaena cylindrica. Biochim. biophys. Acta*, **253**, 104–109.
574. SMITH, T. A. (1968). The biosynthesis of putrescine in higher plants and its relation to potassium nutrition. In *Recent Aspects of Nitrogen Metabolism in Plants*, ed. HEWITT, E. J. and CUTTING, C. V., 139–146. Academic Press, London.

575. SMITH, T. A. (1971). The occurrence, metabolism and functions of amines in plants. *Biol. Rev.*, **46**, 201–241.

576. SODEK, L. and WRIGHT, S. T. C. (1969). The effect of kinetin on ribonuclease, acid phosphatase, lipase and esterase levels in detached wheat leaves. *Phytochem.*, **8**, 1629–1640.

577. SODEK, L. and WILSON, C. M. (1970). Incorporation of leucine-^{14}C and lysine-^{14}C into protein in the developing endosperm of normal and opaque-2 corn. *Archs. Biochem. Biophys.*, **140**, 29–38.

578. SODEK, L. and WILSON, C. M. (1973). Metabolism of lysine and leucine derived from storage protein during the germination of maize. *Biochim. biophys. Acta*, **304**, 353–362.

579. SPENCER, D. and WHITFELD, P. R. (1967). DNA synthesis in isolated chloroplasts. *Biochem. biophys. Res. Commun.*, **28**, 538–542.

580. SPENCER, D., WHITFELD, P. R., BOTTOMLEY, W. and WHEELER, A. M. (1971). The nature of the proteins and nucleic acids synthesized by isolated chloroplasts. In *Autonomy and Biogenesis of Mitochondria and Chloroplasts*, ed. BOARDMAN, N. K., LINNANE, A. W. and SMILLIE, R. M., 372–382. North Holland Publishing Company, Amsterdam.

581. SPENSER, I. D. (1970). Biosynthesis of alkaloids. In *Chemistry of the Alkaloids*, ed. PELLETIER, S. W., 669–718. Van Nostrand–Reinhold Company, New York.

582. SPIRO, R. G. (1970). Glycoproteins. *A. Rev. Biochem.*, **39**, 599–638.

583. SPLITTSTOESSER, W. E. (1969). The appearance of arginine and arginase in pumpkin cotyledons. Characterization of arginase. *Phytochem.*, **8**, 753–758.

584. SPRINZL, M. and CRAMER, F. (1973). Accepting site for amino acylation of tRNA^{phe} from yeast. *Nature, Lond.*, **245**, 3–5.

585. SPRONK, A. M. and COSSINS, C. A. (1972). Folate derivatives of photosynthetic tissues. *Phytochem.*, **11**, 3157–3165.

586. SRIVASTAVA, B. I. S. (1968). Studies on the chromatin of barley leaves during senescence. *Biochem. J.*, **110**, 683–686.

587. SRIVASTAVA, B. I. S. and WARE, G. (1965). The effect of kinetin on nucleic acids and nucleases of excised barley leaves. *Pl. Physiol., Lancaster*, **40**, 62–64.

588. ST. ANGELO, A., ORY, R. L. and HANSEN, H. J. (1969). Localization of an acid proteinase in hempseed. *Phytochem.*, **8**, 1135–1138.

589. STAUB, M. and DENES, G. (1966). Mechanism of arginine biosynthesis in *Chlamydomonas reinhardti*. I. Purification and properties of ornithine acetyltransferase. *Biochim. biophys. Acta*, **128**, 82–91.

590. STERN, H. (1966). The regulation of cell division. *A. Rev. Pl. Physiol.*, **17**, 345–378.

591. STEWARD, F. C. and DURZAN, D. J. (1965). Metabolism of nitrogenous compounds. In *Plant Physiology, A Treatise*, Vol. IVA. *Metabolism: Organic Nutrition and Nitrogen Metabolism*, ed. STEWARD, F. C., 379–686. Academic Press, New York.

592. STEWARD, F. C., LYNDON, R. F. and BARBER, J. T. (1965). Acrylamide gel electrophoresis of soluble plant proteins: a study on pea seedlings in relation to development. *Am. J. Bot.*, **52**, 155–164.

593. STEWARD, F. C. and POLLARD, J. K. (1957). Nitrogen metabolism in plants: ten years in retrospect. *A. Rev. Pl. Physiol.*, **8**, 64–114.

594. STEWART, C. R. (1972). Effects of proline and carbohydrates on the metabolism of exogenous proline by excised bean leaves in the dark. *Pl. Physiol., Lancaster*, **50**, 551–555.

595. STEWART, C. R. and BEEVERS, H. (1967). Gluconeogenesis from amino acids in germinating castor bean endosperm and its role in transport to the embryo. *Pl. Physiol., Lancaster*, **42**, 1587–1595.

596. STEWART, C. R., MORRIS, C. J. and THOMPSON, J. F. (1971). Changes in amino acid content of excised leaves during incubation. II. Role of sugar in accumulation of proline in wilted leaves. *Pl. Physiol., Lancaster*, **41**, 1585–1590.

597. STEWART, W. D. P. (1966). *Nitrogen Fixation in Plants*. Athlone Press, London.

598. STEWART, W. D. P. (1969). Biological and ecological aspects of nitrogen fixation by free-living micro-organisms. *Proc. R. Soc., B*, **172**, 367–388.

599. STEWART, W. D. P., HAYSTEAD, A. and PEARSON, H. W. (1969). Nitrogenase activity in heterocysts of blue-green algae. *Nature, Lond.*, **224**, 226–228.

600. STEWART, W. D. P. and LEX, M. (1970). Nitrogenase activity in the blue-green alga *Plectonema boryanum* strain 594. *Arch. Mikrobiol.*, **73**, 250–260.

601. STOBART, A. K. and THOMAS, D. R. (1968). δ-Aminolaevulinic acid dehydratase in tissue cultures of *Kalanchoë crenata*. *Phytochem.*, **7**, 1313–1316.

602. STOKES, P. (1953). A physiological study of embryo development in *Heracleum sphondylium* L. III. The effect of temperature on metabolism. *Ann. Bot., N.S.*, **17**, 157–173.

603. STONER, C., HODGES, T. K. and HANSON, J. B. (1964). Chloramphenicol as an inhibitor of energy linked functions in maize mitochondria. *Nature, Lond.*, **203**, 258–261.

604. STOUT, E. R. and ARENS, M. Q. (1970). DNA polymerase from maize seedlings. *Biochim. biophys. Acta*, **213**, 90–100.

605. STOUT, E. R. and MANS, R. J. (1967). Partial purification and properties of RNA polymerase from maize. *Biochim. biophys. Acta*, **134**, 327–336.

606. STREET, H. E. and SHEAT, D. E. G. (1958). The absorption and availability of nitrate and ammonia. In *Encyclopedia of Plant Physiology*, Vol. VIII, ed. RUHLAND, W., 150–165. Springer Verlag, Berlin.

607. STREETER, J. G. (1973). *In vivo* and *in vitro* studies of asparagine biosynthesis in soybean seedlings. *Archs. Biochem. Biophys.*, **157**, 613–624.

608. STREETER, J. G. and THOMPSON, J. F. (1972). Anaerobic accumulation of γ-amino butyric acid and alanine in radish leaves (*Raphanus sativus* L.). *Pl. Physiol., Lancaster*, **49**, 572–578.

609. STREETER, J. G. and THOMPSON, J. F. (1972). *In vivo* and *in vitro* studies of γ-amino butyric acid metabolism with the radish plant (*Raphanus sativus* L.). *Pl. Physiol., Lancaster*, **49**, 579–584.

610. STREHLER, B. (1967). The nature of cellular ageing. In *Aspects of the Biology of Ageing. S.E.B. Symposium XXI*, ed. WOOLHOUSE, H. W., 149–178. Academic Press, London.

611. STREICHER, S. L. and VALENTINE, R. C. (1973). Comparative biochemistry of nitrogen fixation. *A. Rev. Biochem.*, **42**, 279–302.

612. SUMIDA, S. and MUDD, J. B. (1970). The structure and biosynthesis of phosphatidyl inositol in cauliflower inflorescence. *Pl. Physiol., Lancaster*, **45**, 712–718.

613. SUMNER, J. B. (1926). The isolation and crystallization of the enzyme urease. Preliminary paper. *J. biol. Chem.*, **69**, 435–441.

614. SUZUKI, N. and IWAI, K. (1970). The occurrence and properties of dihydrofolate reductase in pea seedlings. *Pl. Cell Physiol., Tokyo*, **11**, 199–208.

615. SUZUKI, N. and IWAI, K. (1973). The occurrence and properties of methenyltetrahydrofolate cyclohydrolase in plants. *Pl. Cell Physiol., Tokyo*, **14**, 319–328.

616. SWAIN, R. R. and DEKKER, E. E. (1966). Seed germination studies. II. Pathways for starch degradation in germinating pea seedlings. *Biochim. biophys. Acta*, **122**, 87–100.

617. SYRETT, P. J. (1953). The assimilation of ammonia by nitrogen-starved cells of *Chlorella vulgaris*. II. The assimilation of ammonia to other compounds. *Ann. Bot., N.S.*, **17**, 21–36.

618. TAPPER, B. A. and BUTLER, G. W. (1967). Conversion of oximes to mustard oil glucosides (glucosinolates). *Archs. Biochem. Biophys.*, **120**, 719–721.

619. TAPPER, B. A., CONN, E. E. and BUTLER, G. W. (1967). Conversion of α-ketoisovaleric acid oxime and isobutyraldoxime to linamarin in flax seedlings. *Archs. Biochem. Biophys.*, **119**, 593–595.

620. TAVARES, J. and KENDE, H. (1970). The effect of 6-benzylaminopurine on protein metabolism in senescing corn leaves. *Phytochem.*, **9**, 1763–1770.

621. TEWARI, K. K. and WILDMAN, S. G. (1967). DNA polymerase in isolated tobacco

chloroplasts and nature of the polymerized product. *Proc. natn. Acad. Sci. U.S.A.*, **58**, 689–696.

622. THEIMER, R. R. and BEEVERS, H. (1971). Uricase and allantoinase in glyoxysomes. *Pl. Physiol., Lancaster*, **47**, 246–251.

623. THIMANN, K. V. (1935). On the plant growth hormone produced by *Rhizopus syinus*. *J. biol. Chem.*, **109**, 279–291.

624. THOMAS, J. (1970). Absence of the pigments of photosystem II of photosynthesis in heterocysts of a blue-green alga. *Nature, Lond.*, **228**, 181–183.

625. THOMPSON, J. F. and MOORE, D. P. (1968). Enzymatic synthesis of cysteine and S-methylcysteine in plant extracts. *Biochem. biophys. Res. Commun.*, **31**, 281–286.

626. THOMPSON, J. F., MORRIS, C. J. and GERING, R. K. (1960). The effect of mineral supply on the amino acid composition of plants. *Qualitas Plantarum et Materiae Vegetabiles*, **6**, 261–275.

627. THORNBURG, W. and SIEGEL, A. (1973). Characterization of the rapidly reassociating deoxyribonucleic acid of *Cucurbita pepo* L. and the sequences complementary to ribosomal and transfer ribonucleic acids. *Biochemistry*, **12**, 2759–2765.

628. THRELFALL, D. R. and WHISTANCE, G. R. (1971). Biosynthesis of isoprenoid quinones and chromanols. In *Aspects of Terpenoid Chemistry and Biochemistry*, ed. GOODWIN, T. W., 357–404. Academic Press, London.

629. TING, I. P. and ZSCHOCHE, W. C. (1970). Asparagine biosynthesis by cotton roots. Carbon dioxide fixation and cyanide incorporation. *Pl. Physiol., Lancaster*, **45**, 429–434.

630. TOLBERT, N. E. (1971). Microbodies—peroxisomes and glyoxysomes. *A. Rev. Pl. Physiol.*, **22**, 45–74.

631. TOWILL, L. E. and NOODÉN, L. D. (1973). An electrophoretic analysis of the acid soluble chromosomal proteins from different organs of maize seedlings. *Pl. Cell Physiol., Tokyo*, **14**, 851–863.

632. TRAVIS, R. L., JORDAN, W. R. and HUFFAKER, R. C. (1969). Evidence for an inactivating system of nitrate reductase in *Hordeum vulgare* L. during darkness that requires protein synthesis. *Pl. Physiol., Lancaster*, **44**, 1150–1156.

633. TRAVIS, R. L. and KEY, J. L. (1971). Correlation between polyribosome level and the ability to induce nitrate reductase in dark grown corn seedlings. *Pl. Physiol., Lancaster*, **48**, 617–620.

634. TRAVIS, R. L., LIN, C. Y. and KEY, J. L. (1972). Enhancement by light of the *in vitro* protein synthetic activity of cytoplasmic ribosomes isolated from dark-grown maize seedlings. *Biochim. biophys. Acta*, **277**, 606–614.

635. TREWAVAS, A. (1968). Relationship between plant growth hormones and nucleic acid metabolism. In *Progress in Phytochemistry*, Vol. 1, ed. REINHOLD, L. and LIWSCHITZ, Y., 113–160. Interscience, London.

636. TREWAVAS, A. (1970). The turnover of nucleic acids in *Lemna minor*. *Pl. Physiol., Lancaster*, **45**, 742–751.

637. TREWAVAS, A. (1972). Determination of the rates of protein synthesis and degradation in *Lemna minor*. *Pl. Physiol., Lancaster*, **49**, 40–46.

638. TREWAVAS, A. (1972). Control of the protein turnover rates in *Lemna minor*. *Pl. Physiol., Lancaster*, **49**, 47–51.

639. TREWAVAS, A. (1973). The phosphorylation of ribosomal protein in *Lemna minor*. *Pl. Physiol., Lancaster*, **51**, 760–767.

640. TRIONE, E. J. (1966). Metabolic changes associated with vernalization of wheat. I. Carbohydrate and nitrogen patterns. *Pl. Physiol., Lancaster*, **41**, 277–281.

641. TRIONE, E. J., YOUNG, J. L. and YAMOMOTO, M. (1967). Free amino acid changes associated with vernalization of wheat. *Phytochem.*, **6**, 85–91.

642. TROMP, J. (1970). Storage and mobilization of nitrogenous compounds in apple trees with special reference to arginine. In *Physiology of Tree Crops*, ed. LUCKWILL, L. C. and CUTTING, C. V., 143–159. Academic Press, London and New York.

643. TROMP, J. and OVAA, J. C. (1973). Spring mobilization of protein nitrogen in apple bark. *Physiologia Pl.*, **29**, 1–5.

644. TRONIER, B., ORY, R. L. and HENNINGSEN, K. W. (1971). Characterization of the fine structure and proteins from barley protein bodies. *Phytochem.*, **10**, 1207–1211.

645. TSAI, C. S. and AXELROD, B. (1965). Catabolism of pyrimidines in rape seedlings. *Pl. Physiol., Lancaster*, **40**, 39–44.

646. TURNER, J. F. (1969). Starch synthesis and changes in uridine diphosphate glucose pyrophosphorylase and adenosine diphosphate glucose pyrophosphorylase in the developing wheat grain. *Aust. J. biol. Sci.*, **22**, 1321–1327.

647. TWARDOWSKI, T. and LEGOCKI, A. B. (1973). Purification and some properties of elongation factor 2 from wheat germ. *Biochim. biophys. Acta*, **324**, 171–183.

648. UDVARDY, J. and FARKAS, G. L. (1972). Abscisic acid stimulation of the formation of an ageing-specific nuclease in *Avena* leaves. *J. exp. Bot.*, **23**, 914–920.

649. UNDERHILL, E. N., CHISHOLM, M. D. and WETTER, L. R. (1962). Biosynthesis of mustard oil glucosides. I. Administration of C^{14}-labelled compounds to horseradish, nasturtium and watercress. *Can. J. Biochem. Physiol.*, **40**, 1505–1514.

650. VANDERHOEF, L. N. and KEY, J. L. (1970). The fractionation of transfer ribonucleic acid from roots of pea seedlings. *Pl. Physiol., Lancaster*, **46**, 294–298.

651. VANECKO, S. and VARNER, J. E. (1955). Studies on nitrite metabolism in higher plants. *Pl. Physiol., Lancaster*, **30**, 388–390.

652. VARNER, J. E. (1961). Biochemistry of senescence. *A. Rev. Pl. Physiol.*, **12**, 245–264.

653. VARNER, J. E., BALCE, L. V. and HUANG, R. C. (1963). Senescence of cotyledons of germinating peas. Influence of axis tissue. *Pl. Physiol., Lancaster*, **38**, 89–92.

654. VARNER, J. E. and WEBSTER, G. C. (1955). Studies on the enzymatic synthesis of glutamine. *Pl. Physiol., Lancaster*, **30**, 393–402.

655. VASCONCELOS, A. C. L. and BOGORAD, L. (1971). Proteins of cytoplasmic chloroplast, and mitochondrial ribosomes of some plants. *Biochim. biophys. Acta*, **228**, 492–502.

655a. VERMA, D. P. S., NASH, D. T. and SCHULMAN, H. M. (1974). Isolation and *in vitro* translation of soybean leghaemoglobin mRNA. *Nature, Lond.*, **251**, 74–77.

656. VESSAL, M. and HASSID, W. Z. (1972). Partial purification and properties of L-glutamine D-fructose 6-phosphate amidotransferase from *Phaseolus aureus*. *Pl. Physiol., Lancaster*, **49**, 977–981.

657. VICKERY, H. B., PUCKER, G. W., SCHOENHEIMER, R. and RITTENBERG, D. (1940). The assimilation of ammonia nitrogen by the tobacco plant: a preliminary study with isotopic nitrogen. *J. biol. Chem.*, **135**, 531–539.

658. VIRGIN, H. I. (1955). Protochlorophyll formation and greening in etiolated barley leaves. *Physiologia Pl.*, **8**, 631–643.

659. VOLD, B. S. and SYPHERD, P. S. (1968). Changes in soluble RNA and ribonuclease activity during germination of wheat. *Pl. Physiol., Lancaster*, **43**, 1221–1226.

660. WALBOT, V. (1971). RNA metabolism during embryo development and germination of *Phaseolus vulgaris*. *Devl Biol.*, **26**, 369–379.

661. WALBOT, V., BRADY, T., CLUTTER, M. and SUSSEX, L. (1972). Macromolecular synthesis in *Phaseolus coccineus* embryos and suspensors. *Devl Biol.*, **29**, 104–112.

661a. WALBOT, V., CAPDEVILA, A. and DURE, L. S. (1974). Action of 3′d adenosine (cordycepin) and 3′d cytidine on the translation of stored mRNA of cotton cotyledons. *Biochem. biophys. Res. Commun.*, **60**, 103–110.

662. WALDSCHMIDT-LEITZ, E. and METZNER, P. (1962). Über die Prolamine aus Weizen, Roggen, Mais und Hirse. *Hoppe-Seyler's Z. physiol. Chem.*, **329**, 52–61.

663. WALKER, D. A. (1964). Improved rates of carbon dioxide fixation by illuminated chloroplasts. *Biochem. J.*, **92**, 22c–23c.
664. WALLACE, W. and NICHOLAS, D. J. D. (1969). The biochemistry of nitrifying microorganisms. *Biol. Rev.*, **44**, 359–391.
665. WALLACE, W. and PATE, J. S. (1967). Nitrate assimilation in higher plants with special reference to the cocklebur (*Xanthium pennsylvanicum* Wallr.). *Ann. Bot., N.S.*, **31**, 213–228.
666. WALLER, G. R., YANG, K. S., GHOLSON, R. K., HADWIGER, L. A. and CHAYKIN, S. (1966). The pyridine nucleotide cycle and its role in the biosynthesis of ricinine by *Ricinus communis* L. *J. biol. Chem.*, **241**, 4411–4418.
667. WANG, D. and WAYGOOD, E. R. (1964). Enzymes of synthesis of purine nucleotides. In *Modern Methods of Plant Analysis*, Vol. VII, ed. LINSKENS, H. F., SUNWAL, B. C. and TRACEY, M. V., 421–427. Springer Verlag, Berlin.
668. WANKA, F. and BAUER, F. W. (1967). On phosphorylation and deamination of pyrimidine and deoxypyrimidine nucleosides by enzymes from germinating corn seedlings. *Z. Pflanzenphysiol.*, **58**, 175–186.
669. WANKA, F. and WALBOOMERS, J. M. (1966). Thymidine kinase and uridine kinase in corn seedlings. *Z. Pflanzenphysiol.*, **55**, 458–465.
670. WANKA, F., VASIL, I. K. and STERN, H. (1964). Thymidine kinase: the dissociability and its bearing on the enzyme activity in plant material. *Biochim. biophys. Acta*, **85**, 50–59.
671. WAREING, P. F., HANNEY, C. E. A. and DIGBY, J. (1964). The role of endogenous hormones in cambial activity and xylem differentiation. In *The Formation of Wood in Forest Trees*, ed. ZIMMERMAN, N., 323–344. Academic Press, New York.
672. WAREING, P. F. and SAUNDERS, P. F. (1971). Hormones and dormancy. *A. Rev. Pl. Physiol.*, **22**, 261–288.
673. WEBSTER, B. D. (1970). A morphogenetic study of leaf abscission in *Phaseolus*. *Am. J. Bot.*, **57**, 443–451.
674. WEBSTER, G. C. and VARNER, J. E. (1955). Aspartate metabolism and asparagine synthesis in plant systems. *J. biol. Chem.*, **215**, 91–99.
675. WEEKS, D. P. and MARCUS, A. (1971). Preformed messenger of quiescent wheat embryos. *Biochim. biophys. Acta*, **232**, 671–684.
676. WEEKS, D. P., VERMA, D. P. S., SEAL, S. N. and MARCUS, A. (1972). Role of ribosomal subunits in eukaryotic protein chain initiation. *Nature, Lond.*, **236**, 167–168.
677. WEISSER, C. J. (1970). Cold resistance and injury in woody plants. *Science, N.Y.*, **169**, 1269–1278.
678. WELLS, G. and HAGEMAN, R. H. (1970). The conversion of NADPH to NADH by phosphatase activity in extracts of green leaves. *Pl. Physiol., Lancaster*, **46** (Suppl.), 45.
678a. WELLS, G. N. and HAGEMAN, R. H. (1974). Specificity for nicotinamide adenine dinucleotide by nitrate reductase from leaves. *Pl. Physiol., Lancaster*, **54**, 136–141.
679. WELLS, G. N. and BEEVERS, L. (1973). Protein synthesis in cotyledons of *Pisum sativum* L. II. The requirements for initiation with plant messenger RNA. *Plant Science Letters*, **1**, 281–286.
680. WELLS, G. N. and BEEVERS, L. (1974). Protein synthesis in the cotyledons of *Pisum sativum*. Protein factors involved in the binding of phenylalanyl-transfer ribonucleic acid to ribosomes. *Biochem. J.*, **139**, 61–69.
681. WELLS, G. N. and BEEVERS, L. (1974). Messenger RNA independent formation of a methionyl tRNA initiation complex. *Pl. Physiol., Lancaster*, **53** (Suppl.), 44.
682. WELLS, J. R. E. (1965). Purification and properties of a proteolytic enzyme from French beans. *Biochem. J.*, **97**, 228–235.
683. WELLS, J. R. E. (1969). Characterization of three proteolytic enzymes from French beans. *Biochim. biophys. Acta*, **167**, 388–398.
684. WELLS, R. and BIRNSTIEL, M. (1969). Kinetic complexity of chloroplastal deoxyribonucleic acid and mitochondrial deoxyribonucleic acid from higher plants. *Biochem. J.*, **112**, 777–786.

685. WELLS, R. and INGLE, J. (1970). The constancy of the buoyant density of chloroplast and mitochondrial deoxyribonucleic acids in a range of higher plants. *Pl. Physiol., Lancaster*, **46**, 178–179.

686. WHEELER, A. W. (1968). Changes in auxins in expanding and senescent primary leaves of dwarf French bean (*Phaseolus vulgaris*). *J. exp. Bot.*, **19**, 102–107.

687. WHEELER, C. T. and BOULTER, D. (1967). Nucleic acids of developing seeds of *Vicia faba* L. *J. exp. Bot.*, **18**, 229–240.

688. WHITESIDE, J. A. and THURMAN, D. A. (1971). The degradation of canavanine by Jack Bean cotyledons. *Planta*, **98**, 279–284.

689. WIDHOLM, J. M. (1971). Control of tryptophan biosynthesis in plant tissue cultures: lack of repression of anthranilate and tryptophan synthetases by tryptophan. *Physiologia Pl.*, **25**, 75–79.

690. WIDHOLM, J. M. (1972). Tryptophan biosynthesis in *Nicotiana tabacum* and *Daucus carota* cell cultures: site of action of inhibitory tryptophan analogs. *Biochim. biophys. Acta*, **261**, 44–51.

691. WIDHOLM, J. M. (1973). Measurement of the five enzymes which convert chorismate to tryptophan in cultured *Daucus carota* cell extracts. *Biochim. biophys. Acta*, **320**, 217–226.

692. WIGHTMAN, F. and COHEN, D. (1968). Intermediary steps in the enzymatic conversion of tryptophan to IAA in cell free systems from mung bean seedlings. In *Biochemistry and Physiology of Plant Growth Substances*, ed. WIGHTMAN, F. and SETTERFIELD, G., 273–288. Runge Press, Ottawa.

693. WILLIAMS, G. R. and MASER, J. A. (1973). Plant ribosome-bound proteins: a fraction from leaves that dissociates bean ribosomes. *Biochem. biophys. Res. Commun.*, **53**, 52–58.

694. WILLIAMS, G. R. and NOVELLI, G. D. (1964). Stimulation of an *in vitro* amino acid incorporating system by illumination of dark-grown plants. *Biochem. biophys. Res. Commun.*, **17**, 23–27.

695. WILLIAMS, G. R. and NOVELLI, G. D. (1968). Ribosome changes following illumination of dark grown plants. *Biochim. biophys. Acta*, **155**, 183–192.

696. WILLIAMS, G. R. and WILLIAMS, A. S. (1970). Hybridization of bean leaf leucyl-tRNA with nuclear DNA and with chloroplast DNA. *Biochem. biophys. Res. Commun.*, **39**, 858–863.

697. WILLIS, A. J. and YEMM, E. W. (1955). The respiration of barley plants. VIII. Nitrogen assimilation and the respiration of the root system. *New Phytol.*, **54**, 163–181.

698. WILSON, C. M. (1966). Bacteria, antibiotics and amino acid incorporation into maize endosperm protein bodies. *Pl. Physiol., Lancaster*, **41**, 325–327.

699. WILSON, C. M. (1968). Plant nucleases. II. Properties of corn ribonucleases I and II and corn nuclease I. *Pl. Physiol., Lancaster*, **43**, 1339–1346.

700. WILSON, C. M. (1971). Plant nucleases. III. Polyacrylamide gel electrophoresis of corn ribonuclease isoenzymes. *Pl. Physiol., Lancaster*, **48**, 64–68.

701. WILSON, P. W. (1969). First steps in biological nitrogen fixation. *Proc. R. Soc., B*, **172**, 319–325.

702. WOLCOTT, J. H. and ROSS, C. (1967). Orotidine-5′-phosphate decarboxylase and pyrophosphorylase of bean leaves. *Pl. Physiol., Lancaster*, **42**, 275–279.

703. WOLLGIEHN, R. (1967). Nucleic acid and protein metabolism of excised leaves. In *Aspects of the Biology of Ageing. S. E. B. Symposium* XXI, ed. WOOLHOUSE, H. W., 231–248. Academic Press, New York.

704. WONG, K. F. and DENNIS, D. T. (1973). Aspartokinase from wheat germ. Isolation, characterization and regulation. *Pl. Physiol., Lancaster*, **51**, 322–326.

705. WONG, K. F. and DENNIS, D. T. (1973). Aspartokinase in *Lemna minor* L. Studies on the *in vivo* and *in vitro* regulation of the enzymes. *Pl. Physiol., Lancaster*, **51**, 327–331.

706. WONG, P. P. and EVANS, H. J. (1971). Poly-β-hydroxybutyrate utilization by soybean (*Glycine max* Merr.) nodules and assessment of its role in maintenance of nitrogenase activity. *Pl. Physiol., Lancaster*, **47**, 750–755.

707. WONG, P. P., EVANS, H. J., KLUCAS, R. and RUSSELL, S. (1971). Investigations into

the pathway of electron transport to the nitrogenase from nodule bacteroids. *Plant and Soil*, Special Volume, 525–543.

708. WOODCOCK, C. F. and BOGORAD, L. (1973). Nucleic acid and information processing in chloroplasts. In *Structure and Function of Chloroplasts*, ed. GIBBS, M., 89–128. Springer Verlag, New York.

709. WRAY, J. L. and FILNER, P. (1970). Structural and functional relationships of enzyme activities induced by nitrate in barley. *Biochem. J.*, **119**, 715–725.

710. WRIGHT, S. T. C. and HIRON, R. W. P. (1969). (+)-Abscisic acid, the growth inhibitor induced in detached wheat leaves by a period of wilting. *Nature, Lond.*, **224**, 719–720.

711. WYEN, N. V., ERDEI, S. and FARKAS, G. L. (1971). Isolation from *Avena* leaf tissues of a nuclease with the same type of specificity towards RNA and DNA. Accumulation of the enzyme during leaf senescence. *Biochim. biophys. Acta*, **232**, 472–483.

712. YANG, S. F. (1968). Biosynthesis of ethylene. In *Biochemistry and Physiology of Plant Growth Substances*, ed. WIGHTMAN, F. and SETTERFIELD, G., 1217–1226. Runge Press, Ottawa.

713. YARWOOD, A., BOULTER, D. and YARWOOD, T. N. (1971). Methionyl-tRNAs and initiation of protein synthesis in *Vicia faba* L. *Biochem. biophys. Res. Commun.*, **44**, 353–361.

714. YATSU, L. Y. and JACKS, T. Y. (1968). Association of lysosomal activity with aleurone grains in plant seeds. *Archs. Biochem. Biophys.*, **124**, 466–471.

715. YEMM, E. W. (1950). Respiration of barley plants. IV. Protein catabolism and the formation of amides in starving leaves. *Proc. R. Soc., B*, **136**, 632–649.

716. YEMM, E. W. and WILLIS, A. J. (1956). The respiration of barley plants. IX. The metabolism of roots during the assimilation of nitrogen. *New Phytol.*, **55**, 229–252.

717. YOMO, H. and TAYLOR, M. P. (1973). Histochemical studies on protease formation in the cotyledons of germinating bean seeds. *Planta*, **112**, 35–43.

718. YOUNG, J. L. and VARNER, J. E. (1959). Enzyme synthesis in the cotyledons of germinating seeds. *Archs. Biochem. Biophys.*, **84**, 71–78.

719. ZASLOFF, M. and OCHOA, S. (1973). Polypeptide chain initiation in eukaryotes. IV. Purification and properties of supernatant initiation factor from *Artemia salina* embryos. *J. molec. Biol.*, **73**, 65–76.

720. ZECH, A. C. and PAULI, A. W. (1962). Changes in total free amino nitrogen, free amino acids and amides of winter wheat during cold hardening. *Crop Sci.*, **2**, 421–423.

721. ZUBER, H. (1968). Reinigung und Eigenschaften der Carboxypeptidase aus Citrusfrüchten. *Hoppe-Seyler's Z. physiol. Chem.*, **349**, 1337–1352.

722. ZUCKER, M. (1968). Sequential induction of phenylalanine ammonia-lyase and lyase-inactivating system in potato tuber disks. *Pl. Physiol., Lancaster*, **43**, 365–374.

723. ZWAR, J. A. and JACOBSEN, J. V. (1972). A correlation between a ribonucleic acid fraction selectively labelled in the presence of gibberellic acid and amylase synthesis in barley aleurone layers. *Pl. Physiol., Lancaster*, **49**, 1000–1006.

Index